Lateral DNA Transfer

Mechanisms and Consequences

Lateral DNA Transfer

Mechanisms and Consequences

Frederic Bushman

The Salk Institute
La Jolla, California

COLD SPRING HARBOR LABORATORY PRESS
Cold Spring Harbor, New York

Lateral DNA Transfer
Mechanisms and Consequences

Developmental Editor: Judy Cuddihy
Project Coordinator: Inez Sialiano
Production Editor: Patricia Barker
Desktop Editor: Susan Schaefer
Book Designer: Denise Weiss
Cover Designer: Ed Atkeson (Berg Design)

Front cover (paperback edition): Illustration rendered by Witek Kwiatkowski, Jamie Simon, and Frederic Bushman from coordinates reported by Larsen et al. (*Biochemistry 30:* 4443–4449 [1991].)

Library of Congress Cataloging-in-Publication Data

Bushman, Frederic.
    Lateral DNA transfer : mechanisms and consequences /
Frederic Bushman.
      p. cm.
    Includes bibliographical references and index.
    ISBN 0-87969-603-6 (case : alk. paper) — ISBN 0-87969-621-4 (pbk. :
alk. paper)
    1. Mobile genetic elements. 2. Transposons. 3. Transfection. I. Title.
    QH452.3 .B87 2001
    572.8´69—dc21

                            2001042431

10 9 8 7 6 5 4 3 2 1

To my family

Contents

Preface

THIS BOOK GREW OUT OF A FASCINATION with the way DNA changes over time and the consequences of these changes for life on earth. The amount of new information in this area has ballooned in recent years, in large measure due to the flood of data from the complete genome sequences. This year the human genome sequence has been added to the list. These genome sequences allow the full extent of DNA transfer between organisms to be appreciated for the first time. Our own DNA is a complex composite of imported sequences and mobile genetic parasites, and the dynamic remodeling of our genome continues to this day. The same is true for all of the other organisms for which genome sequences are available.

Work in areas as diverse as medical microbiology and DNA enzymology has also revealed the importance of DNA transfer between organisms, but the extent and diversity of the data present an ongoing challenge. The volume of new information has made keeping up with the subject difficult or impossible for busy students and researchers. Even specialists often have trouble staying current with work in related areas. This book is intended to help a diverse community appreciate the full picture. I have struggled to make the presentation inviting—highlighting surprising and odd discoveries where possible—while still covering the subject systematically.

No writer could be an expert in all the areas covered in this book, so the risk of incompleteness or inaccuracy is high. I can only hope that readers feel that a broad overview of lateral transfer is useful enough to make any lapses forgivable.

This book would not have been possible without the help of an outstanding group of colleagues. I am extremely grateful to all those who read sections of the book and provided invaluable comments and suggestions: Jonathon Eisen (Chapters 5 and 10), Alex Gann (Chapters 1–14), Erik Johnson (Chapters 1–14), Renee Ketting (Chapters 9, 10, and 13), Ned Landau (Chapters 6 and 7), Sandy Martin (Chapters 8, 9, and 14), Matt Meselson (Chapters 1–7), Leslie Orgel (Chapters 1–14), Murcia Podar (Chapters 1–14), William Reznikoff (Chapters 1–5), Suzanne Sandemeyer (Chapter 8), David Schatz (Chapter 11), Astrid Schroeder (Chapters 1–14), and Anca Segal (Chapters 1–5), and Detlief Weigel (Chapters 1–3).

I thank my family for their support and patience throughout the project.

Judy Cuddihy, my developmental editor, provided crucial help at many points in the project. She edited the text extensively, making the final work much more readable. Her contribution approaches coauthor status. Inez Sialiano and Donna Rush provided excellent editorial help as well. I also thank Pat Barker and Susan Schaefer, who handled production issues, and Jan Argentine and Denise Weiss, Editorial Development and Production Managers, respectively.

Finally, I am particularly grateful to three outstanding men who have greatly influenced my growth as a scientist: my graduate thesis advisor Mark Ptashne, my postdoctoral supervisor Kiyoshi Mizuuchi, and my present next-door neighbor at Salk, Leslie Orgel. All three are distinguished by their extremely wide knowledge of molecular mechanisms and exceptional common sense. All three have thought deeply about the mechanisms by which DNA changes over time. The ideas I absorbed from them are reflected in the book in too many places to list specifically—much of what is best in the work can be attributed to Mark, Kiyoshi, and Leslie.

Lateral DNA Transfer

Mechanisms and Consequences

1

Introduction

DNA IS WELL KNOWN AS THE MOLECULAR REPOSITORY of the program of life. What is not as well known is that the DNA composition of organisms can be remarkably fluid. Surprisingly often, DNA is transferred from one organism to another, and that DNA can become stably incorporated in the recipient, permanently changing its genetic composition. This process is called "lateral transfer" or "horizontal transfer"; this is in contrast to the inheritance of genes by descent from one's parents—that process is termed "vertical" inheritance.

As an example of the scale of lateral transfer between organisms, consider transfer by marine bacteriophages in the wild. Ocean waters typically contain enormous numbers of viruses—usually more than ten million per milliliter of seawater. Most of these viruses are probably bacteriophages (also called phages), viruses that infect bacteria. Only a few marine phages have been isolated and studied in the laboratory, thus little is known about this vast population (Fig. 1.1). During their growth, some of the marine phages pick up genes of the host cell and transfer them by infection to a new cell, where the transferred sequences sometimes become stably incorporated, a process termed transduction. Biologists have measured the rate of transduction between several pop-

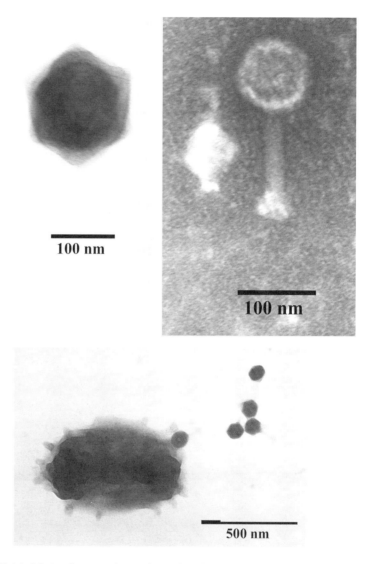

FIGURE 1.1. Marine bacteriophage shown by electron microscopy. Photography courtesy of Dr. Greig Steward.

ulations of marine bacteria, revealing that gene transfer takes place about once in every 10^8 phage infections. These frequencies may not seem terribly high, but given the very high concentrations of bacteria and viruses in seawater, and the tremendous volume of water in the ocean, it follows that gene transfer between organisms takes place about 20 million billion times *per second* in the oceans. Although there are several rough estimates in this calculation, there is no doubt that the rate of gene transduction worldwide is extremely high.

This book is about how lateral transfer occurs, the extent to which it happens, and the implications it has for our understanding of everything from evolution to the emergence of new medical threats. It has recently become possible to sequence the entire DNA content (genome) of essentially any organism. Comparative analysis of these completed genomes has documented extensive lateral transfer, even in the human genome. Meanwhile, studies of the molecular machinery used to transfer DNA—in viruses, transposons, and conjugative plasmids—have produced a detailed picture of how this is achieved. But that rich area of research is often hard for outsiders to fathom. At the same time, the study of medically important infectious agents has documented many examples of lateral transfer that are important to human health.

Although related, these fields are generally studied by separate communities of scientists, each with their own jargon and objectives. The goal of this book is to provide an overview of both the DNA transfer machinery and the biological consequences of lateral transfer, and to present it in a way that is accessible to a wide range of readers.

This book focuses primarily on DNA transfer between organisms—of the same or different species—since contemporary research has yielded particularly striking new discoveries in this area with very wide implications. Quite often, related machinery mediates the rearrangement of DNA segments within a cell and transfer of DNA between cells, as well as transfer of DNA between organisms. In many cases, it would therefore be artificial to ignore related systems that transfer DNA within or between cells, and so they are often included to provide a wider picture.

Many remarkable stories have emerged from studies of lateral gene transfer in addition to that of the marine phages. For example, hybrid wallabies have been identified that contain chromosomes different from either parent, apparently because an interspecific cross activated an endogenous retrovirus that proliferated to great numbers in the hybrid animal, thereby expanding chromosome sizes (Chapter 13). In another case, it has been proposed that the predatory mite *Proctolaelaps regalis* has probably transferred DNAs between insects during messy "hit-and-run" feeding bouts on fruit fly eggs, leading to new mobile element infestations in flies (Chapter 9). Even the evolution of sex appears to be linked to proliferation of mobile elements (Chapter 14). Where possible, the denser studies of DNA transfer mechanisms are interspersed with colorful examples of the biological consequences.

Phage were discovered independently in 1915 by Frederick Twort and in 1917 by Felix d'Herelle. The name "bacteriophage" is derived from the Greek phagos, meaning "to eat," as the growth of bacteriophage often kills the host bacterial cell.

BOOK SUMMARY

This book is organized in a modular fashion that allows it to be read from start to finish or read selectively for information on specific areas. The chapter titles and subheadings provide a road map of the contents of each section, and references to other sections within the chapters point readers to related material.

This book is aimed at an audience with some background in biology but no special knowledge of mobile DNA or lateral transfer. Chapter 2 begins with a review of molecular biology basics for readers in need of a refresher course. This can be skipped by those who are familiar with the material.

Chapters 3–5 survey lateral DNA transfer mechanisms of prokaryotes. Chapter 3 covers the rise of antibiotic resistance in microbial pathogens, which is largely a consequence of lateral transfer of resistance genes. Conversion of bacteria to a pathogenic state is also often a consequence of DNA transfer, in this case blocks of genes called pathogenicity islands. Chapter 4 covers transfer of pathogenicity islands, together with the replication mechanisms of the bacteriophages that often transfer these islands by transduction. The idea that lateral transfer is a major source of prokaryotic genetic change was highlighted by the complete microbial genome sequences that began to appear in 1995. Chapter 5 covers the prokaryotic genome sequences and the evidence they provide for widespread lateral transfer.

Chapters 6–11 cover lateral DNA transfer in eukaryotes. Many forms of lateral transfer are less frequent in eukaryotes than in prokaryotes, but lateral transfer is clearly an important source of genetic novelty in eukaryotes as well. Chapter 6 covers one of the best-understood eukaryotic DNA transfer vehicles, the retroviruses. Retroviruses or mutated remnants have been found in the genomes of most of the organisms in which they have been sought, emphasizing their ubiquity. Retroviruses can modify the host cell genome not only by integrating their genomes into a host chromosome, but also by transporting cellular genes between cells.

In many cases, infectious agents evolve to be relatively benign, probably in part because they themselves can leave more offspring during infection of a relatively healthy host. Infectious agents newly introduced into the human population, in contrast, are often much more pathogenic. Chapter 7 reviews the biology of HIV, a retrovirus recently introduced into the human population that is the cause of the devastating AIDS epidemic. To replicate, HIV must integrate a cDNA copy of its genome into a chromosome of the host, a life-style that has many implications for disease progression and therapy.

Chapters 8 and 9 cover two types of mobile elements in eukaryotes, the retrotransposons, which form new copies by using an RNA intermediate, and the DNA transposons, which move via DNA-mediated pathways. Many of these elements are known to have been transferred between organisms and so are mobile both within and between cells.

Five eukaryotic genomes are completely sequenced or nearly so. Chapter 10 surveys these sequences and the record of lateral DNA transfer they provide.

Chapter 11 surveys the recombination machinery responsible for constructing the antigen-binding proteins such as antibodies.

Prokaryotes are organisms lacking nuclei and membrane-enclosed compartments within their cells. Eukaryotes contain membrane-enclosed nuclei and a variety of internal membrane-enclosed structures. Two of the three domains of life are prokaryotic, the Bacteria and Archaea. Eukaryotes—the animals, plants, fungi, and protists—comprise the third domain.

Remarkably, there is now good evidence that this system arose from a captured DNA transposon. It is not known whether the transposon was first introduced by lateral transfer, although this is a reasonable guess.

A number of different mechanisms transfer DNA between species that belong to different domains of life—the Bacteria, Archaea, and Eukarya—the most divergent phylogenetic taxa. These long-range transfers are the topic of Chapter 12.

Cells have gone to great lengths to regulate the activity of mobile DNA elements. In some cases, cells seek to suppress harmful elements, such as selfish genomic parasites or lytic viruses. In other cases, cells have evolved to take up and incorporate DNA, as in inducible transformation in some bacteria. Chapter 13 surveys the regulation of lateral DNA transfer, both by cells and by mobile elements themselves.

The book ends with a review of some main themes from earlier chapters and speculative extrapolations from these ideas. For example, lateral transfer has been proposed to be important in the early evolution of modern organisms and in the origin of sex. These and other ideas are covered in Chapter 14.

Sets of chapters can be read for overviews of particular areas. Chapters 3–5 cover lateral transfer in prokaryotes, and Chapters 6–11 cover eukaryotes. Lateral transfer and the complete genome sequences are presented in Chapters 5 and 10. The most medically important examples can be found in Chapters 3, 4, 7, and 11. Sections of greater interest to specialized readers are set off from the main text in shaded boxes, and may be skipped by the more casual reader.

References to recent reviews and a few key papers are cited under each subheading. In this relatively short book, it is not possible to include complete references to the primary literature. The indicated review articles provide more thorough citations. In most cases, equally good primary papers could have been chosen instead of those presented, and I apologize to all those whose work was not specifically mentioned.

This book is about natural mechanisms of DNA transfer, so technological applications are not treated in detail. Uses of lateral transfer machinery in biotechnology, a very broad topic, are summarized briefly in shaded boxes outside the main text. The interested reader can learn about these methods from the cited technical material. The history of lateral DNA transfer is also treated, but relatively briefly. Boxed text presents key discoveries in the emerging study of lateral DNA transfer and accounts of the people involved.

We begin our survey of lateral gene transfer by introducing some of the main systems that mobilize DNA molecules. Specific systems and the consequences of their action are described in more detail in subsequent chapters. This chapter concludes by introducing a few of the main themes that run through many of the chapters.

THE MACHINERY OF LATERAL DNA TRANSFER (1, 2)

For a DNA molecule to move between cells, it must exit across the membrane of the donor and enter through the membrane of the recipient. The transferred DNA sequence must then achieve a state in the recipient cell that allows it to persist. Each of these steps is typically carried out by proteins dedicated to DNA transfer and DNA assimilation.

Three broad mechanisms mediate efficient movement of DNA between cells—transduction, conjugation, and transformation. These mechanisms are introduced below and considered in detail in the following chapters.

Transduction

Transduction refers to the transfer of a DNA sequence from one cell to another by a virus (Fig. 1.2). Transduction mechanisms can take many forms. Some viruses transfer the same host cell DNA sequences every time, a mechanism called specialized transduction. Other viruses can transfer essentially any host sequence, a mechanism called generalized transduction. Transduction can take place as a normal part of the viral life cycle, or abnormally as a side product of the replication mechanism. Transduction is discussed in detail in Chapter 4.

Conjugation

Conjugation is the directed transfer of DNA from one cell to another (Fig. 1.3). Conjugation historically refers to the transfer of DNAs between bacterial cells, although, as discussed in Chapter 12, *Agrobacteria* uses an essentially indistinguishable mechanism to transfer DNA to eukaryotic cells. Conjugation involves a special apparatus called the pilus for binding the donor and recipient cells, bringing them into direct cell-to-cell contact. A DNA strand from the donor cell is extruded into the recipient cell. If the newly transferred DNA becomes stably associated with the new cell, the recipient can take on new characters encoded in the transferred DNA. The genes for the conjugation apparatus are commonly found on the DNA that becomes transferred, usually extrachromosomal DNA circles such as plasmids.

Terminology differs in the field of cancer biology, where cellular "transformation" means conversion of a normal animal cell to a state of unregulated growth.

Transformation

Transformation involves the simple uptake by a cell of DNA from the surrounding environment and expression of that DNA in the recipient cell (Fig. 1.4). Both prokaryotic and eukaryotic cells can be induced to take up DNA in the laboratory, but only in bacteria do we see cells that are highly transformable under normal growth condi-

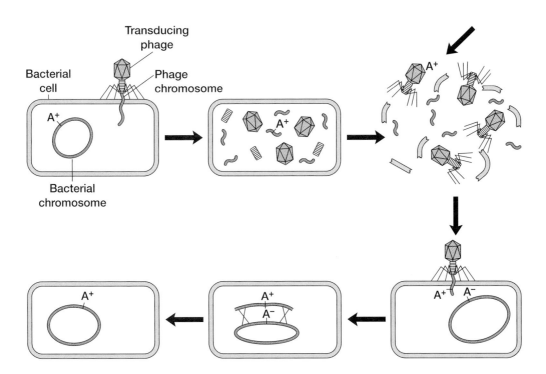

FIGURE 1.2. Gene transduction. Starting from the upper left, a bacteriophage binds to a bacterial cell and injects its genome. Replicating bacteriophages typically package their own chromosomes, but on rare occasions host cell DNA is packaged instead (*arrow, top right*). A phage containing cellular DNA can then inject it into a new host cell (*bottom right*). Recombination with the endogenous chromosome (*bottom middle*) can then install the new gene. In the example, A+ genetic information from the original cell is transferred to an A− bacterium, converting it to A+ (*bottom left*). Details can vary with different phages and transduction pathways (Chapter 4). (Modified, with permission, from Griffiths et al. 1993 [©.W.H. Freeman]. [6])

tions. For some bacteria, transformation is induced by specific environmental conditions, whereas others are always "competent" for transformation. Transformation is typically carried out by dedicated cellular machinery that binds the extracellular DNA and transfers it across the cellular membrane. Transformation is discussed in more detail in Chapter 3.

Plasmids are typically circular DNA molecules that replicate independently of the cellular chromosome.

Transposition

Another prominent mechanism that moves DNA molecules to new locations is transposition, in which colinear DNA segments move from one location in the genome to another in that same cell. Transposons typically encode the enzymes that direct DNA breaking and joining reactions involved in transposition, and the sites of action for these enzymes. Transposons are known in prokaryotes and eukaryotes, although the types differ in each.

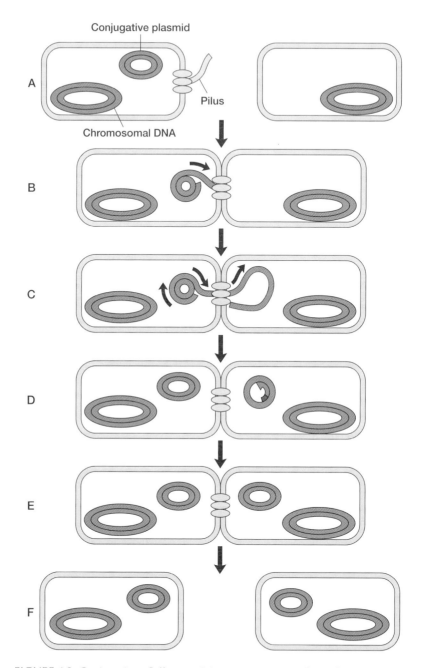

FIGURE 1.3. Conjugation. Cells containing a conjugative plasmid synthesize the pilus/mating pore structure (*A*). The pilus binds cells together. Nicking of the oriT origin of transfer and DNA synthesis begins the transfer of a single strand from the donor to the recipient cell (*B*). Replication and transfer proceed (*C*) until a complete plasmid is transferred. Transfer is terminated by circularization of the newly transferred plasmid in the recipient cell (*D*). Replication of the second strand in the recipient completes formation of the transferred plasmid (*E*), after which the mating pair separates (*F*). (Modified, with permission, from Griffiths et al. 1993 [© W.H. Freeman]. [6])

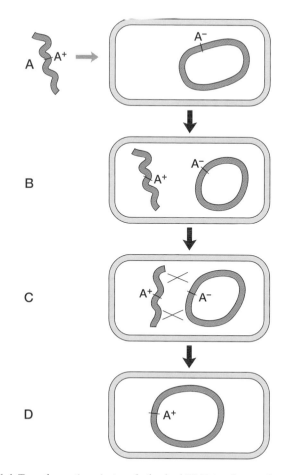

FIGURE 1.4. Transformation. A strand of naked DNA is taken up by a "competent" cell (*A*). The DNA strand encodes the A^+ phenotype. In the example, the recipient cell is A^- (*A,B*). Recombination between the incoming DNA and the resident chromosome (*C*) installs the new sequence in the recipient cell. The net effect is to convert the cell from A^- to A^+.

Transposition is not itself a mechanism of lateral DNA transfer, because there is no obligate exiting of one cell and entry into another. However, in a few cases, transposition reactions are directly associated with lateral transfer per se. More generally, transposition can link cellular sequences to vehicles for lateral transfer between cells, such as conjugative plasmids or viruses, resulting in their mobilization. Transposon function and their effects on lateral transfer are discussed in Chapters 3 and 5 (prokaryotes) and Chapters 8–10 (eukaryotes).

Incorporating New DNA after Lateral Transfer

To persist, newly introduced DNA must be able to replicate in the recipient cell and segregate to daughter cells during division.

Otherwise, division of the host cell will eventually dilute away the newly introduced DNA. Some DNA sequences such as plasmids encode their own machinery for replication. Once introduced into a permissive cell by any of the above three means, a plasmid can replicate independently of the host chromosome. However, those DNA sequences that cannot replicate independently must become stably linked to the host cell DNA to persist during replication. Following are some of the mechanisms that influence this process.

Homologous Recombination

All cells encode systems that can break identical DNA segments and rejoin them in new arrangements (Fig. 1.5). This process is known as "homologous recombination," emphasizing the requirement for DNA sequence matches in the two interacting DNA molecules. Depending on the arrangements of the matching sites in the two recombination partners, unmatched sequences in the transduced DNA can sometimes become joined to recipient cell DNA if flanked by regions of homology. Homologous recombination is a major mechanism for incorporating newly transferred DNA into the recipient cell chromosome after transduction, transformation, and conjugation.

Homologous recombination evolved at least in part to aid in the repair of damaged DNA, particularly the reactivation of stalled DNA replication intermediates. It is estimated that replication of the *E. coli* chromosome stalls 15–50% of the time (results vary with the assay used), generating potentially toxic double-stranded DNA breaks. Recombination-mediated reactivation is required to restart DNA synthesis. Homologous recombination also performs many other functions in DNA maintenance, for example, repairing double-stranded breaks in DNA.

Site-specific Recombination

A second type of recombination, called "site-specific recombination," also figures prominently in our story. In this case, dedicated recombination proteins recognize specific DNA sequences on each of two DNA molecules. The sites need not be identical in sequence, distinguishing site-specific recombination from homologous recombination (Fig. 1.6). The resulting protein–DNA complexes then carry out the orderly exchange of DNA strands. As shown in Chapter 4, the phage integration machinery introduced in Chapter 1 is an example of a site-specific recombination system.

DNA Repair

Cells encode elaborate machinery for repair of damage to DNA, which can be caused by exposure to UV light, X-rays, or chemical mutagens, in each case changing the chemical structure of DNA. DNA is the only cellular macromolecule that is repaired when damaged, emphasizing

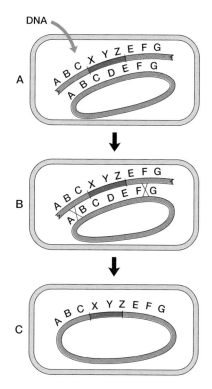

FIGURE 1.5. A closer look at homologous recombination. In this example, a sequence containing information for XYZ enters a cell flanked by regions of homology with the host chromosome (*top*). Recombination within the regions of homologyABC and EFG (*middle*) result in incorporation of XYZ information and replacement of CD information (*bottom*).

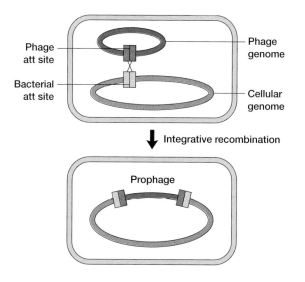

FIGURE 1.6. Site-specific recombination, taking phage integration as an example. A phage genome is injected into a recipient bacterial cell and becomes circularized (*top*). The phage and bacterial chromosomes contain specific DNA sequences called attachment sites (att) where the DNA cleavage and joining reactions take place. Completion of site-specific recombination yields integrated phage (*bottom*). Each host–virus DNA junction is composed of a hybrid attachment site derived from both phage and bacterial DNA.

its role as the central informational molecule in cells. Other damaged macromolecules are simply degraded and resynthesized, with no attempt at repair.

DNA becomes damaged at a remarkable rate. It has been estimated that a typical cell needs to repair some 10,000 sites of DNA damage per day, an incredible number. Not surprisingly, cells devote considerable machinery to the job. Even the minimal *Mycobacteria genitalium* genome (Plate 1) devotes a handful of its 470 genes to DNA repair. At the other extreme, the bacterium *Deinococcus radiodurans* devotes fully 30–40 genes to DNA repair, explaining at least in part this bacterium's extreme resistance to radiation.

Lateral transfer systems regularly hijack these repair enzymes for their own purposes. Often the mobile DNA elements involved carry out only the early steps of replicating their DNA, acting only up to the point where the host cannot easily survive without finishing the job. Thus, many mobile elements stick host DNA repair systems with the job of completing the element replication cycle. Examples come up in most of the chapters of the book.

MOBILE DNA AND EVOLUTION (9)

Lateral DNA transfer is a key contributor to evolutionary change. Much of the speciation of bacteria, for example, has probably been mediated by lateral DNA transfer events (discussed in Chapter 5). There are certainly mechanisms for genetic change in addition to lateral transfer, such as accumulation of base pair substitutions into DNA, but for many traits, exchange of blocks of DNA is probably the most important source of novelty, particularly in prokaryotes. Many laboratory experiments have documented the ability of DNA transfer to change the phenotype of cells. Diverse changes in organisms are also known to arise by lateral transfer in natural settings. The newly completed genome sequences provide a rich record of probable lateral transfer events. In this section, we briefly introduce several of the broad issues in evolutionary biology raised by studies of lateral DNA transfer, setting the stage for specific examples in later chapters.

The terminology describing the relationships between a host organism and an invader can also be applied to newly acquired DNA. Organisms (or sequences) that benefit the host are "mutualists," those that harm the host are "parasites," whereas those that are harmless to the host (but benefit themselves) are "commensals." Mutualists have also been called "symbionts," but recently "symbiosis" has come to refer to any close physical relationship between two organisms, either helpful or harmful, so the more specific terms are favored where possible. Newly acquired DNA sequences can be mutualists, parasites, or commensals.

Selfish DNA (3–5, 7, 8)

Many mobile DNA elements do not contribute to the long-term welfare of their host cell, existing instead as "selfish" DNA. Selfish DNA has been defined as sequences that make no positive contribution to the phenotype of the host but can form additional copies of themselves in the genome, thereby assuring their persistence. Such sequences are commensals or parasites because they themselves benefit but do not contribute positively and sometimes oppose the welfare of the host. Today it is clear that many of the abundant repeated DNA sequences in many organisms are composed of copies of mobile DNA elements. In humans, mobile DNA elements or their relics constitute over 40% of the genome (Chapter 10).

Most of the genetic systems involved in DNA transfer between cells qualify as selfish, although many are more autonomous than the repeated genomic sequences that inspired the original definition. For example, integrating viruses and conjugative plasmids qualify in a broad sense as selfish DNA elements. For these, the genes for DNA transfer are themselves encoded on the transferred elements, allowing plasmids or viruses to exit the host cell if conditions are bad and seek better chances elsewhere. Many such elements are not just commensal, but actively parasitic or even lethal. Plasmids may consume synthetic precursors and chemical energy. Viruses often exhaust and break open the host cell in the course of producing progeny. Such elements are plainly selfish—they exist for their own good and often at the expense of the host.

Pathogenesis and Mutualism

Long evolution of the host and parasite together often results in a more benign relationship between the two. It is well known from studies of infectious diseases that microbes newly introduced into humans from other species (zoonosis, in the medical jargon) are often particularly pathogenic. Microbes that have experienced long coevolution with humans, in contrast, are often commensal or nearly so, and can be hard to detect at all. Pathogenesis is seen with some mobile DNA elements, such as the P elements of fruit flies, which appear to have been recently introduced into flies. Related elements that have coexisted for prolonged periods with their hosts, in contrast, are often down-regulated in activity so as to minimize their burden. P elements and related mobile sequences are discussed in Chapter 9.

Although many mobile DNA elements are evolved to be selfish, they can be beneficial mutualists as well. For example, as described in Chapter 3, antibiotic-resistance genes in mobile elements confer a competitive advantage on both the host and the mobile elements transporting them. Mobile elements occasionally provide another more direct benefit. The draft human genome sequences reveal many examples in which genes for mobile element machinery have themselves been coopted by cells for their own uses. For example, as is described in Chapter 11, transposon sequences are thought to be evolutionary precursors of genes encoding key enzymes of the human immune system.

Borrowing from the Collection of World Genes

Many mobile DNA elements are capable of capturing cellular sequences and moving them between hosts. This means that over evolutionary time, the collection of genes present in one species is potentially available to many other species. Of course, in practice, there are limitations to the likelihood of transfer, and the process is generally more prominent in prokaryotes than in eukaryotes. Nevertheless, as discussed in Chapter 12, there are many documented cases of DNA transfer between Eubacteria, Archaea, and Eukarya, the three domains of life. Gene transfer by mobile DNA systems greatly broadens the sequence repertoire available to organisms over evolutionary time.

These evolutionary themes recur in the following chapters in diverse forms. The consequences of lateral transfer will be evident in many examples, from bacteria to humans. As we shall see, mobile DNA systems are central to many aspects of evolutionary change.

REFERENCES

1. Berg D.E. and Howe M.M., eds. 1989. *Mobile DNA*. American Society for Microbiology, Washington, D.C.
2. Cox M.M., Goodman M.F., Kreuzer K.N., Sherratt D.J., Sandler S.J., and Marians K.J. 2000. The importance of repairing stalled replication forks. *Nature* **404:** 37–41.
3. Crick F.H.C. 1979. Split genes and RNA splicing. *Science* **204:** 264–271.
4. Dawkins R. 1976. *The selfish gene.* Oxford University Press, Oxford, United Kingdom.
5. Doolittle W.F. and Sapienza C. 1980. Selfish genes, the phenotype paradigm and genome evolution. *Nature* **284:** 601–603.
6. Griffiths A.J.F., Miller J.H., Suzuki D.T., Lewontin R.C., and Gelbart W.M. 1993. *An introduction to genetic analysis.* W.H. Freeman, New York.
7. Orgel L.E. and Crick F.H.C. 1980. Selfish DNA: The ultimate parasite. *Nature* **284:** 604–607.
8. Orgel L.E., Crick F.H.C., and Sapienza C. 1980. Selfish DNA. *Nature* **288:** 645–646.
9. Syvanen M. 1994. Horizontal gene transfer: Evidence and possible consequences. *Annu. Rev. Genet.* **28:** 237–261.

CHAPTER 2

DNA and Lateral Transfer

SYSTEMS FOR LATERAL DNA TRANSFER CAN TAKE MANY FORMS and some-times generate complex DNA products, but much of the story involves two simple themes. The first is the structure of the DNA molecule. The nature of DNA constrains the types of reactions that can break and rejoin molecules in new arrangements. The chemistry of the DNA sequence specifies the sites at which other molecules bind to the DNA polymer. The physical nature of DNA, such as its resistance to bending and twisting, dictates the conformations of larger protein–DNA complexes that typically mediate DNA transfer.

The second theme is the nature of the proteins involved in the lateral transfer of DNA. Again and again in the field, previously disparate observations have come to be explained by related mechanisms. Many recent studies emphasize that much of the transfer machinery falls into a few broad classes.

This chapter introduces the structure of DNA, then presents a brief review of molecular biology basics. The chapter concludes with a brief summary of the main points. Readers with a background in molecular biology may wish to skip the review section. Chapter 3 begins the survey of specific systems with the transfer of antibiotic-resistance genes by bacterial conjugation.

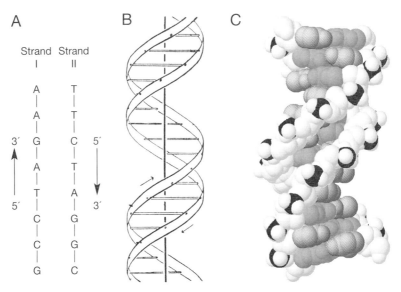

A
Strand   Strand
  I   II

   A ——— T
   A ——— T
3′ G ——— C 5′
   A ——— T
   T ——— A
5′ C ——— G 3′
   C ——— G
   G ——— C

FIGURE 2.1. Three views of the DNA helix. (*A*) Schematic diagram of the DNA helix, illustrating pairing of A with T and G with C, and also showing the antiparallel directions of the two strands. (*B*) The DNA double helix. The vertical line indicates the long axis of the DNA helix. The "ribbons" indicate the phosphate backbone of each DNA chain, the vertical lines the DNA base pairs. (Reprinted, with permission, from Watson and Crick 1953 [©Macmillan Magazines Ltd.].[12]) (*C*) A "space-filling" model of the DNA helix showing all the component atoms. Phosphorus atoms are shown in black, purine bases in gray, and pyrimidines in green. (Image rendered using coordinates published by Larsen et al. 1991. [8])

THE STRUCTURE OF DNA (12)

Deoxyribonucleic acid, or DNA, is the chemical repository of genetic information. Three views of the DNA helix are shown in Figure 2.1. The leftmost diagram (Fig. 2.1A) shows a schematic picture of DNA, with the sugar-phosphate backbones of the two strands shown as lines connecting the DNA bases adenine (A), thymine (T), guanine (G), and cytosine (C). The two strands are linked by the hydrogen bonds between the paired DNA bases—A pairing with T and G pairing with C (Fig. 2.2). Because each strand pairs with a partner containing matched, or "complementary," bases, the unique sequence of the bases can be duplicated (see section on DNA replication below) and passed on to daughter cells. The DNA backbone is composed of sugar units linked in an asymmetric fashion by phosphate groups, giving each of the two strands a direction. In the double helix, the backbones of the two strands run in opposite or "antiparallel" directions. (Fig. 2.1A).

The deoxynucleotide building blocks that comprise DNA are shown in Figure 2.3. Each DNA base is attached to a sugar molecule, the deoxyribose. The deoxyribose sugar together with a base is called a nucleoside. The nucleoside sugar can be linked to a phosphate molecule, to yield a deoxynucleoside 5′ phosphate (the numerical desig-

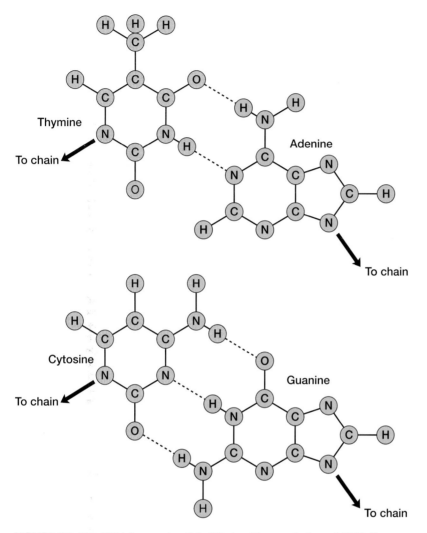

FIGURE 2.2. The DNA base pairs. (Modified, with permission, of W.H. Freeman from Lehninger 1979 [©Worth Publishers].[9])

nations indicate the atoms on the sugar ring). A deoxynucleoside linked to phosphate is also called a nucleotide. The DNA chain is linked up by joining the 5′ phosphate to a 3′ position of the deoxyribose sugar in the next nucleotide. In the double helix, going from bottom to top, one strand runs in the 5′ to 3′ direction and the other runs in the 3′ to 5′ direction, yielding the antiparallel pairing.

Figure 2.1B shows another schematic diagram of the DNA chain, this time twisted into the familiar DNA double helix. This diagram was first presented in the famous 1953 paper by Francis Crick and James Watson describing the double-helical structure of DNA. The constraints of forming the base pairs and accommodating the chemical bonds holding the bases together require winding the DNA "ladder" into a right-handed double-helical spiral. Figure 2.1C shows all the atoms of the double helix in a precise "space-filling" model.

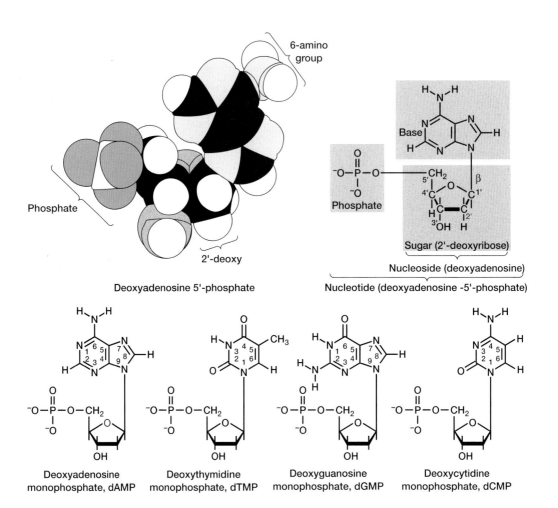

FIGURE 2.3. Detailed picture of the bases in DNA. (Redrawn, with permission, from Kornberg and Baker 1992 [© W.H. Freeman]. [7])

A BRIEF OVERVIEW OF MOLECULAR BIOLOGY

The Central Dogma (1, 2, 4, 7)

The process by which information contained in the sequence of the DNA base pairs is used to construct proteins is described by "The Central Dogma" of molecular biology, so named by Francis Crick (Fig. 2.4). Actually, Crick later admitted that he gave the theory its name at a time when he was somewhat hazy on what the word "dogma" meant! The name was intended to convey "central idea" but without the overtones of "belief that cannot be doubted" inherent in the word "dogma."

In any event, the Central Dogma states that information in nucleic acid sequence guides the formation of a sequence in proteins.

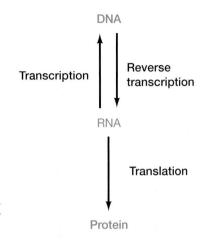

DNA

Transcription | Reverse transcription

RNA

Translation

Protein

FIGURE 2.4. Diagram of the Central Dogma and its reversal by reverse transcription.

Typically, the DNA sequence is copied into the related polymer ribonucleic acid, or RNA; information in the RNA sequence specifies the sequence of a protein. RNA is closely related in structure to DNA, the difference being in the presence of a slightly different sugar in the backbone, and the substitution of the base uracil (U) for thymine (T). RNA molecules are copied or "transcribed" from DNA using the same rules for base-pairing that hold together the two strands of the double helix. Thus, the sequence of an RNA molecule preserves the information originally present in the DNA sequence that served as the template. RNA molecules that code for proteins are called messenger RNAs or mRNAs because they carry the information for synthesizing proteins from the DNA to ribosomes, the machinery that synthesizes proteins. Reading of the RNA information to make protein is called "translation."

Proteins are composed of amino acid building blocks. There are 20 amino acids typically found in proteins, and they are arranged in different linear sequences to form specific protein chains. Proteins range in size from only a few amino acids to thousands. RNA sequence is mapped onto amino acid sequence by means of the genetic code. The RNA sequence is read in sets of three bases, with most base triplets encoding an amino acid; for example, AAA encodes lysine, GGC encodes glycine. Several specific triplets act as translation start or stop signals. The triplet AUG encodes methionine, but also acts as a translation start signal. The triplets UGA, UAG, and UAA all signal translation stop. The genetic code is almost universal, although as is usually the case in biology, there are a few exceptions.

Proteins, once synthesized, have the remarkable property of spontaneously folding so that the newly synthesized linear peptide chain adopts a defined three-dimensional structure, the functional protein. One key experiment, by Christian Anfinsen, showed that a protein capable of degrading RNA molecules could be unfolded, but that the protein could spontaneously refold and so regain the ability to degrade RNA. Later studies have shown that many proteins have this ability to fold spontaneously, although cells also synthesize machinery to speed and facilitate the process. Thus, the information contained in the DNA sequence can be transcribed into RNA and translated to make a protein; the protein will then fold and assume its role in cells.

Transcription and translation are each carried out by a dedicated set of enzymes. Transcription is carried out by the enzyme RNA polymerase, which synthesizes RNA chains by copying a DNA template. Translation is carried out by the ribosome, a complex machine with both RNA and protein components. Ribosomes direct synthesis of protein chains guided by the mRNAs. Many other proteins contribute to the regulation and execution of transcription and translation.

The Central Dogma, like most biological laws, has exceptions. Several classes of mobile elements violate its proscriptions by reversing the transcription step, so that RNA of the mobile element is "reverse transcribed" to make a DNA copy. For example, the retroviruses, the topics of Chapters 6 and 7, carry out reverse transcription as a required step in the retroviral replication cycle. (To be fair, Crick did allow for exceptions in the original article on the Central Dogma.)

DNA Replication (7)

The structure of DNA suggested to Crick and Watson a simple mechanism for its replication. If the two DNA strands are pulled apart, each contains the information necessary for copying the other. An A on one unpaired strand, for example, can bind to a single T nucleotide, a G to a C, and so forth. Connecting the individual nucleotides yields a new and complementary DNA strand. Complete copying of the two DNA strands produces two new DNA duplexes (Fig. 2.5), each identical to the parent. This replication reaction is carried out by a cell every time it divides, thereby ensuring that each daughter cell receives the full complement of DNA sequences.

DNA replication is carried out in diverse ways in different organisms, but a related set of mechanisms is used in most cases (Fig. 2.6). Most cells initiate DNA replication at defined locations on the chromosome, named origins of replication. *Escherichia coli*, the widely studied intestinal bacterium, has a single origin of replication on its one chromosome. Initiation differs in eukaryotes, which have many origins of replication in their much larger chromosomes. Origins serve to assemble the replication enzymes and regulatory proteins that carry out the copying of each DNA strand.

The central enzyme of DNA replication is DNA polymerase, the protein that links activated DNA nucleotides together to form new chains. Before DNA polymerase can be recruited to a replication origin, regulatory proteins bind to the origin sequences and unpair the DNA strands. This allows the polymerase and other factors at the "replication bubble" to have access to the single-stranded DNA. The primer required to initiate synthesis is provided by short RNA chains synthesized by the enzyme primase. Polymerase, primase, and other proteins assemble at the replication fork, the point of new DNA synthesis. DNA polymerase adds nucleotides sequentially to the growing DNA chain, moving along the DNA template with each cycle of addition. Each nucleotide chosen to extend the chain is paired with the complementary base in the single-stranded DNA template, using the rules of Watson-Crick base-pairing.

Replication forks proceed along the chromosome in a bidirectional fashion away from the origin of replication. Synthesis terminates when two replication forks proceeding in opposite directions meet, thereby completing synthesis of the DNA segment between the origins.

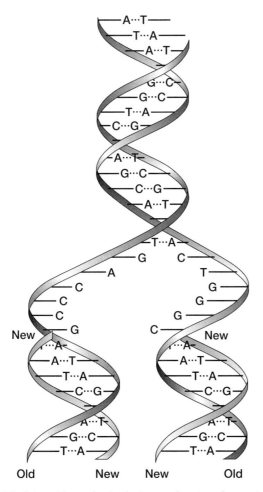

FIGURE 2.5. Nucleic acid synthesis during replication. Separation of the DNA strands allows each to serve as template for synthesis of a new DNA chain. Note how base-pairing directs incorporation of the correct nucleotide in the new chain. (Redrawn from Watson 1965 by permission of Addison Wesley Longman Publishers, Inc. [11])

Replication origin

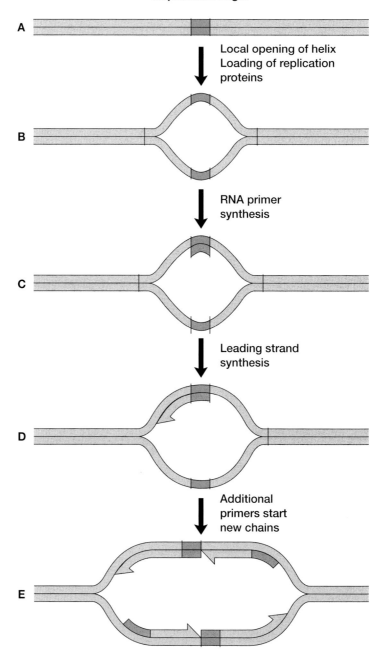

FIGURE 2.6. Pathway of DNA replication initiated at a cellular origin. To initiate DNA synthesis, a segment of DNA containing a replication origin (*A*) becomes unpaired within the origin sequence (*B*), allowing loading of replication proteins that initiate DNA synthesis (*C*). As each replication fork migrates away from the origin (*D*), new strands are initiated (*E*, lagging strand synthesis). Short RNA strands are synthesized to prime replication (shown dark gray). (Modified from Alberts et al. 1994. by permission of Routledge, Inc., part of the Taylor Francis Group. [1])

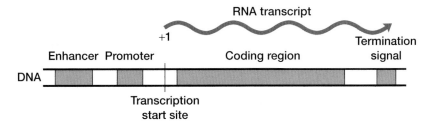

FIGURE 2.7. An idealized gene.

The Nature of Genes (1, 5, 10)

Much of the cellular DNA is organized into functional units called genes. Genes are composed of DNA sequences, a portion of which is copied to make RNA molecules. Many genes encode mRNAs, which are subsequently translated to make proteins. Other genes encode RNAs that carry out specific functions, such as the RNA components of the ribosome.

Genes vary greatly in size. Some are encoded by only a few hundred base pairs, such as the genes for some small RNAs involved in translation. Others are enormous. For example, on human chromosome 22, the smallest protein coding gene is about 1000 bases (a kilobase, usually abbreviated kb), whereas the largest is 583 kb.

All genes contain signals for starting and stopping transcription, the process by which DNA information is converted into RNA. The mechanisms for doing so are diverse and differ between prokaryotes and eukaryotes. Some common features are summarized with a simplified gene, shown in Figure 2.7. Upstream of the gene is a binding site for RNA polymerase, the enzyme responsible for RNA synthesis. The frequency of initiation is controlled by the affinity of polymerase for local DNA sequences and regulatory DNA-binding proteins, some of which bind to specific sequences near the polymerase-binding site. This loosely defined set of sequences is called the gene promoter. Other sequences, often present at more distant locations on the DNA, can also influence the rate of transcription initiation. These sequences are called enhancers or silencers, depending on whether they act positively or negatively. Transcription termination differs among various kinds of genes and organisms. In many cases, there are also sequences that signal the termination of transcription or cleave the RNA chain at its end, in either case yielding a defined end to the transcribed RNA.

Many genes are not colinear in the chromosome but are broken up into pieces. The spacer sequences are called introns, and the coding segments exons. Both introns and exons are present in the initial RNA transcript, but the introns are removed by a process called RNA splicing, yielding continuous coding regions that serve as templates for translation. Each human gene contains on average 9 introns. Introns are rare in bacteria but not unknown. The origin of introns is obscure, but it has been suggested that there may be a connection with mobile DNA elements (Chapters 8 and 14). Some RNAs undergo additional types of chemical modifications before assuming their roles in cells.

Organization of Genes into Chromosomes (1, 3, 5, 6)

Most of the cellular genes are organized into larger DNA structures called chromosomes. In some bacteria, such as *E. coli*, the genome consists of a single chromosome linked to form a large DNA circle. The number of chromosomes in each cell is species-specific. For example, in humans there are 23 linear chromosomes; dogs have 39, corn has 10, and the fruit fly has 4. Most human cells contain two copies of each chromosome for a total of 46; this is called the diploid set of chromosomes, which has two copies of each gene or genetic locus. Each germ cell contains a haploid set or single copy of each chromosome. The exception is the sex chromosomes X and Y. In women, 2 of the 46 chromosomes are X chromosomes, and no Y chromosomes are present. Men possess one X and one Y chromosome. The human body is composed of roughly 50 trillion cells, and so contains 50 trillion sets of chromosomes.

Almost all cells in an organism contain exactly the same complement of DNA sequences. In multicellular organisms, different genes are activated in different cell types to produce the characteristic proteins needed by each. One of the rare cases where DNA sequences are rearranged during development, in the construction of the vertebrate immune system, is discussed in Chapter 11. DNA sequences are commonly modified by addition of methyl groups, which aid in gene regulation and DNA repair (Chapter 13), but methylation does not permanently alter the informational content of DNA.

The number of genes differs greatly among organisms. In *Mycoplasma genitalium*, the bacteria with the smallest genome yet sequenced, there are only about 470 genes in a chromosome of 580,070 base pairs (Plate 1). These numbers of genes are only estimates, because it is not always easy to recognize genes with certainty in the primary DNA sequence. *E. coli* contains an estimated 4,288 protein-coding genes in a chromosome of 4,639,221 base pairs. Humans, in contrast, possess around 37,000 genes in 3.4 billion base pairs of DNA. Despite the completion of a draft of the human genome sequence, this gene number is still a matter of debate due to the difficulty of identifying genes.

Inside cells, each of the chromosomes is bound by cellular proteins that organize and condense the DNA. If the DNA from one human chromosome were stretched out, it could extend 8.5 centimeters! Chromosomes of all organisms are wound with proteins to reduce this length. The chromosomes are contained within the cell nuclei in eukaryotes and exist in looser nucleoid bodies in prokaryotes.

Although most of the cellular DNA is contained within the chromosomes, there are also DNA molecules that replicate independently. The mitochondria, the energy-producing "factory" within eukaryotic cells, are found outside the nucleus in the cytoplasm. The mitochondria contain a small chromosome that encodes a few of the mitochondrial proteins. Plant cells contain mitochondria and chloroplasts, light-harvesting organelles that also contain their own chromosomes. These organelles were almost certainly derived from mutualistic bacteria acquired during eukaryotic evolution. Transfer of genes from organelle chromosomes to the nuclear genome represents an important mechanism of lateral transfer in higher cells (Chapter 12).

Other forms of DNA that replicate independently of the genome are called plasmids. Plasmids range in size from less than 1 kilobase to more than 300 kilobases. Plasmids are usually dispensable for the replication of the host genome; however, genes carried on plasmids can influence the welfare of the host in diverse ways. In bacteria, most plasmids are rings of DNA that contain an origin of replication and often various genes, but they can also be linear DNA molecules. Plasmids have also been found in a few cases in eukaryotic cells. Plasmids play an important role in several forms of lateral DNA transfer.

Summary

The central player in our story is deoxyribonucleic acid, or DNA. DNA is composed of the A, T, G, and C nucleotides linked together by the sugar-phosphate backbone. Two chains pair with each other by interactions between the nucleotide bases. A pairs with T, and G with C, thereby linking up the two strands. The annealed DNA strands wind up into the familiar twisted ladder shape of the DNA double helix. Biological information is encoded in the sequence of the nucleotide pairs, the rungs of the ladder in the double helix.

The Central Dogma holds that information in nucleic acids sequence programs protein synthesis. The genetic code relates the sequence of DNA triplets to the amino acid sequence of proteins. The double-helical structure of DNA also allows each of the two DNA strands to act as a template for new DNA synthesis, providing a mechanism for DNA replication. Replication initiates at specialized sites, called origins, which load proteins responsible for DNA synthesis.

Genes are DNA segments that, together with their regulatory signals, direct the synthesis of mRNAs or other stable RNAs (ribosomal RNAs, tRNAs, etc.). Genes are arranged in larger structures called chromosomes. The full complement of chromosomes of an organism is called the genome.

REFERENCES

1. Alberts B., Bray D., Lewis J., Raff M., Roberts K., and Watson J.D. 1994. *Molecular biology of the cell.* Garland Publishing, New York.
2. Anfinsen C.B. 1973. Principles that govern the folding of protein chains. *Science* **181:** 223–230.
3. Blattner F.R., Plunkett G.R., Bloch C.A., Perna N.T., Burland V., Riley M., Collado-Vides J., Glasner J.D., Rode C.K., Mayhew G.F., Gregor J., Davis N.W., Kirkpatrick H.A., Goeden M.A., Rose D.J., Mau B., and Shao Y. 1997.

The complete genome sequence of *Escherichia coli* K-12. *Science* **277:** 1453–1474.

4. Crick F.H.C. 1988. *What mad pursuit: A personal view of scientific discovery.* Basic Books, New York.

5. Dunham I., Shimizu N., Roe B.A., Chissoe S., Hunt A.R., Collins J.E., Bruskiewich R., Beare D.M., Clamp M., Smink L.J., et al. 1999. The DNA sequence of human chromosome 22. *Nature* **402:** 489–495.

6. Fraser C.M., Gocayne J.D., White O., Adams M.D., Clayton R.A., Fleischmann R.D., Bult C.J., Kerlavage A.R., Sutton G., Kelley J.M., et al. 1995. The minimal gene complement of *Mycoplasma genitalium*. *Science* **270:** 397–403.

7. Kornberg A. and Baker T. 1992. *DNA replication*, 2nd edition. W.H. Freeman, New York.

8. Larsen T.A., Kopka M.L., and Dickerson R.E. 1991. Crystal structure analysis of the B-DNA dodecamer CGTGAATTCACG. *Biochemistry* **30:** 4443–4449.

9. Lehninger A. 1979. *Biochemistry*. Worth Publishers, New York, New York.

10. Ptashne M. 1992. *A genetic switch*, 2nd edition: *Phage λ and higher organisms*. Cell Press and Blackwell Scientific, Cambridge, Massachusetts.

11. Watson J.D. 1965. *Molecular biology of the gene*. W.A. Benjamin, New York.

12. Watson J.D. and Crick F.H.C. 1953. Molecular structure of the nucleic acids. A structure for deoxyribose nucleic acid. *Nature* **171:** 737–738.

CHAPTER 3

Conjugation, Transposition, and Antibiotic Resistance

WE HAVE COME TO EXPECT that diseases caused by bacteria can be cured with antibiotics, but the rapid spread of antibiotic resistance threatens this comfortable view. It is clear that genes conferring resistance to antibiotics are widespread in nature and that these genes are frequently swapped among pathogens. The wide use of antibiotics provides a powerful selection pressure favoring the growth of resistant bacteria, causing the fraction of bacteria containing resistance genes to increase over time. Largely as a consequence of this, resistance to a new antibiotic typically arises 2–5 years after its introduction. Resistant pathogens are often most abundant in hospitals, where selection by antibiotic use is strongest, but people are most vulnerable.

In this chapter, we first introduce mechanisms of antibiotic action and antibiotic resistance, then examine the mobile DNA systems that transfer resistance genes between cells. The machinery for lateral transfer is considered in order of increasing complexity, from the simplest mobile DNA elements up to the complex composites conferring resistance to many antibiotics at once that are often found in clinical settings.

ANTIBIOTIC-RESISTANT INFECTIONS (26, 37, 46, 55, 61, 66)

As an example of the danger of antibiotic-resistant microbes, consider *Staphylococcus aureus,* which derives its name from the Greek *staphyle,* meaning bunch of grapes, referring to the clustering of the cocci (round bacteria) during growth (Fig. 3.1). *S. aureus* looks even more like grapes when stained with Gram's stain, which colors it purple, thus defining *S. aureus* as a "gram-positive" bacteria. The *"aureus"* derives from the fact that colonies grown on solid media in the laboratory often have a golden hue. *S. aureus* is found on the skin and mucous membranes of healthy humans and animals, and is not usually dangerous, comprising one of the 500–600 species of bacteria that inhabit the body of a healthy person. However, a variety of conditions can cause *S. aureus* to become a life-threatening pathogen. One means for conversion involves the capture of genes for pathogenicity from infecting bacteriophage, an aspect of lateral DNA transfer covered in Chapter 4.

FIGURE 3.1. Scanning electron micrograph of a cluster of *Staphylococcus aureus* cells. (Courtesy Robert Apkarian, Emory University, Atlanta, Georgia.)

Pathogenic *S. aureus* can cause a wide spectrum of diseases. Introduction of *S. aureus* into wounds, particularly associated with foreign objects such as stitches or splinters, can lead to dangerous infections. Some strains of *S. aureus* also secrete toxins, which can kill cells directly or provoke hyperactivation of the immune system. In staphylococcal scalded skin syndrome, disease begins as inflammation and redness around the mouth. Gentle pressure on the affected areas can displace the upper layer of skin, resulting in appearance of blisters. Such blisters can eventually cover large surface areas of skin. The syndrome is not generally fatal, but death can occur due to secondary bacterial infection of the damaged skin. More lethal is toxic shock syndrome, which results from localized growth of *S. aureus* and production of a secreted toxin. The initially high fatality rate seen with this syndrome has been reduced, at least for now, with modern antibiotic treatment. Harder to treat is endocarditis resulting from infection of heart tissue. Initial symptoms include fever, chills, and chest pain. Subsequently, the infection can cause embolism of the heart and other complications, and mortality can reach 50%.

S. aureus was initially treated with penicillin, but today less than 10% of strains are still sensitive. DNA elements conferring resistance have spread through the population worldwide. More recently, *S. aureus* has been treated with second-generation penicillins such as methicillin, drugs that are not degraded by the bacterial enzymes that initially conferred penicillin resistance. However, 30–50% of strains are now resistant to methicillin as well. Such strains are called MRSAs, for methicillin-resistant *Staphylococcus aureus*. The name is not very accurate, however, because strains resistant to methicillin are usually also resistant to many other antibiotics as well. Vancomycin is the only antibiotic left for use against the MRSA strains, but resistance to vancomycin is also spreading under the pressure of increased use. In 1999, the first cases of vancomycin-resistant MRSA strains were reported. For the first time since the introduction of antibiotics, some *S. aureus* infections may be untreatable. For other pathogens as well, such as some strains of *Mycobacterium tuberculosis*, we are fast reaching a state where no antibiotic treatments are effective.

THE ANTIBIOTIC ERA (1, 19, 24, 30, 69)

In the 19th century, prior to the introduction of antibiotics and antiseptic treatments, more than half of all surgical patients developed infections. Expression of large quantities of pus from wounds was considered part of the normal healing process. Crucial studies by Joseph Lister, Robert Koch, and Louis Pasteur established that many diseases are caused by infecting microbes, and as a result of these studies, antiseptic treatments used during surgery greatly reduced the rate of infection. Later, chemical treatments were developed that specifically poisoned infecting microbes. Such "magic bullets" first came into use through the work of Paul Ehrlich and coworkers, who developed arsenical compounds for the treatment of syphilis. Their successes set the stage for widespread efforts to identify chemical treatments for infectious diseases.

A key breakthrough came with the development of penicillin by Alexander Fleming, Ernst Boris Chain, and Howard Florey. Upon returning from a vacation, Alexander Fleming noticed that a mold accidentally contaminating a bacterial plate in his lab was surrounded by a region of reduced bacterial growth. Fleming had the crucial insight that the mold, *Penicillium notatum*, might have secreted an antibacterial substance. Subsequent study revealed that media from *Penicillium* cultures contained the antibacterial activity. Implementing this discovery for medical use required isolation of the agent in pure form and large-scale manufacturing. These goals were achieved by Chain and Florey, involving, among other steps, large-scale mold cultures in stacks of hospital bedpans! These seminal developments allowed the introduction of penicillin into clinical practice.

These findings, together with the discoveries of sulfa drugs by Gerhard Domagk and streptomycin by Selman Waksman, ushered in the era of antibiotics. Importantly, it was recognized that certain types of bacteria, particularly actinomycetes and streptomycetes, often produced compounds with antibiotic properties. This led to systematic efforts to isolate bacteria from the environment and test the small molecules they produce for antibiotic activity. It has emerged that microbes in natural settings probably use antibiotics at least in part to poison potential competitors, explaining their wide natural occurrence.

The first antibiotic-resistant mutants of pneumococcus were discoved in 1951 by Roland Hotchkiss. He found that penicillin resistance and streptomycin resistance could be induced in previously sensitive strains and subsequently transferred between strains, helping establish their character as genetic traits. Around this time, R factors, transmissible plasmids sometimes encoding several antibiotic-resistant determinants, were identified in outbreaks of antibiotic-resistant *Shigella* in Japan. We discuss R factors later in this chapter.

An important corollary of natural antibiotic production took longer to appreciate. If microbes are producing antibiotics normally, they must have means for avoiding poisoning themselves with their own products. Consequently, genes for antibiotic resistance must also be present in bacterial communities, and must have coevolved with systems for antibiotic synthesis.

An alternative explanation is that medical use alone has driven the development of antibiotic-resistance genes. However, several studies support the idea that resistance genes were common before the antibiotic era. In 1970, a study of a remote region of Borneo revealed that bacteria harboring genes for resistance could be found in the guts of native inhabitants. Since it is thought to be unlikely that antibiotic drugs had been used extensively in this region, it was inferred that the evolution of resistance genes predated medical antibiotic use. A similar study of isolated populations in the Solomon Islands reached the same conclusion. It seems likely that resistance genes from this preexisting pool have been transferred to human pathogens under the pressure of contemporary medical use.

MECHANISMS OF ANTIBIOTIC ACTION (1, 26, 46, 55, 68)

To be useful, agents for treating infectious diseases must succeed in poisoning the infecting microbe without poisoning the patient undergoing treatment. Antibiotic drugs were discovered by screening metabolites from cultures of diverse microbes for their ability to poison pathogens. Of these, a subset was later found to be sufficiently nontoxic for use in humans as drugs. Analysis of the targets of these antibiotic agents revealed that they typically act against cellular systems unique to bacteria, thus explaining their lack of toxicity to humans.

Penicillin acts by disrupting the synthesis of the bacterial cell wall (Fig. 3.2). The outer support system of a cell differs between bacterial and human cells. Most bacteria maintain higher concentrations of salt inside their cells than exists in the external medium. This presents a challenge to the cell. Because the membrane enclosing the cell has pores, water tends to rush in to equalize the salt concentration inside and out. This would cause bacterial cells to swell up and burst if not for the mechanical strength provided by the bacterial cell wall. The cell wall is formed of an elaborate cross-linked material called proteoglycan. Sugar polymers extend along the surface of the cell, and these chains are cross-linked together by short protein chains. The construction of the cell wall differs between two of the largest groups of bacteria, the gram-positive and gram-negative bacteria. In fact, it is the cell wall itself that binds the Gram's stain and allows the characteristic purple color to appear in stained gram-positive cells.

Penicillin is a β-lactam antibiotic. This group of antibiotics binds to enzymes called penicillin-binding proteins (PBPs), which are important for cell wall biosynthesis (Fig. 3.3). Binding of β-lactam antibiotics inhibits PBPs, disrupting cross-linking of the cell wall and so causing cells to burst due to osmotic pressure.

Vancomycin also disrupts bacterial cell wall biosynthesis, but by a different mechanism. The sugar chains comprising the cell wall of many gram-positive bacteria are cross-linked by an unusual chemical unit, a dimer of D-alanine molecules. Vancomycin binds the termini of the D-alanine–D-alanine unit and prevents formation of the cross-link. The cell wall lacking these cross-links is weakened, leading to rupture

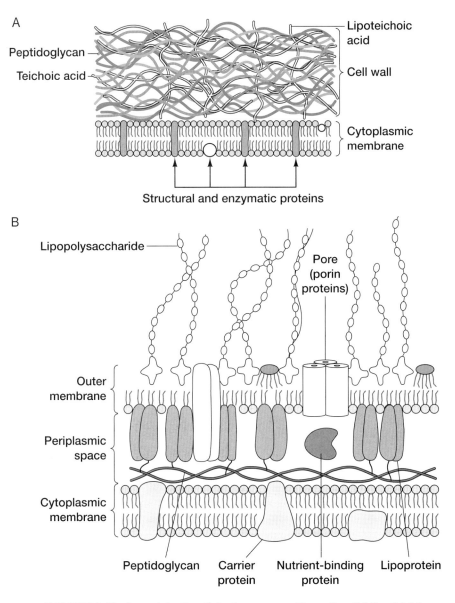

A

Peptidoglycan

Teichoic acid

Lipoteichoic acid

Cell wall

Cytoplasmic membrane

Structural and enzymatic proteins

B

Lipopolysaccharide

Pore (porin proteins)

Outer membrane

Periplasmic space

Cytoplasmic membrane

Peptidoglycan   Carrier protein   Nutrient-binding protein   Lipoprotein

FIGURE 3.2. The bacterial cell wall. In the gram-positive cell wall (*A*), a thick layer of proteoglycan lies above a single lipid bilayer. In the gram-negative cell wall (*B*), two lipid bilayers enclose the periplasmic space. Peptidoglycan strands are present in the periplasmic space. The outer surface is coated by lipopolysaccharide chains. (Redrawn, with permission, from Murray et al. 1998.[46])

of the cell membrane. Many other antibiotics act to disrupt the integrity of the bacterial cell wall, such as cephalosporins, bacitracin, and cycloserine.

The protein synthetic machinery also differs between humans and bacteria (Fig. 3.4). The sizes and sequences of the ribosomal RNAs differ, and many of the steps of translation differ in detail. Bacterial trans-

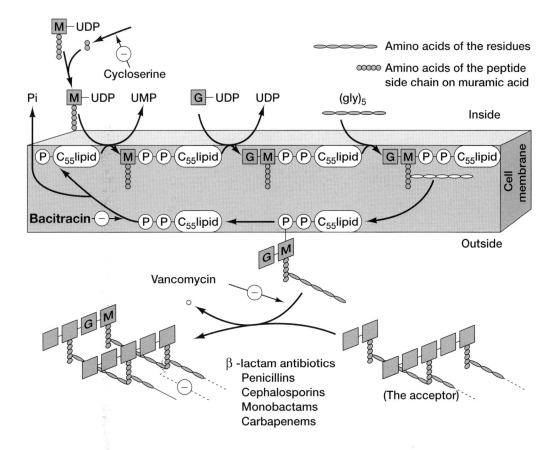

FIGURE 3.3. Cell wall synthesis and the action of antibiotics. The interior of the bacterium is to the top, the outside beneath. A unit consisting of two sugars (M = *N*-acetylmuramic acid; G = *N*-acetylglucosamine) linked to five amino acid units is synthesized attached to the C55 carrier by a diphosphate group. Synthesis takes place inside the bacterium; the C55 lipid flips across the membrane to deliver its cargo to the outer surface. There the sugars are linked to sugar units in the peptidoglycan and the peptides are linked to three peptide units by penta-glycine units in a transpeptidation reaction. The C55 carrier is then recycled back into the bacterium after loss of the diphosphate group. The steps blocked by various antibiotics are as indicated. (Redrawn, with permission, from Rang et al. 1995 [© Churchill Livingstone].[51])

lation can thus be selectively targeted by antibiotics. Tetracycline, for example, binds to the ribosomal site that normally binds the tRNAs, the unit that brings activated amino acid precursors to the ribosomal site of protein synthesis. Tetracycline, as its name implies, is composed of four chemically linked rings. The compound fits into the site normally occupied by tRNA and so prevents further growth of the protein chain. Many other familiar antibiotics act against bacterial protein synthesis, such as aminoglycosides (kanamycin, streptomcyin, neomycin), macrolides (erythromycin, clarithromycin, spiramycin), and puromycin.

Other antibiotics disrupt additional systems unique to bacteria. For example, the quinolone antibiotics disrupt the function of DNA

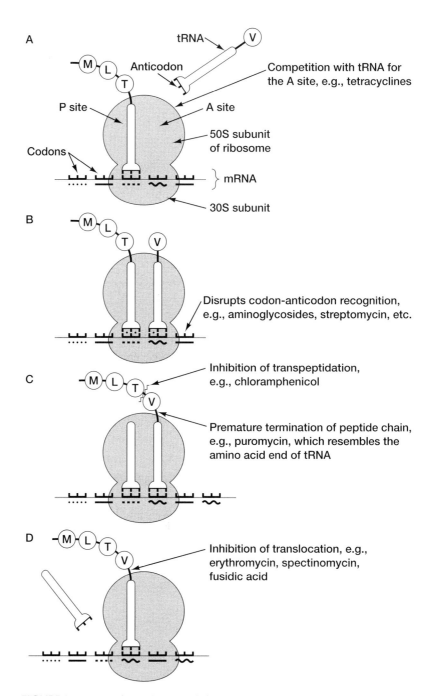

FIGURE 3.4. Bacterial translation and the action of antibiotics. The prokaryotic ribo-some is shown in green. An mRNA bound to the small subunit directs amino acid incorporation. The cycle starts (*A*) with a tRNA containing the nascent peptide chain in the P (polymerization) site. A tRNA charged with the amino acid to be added enters and binds to the A (acceptor) site (*B*). Growth of the polypeptide proceeds with transfer of the chain from the tRNA at the P site to the A site (*C*). The cycle is com-pleted by release of the tRNA at the P site and translocation of the tRNA bound at the A site to the P site. The mode of action of antibiotics that interfere with translation is summarized at the right. (Redrawn, with permission, from Rang et al. 1995 [© Churchill Livingstone].[51])

gyrase, a prokaryotic enzyme that controls the degree to which the DNA ribbon is wound up around itself (supercoiling). Rifampicin inhibits bacterial RNA polymerase, which differs considerably from the human RNA polymerase and so presents a selective target.

The above examples illustrate some of the types of antibiotics and provide the point of departure for considering mechanisms of antibiotic resistance and resistance gene dispersal.

TYPES AND ORIGIN OF ANTIBIOTIC RESISTANCE (26, 40, 46, 55, 74)

Three broad mechanisms account for the appearance of antibiotic resistance. Genes for each type of resistance function can be encoded on DNA elements, such as plasmids and transposons, capable of lateral DNA transfer. In the first mechanism, antibiotic resistance arises when an organism produces an enzyme that modifies the antibiotic molecule and renders it inactive. An example of this is the β-lactamase (*bla*) genes, which confer resistance against penicillin and the other β-lactam antibiotics by producing an enzyme that cleaves the β-lactam ring. In another, some forms of erythromycin resistance arise when bacteria produce enzymes that add chemical groups (methyl groups) to erythromycin and thereby render it inactive.

Another form of antibiotic resistance occurs when the target for the antibiotic becomes modified. For the case of vancomycin, some bacterial strains are resistant because they have replaced the D-alanine–D-alanine cross-link in their cell walls with a D-alanine–D-lactate linkage, which no longer binds vancomycin, thereby conferring resistance. This type of resistance occurs via transfer of genes specifying D-alanine–D-lactate synthesis, the *van* gene cluster, to host species previously making D-alanine–D-alanine.

The third form of antibiotic resistance arises by alterations in the bacteria that reduce the concentration of antibiotic in a cell. This can occur through expression of pump proteins that reside in the bacterial membrane and actively pump out antibiotics. Some types of resistance to tetracycline are an example. In a related form of resistance, entry of antibiotics into cells is blocked. The β-lactam antibiotics enter cells through protein pores, or channels, in the cell wall. Mutations in the *porin* genes encoding these channels block β-lactam entry, rendering cells insensitive.

In many cases, the species of origin of antibiotic-resistance genes is unclear, but there are good candidates in a few cases for the precursors of clinically important genes. One example is vancomycin, which is produced naturally by the bacteria *Amycolatopsis orientalis*. This species synthesizes a cell wall that is cross-linked by D-alanine–D-lactate and encodes a cluster of genes resembling the *van* gene cluster, genes that allow *A. orientalis* to elude the effects of its own antibiotic. Transfer of the *van* gene cluster from *A. orientalis* or a related species to a pathogenic bacterium resulted in clinical vancomycin resistance. In fact, this may have happened twice, because there are two distinct gene clusters, *van A* and *van B*, that both confer medically important resistance, which appears to have arisen independently.

In another example, methicillin resistance in *S. aureus* arose by the substitution of a normal *S. aureus* penicillin-binding protein (PP2a) for one with a low affinity for the β-lactam antibiotics. The gene encoding the modified protein, *mecA*, is 80% identical to a gene from *Staphylococcus sciuri*, a bacterial species found in animals. *mecA* appears to have undergone considerable evolution during its passage between species, including association with mobile elements and the acquisition of methicillin resistance in the encoded protein. Today *mecA* is widespread in pathogenic *S. aureus*.

THE SPREAD OF RESISTANCE GENES

Bacteria have evolved mechanisms that enable them to be resistant to all antibiotics yet devised. In some cases, changes as simple as base pair substitution mutations in the gene encoding the antibiotic target are sufficient to confer resistance. For example, resistance to rifampicin can arise due to a mutation in the gene encoding a subunit of the target, RNA polymerase. Frequently, however, resistance arises when a sensitive bacterium acquires new DNA sequences by lateral transfer from another organism. In some cases, these two mechanisms operate together. Penicillin-binding proteins insensitive to β-lactam antibiotics can arise de novo by mutation of the PBP genes. These genes can then be transferred laterally, for example, linked to an R plasmid, conferring resistance on a formerly sensitive host. The rapid spread of clinical antibiotic resistance is often due more to lateral transfer than to repeated generation of new base substitutions in DNA, but this varies from case to case. Drug resistance in *Mycoplasma tuberculosis*, the causative agent of tuberculosis, for example, is generally due to de novo mutation, although for most pathogens the opposite is true.

In the following sections, we investigate the remarkable array of mechanisms underlying DNA mobility in antibiotic resistance. We begin with conjugative transfer of plasmids, one of the most common means for transferring resistance genes between cells.

Plasmids (35, 48, 55, 63)

In many cases, the genes conferring resistance to antibiotics are found not on the bacterial chromosome, but on extrachromosomal plasmids. There is considerable diversity in the types of plasmids known. In what follows, we focus on the circular double-stranded DNA plasmids found in gram-positive and gram-negative bacteria that commonly mediate antibiotic resistance.

Plasmids replicate independently of the host chromosome. To achieve this, each plasmid encodes an origin of replication, a site where DNA replication proteins assemble and initiate DNA synthesis. In plasmids these sites are typically not larger than a few hundred base pairs. In some cases, plasmids also encode proteins important for their replication. Many plasmids encode a Rep protein, which binds at the origin DNA and facilitates loading of replication factors. Most of the subsequent steps of DNA replication are typically carried out by the host-encoded enzymes for DNA synthesis.

Plasmids replicate to characteristic numbers of copies per host cell. Some plasmids, such as the *Escherichia coli* F plasmid, have their replication tightly linked to the growth cycle of the host. Only 1 or 2 copies of F accumulate per bacterial chromosome. Other plasmids such as *E. coli* ColE1 can reach 10–30 copies per bacterial genome.

Any type of DNA sequence can be carried on a plasmid provided the sequence is not toxic to the plasmid or to the host cell. Examples from natural populations include genes for resistance to antibiotics and heavy metals (such as mercury), diverse toxins, and virulence factors. Genetic maps of two simple plasmids are shown in Figure 3.5.

Naturally occurring bacteria frequently contain plasmids. In a sample of more than 1000 bacteria from coastal marine sediments, about 25% contained plasmids. The majority of bacteria isolated from humans also appear to contain plasmids. Determination of the complete DNA sequences of bacterial genomes has revealed naturally occurring plasmids in several species (discussed in Chapter 5). Thus, mechanisms for lateral transfer in bacteria that involve plasmids invoke widespread machinery. Once an antibiotic-resistant determinant becomes associated with a plasmid, the resistance gene can increase greatly in abundance by plasmid replication.

The finding that plasmids can incorporate most DNA sequences has led to their use as vehicles for recombinant DNA manipulation in the laboratory. A typical step in many protocols for DNA cloning involves the attachment of the gene of interest to a plasmid. This permits the synthesis of many copies of that gene by simply growing large amounts of plasmid-containing bacteria.

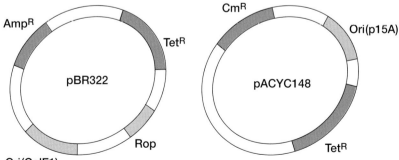

FIGURE 3.5. Structures of some prokaryotic plasmids. Antibiotic-resistance genes are shown in green, origins of replication and replication proteins in gray.

Conjugative Plasmids (11, 20, 35, 36, 38, 48, 55)

Some plasmids encode not only sequences for their replication, but also machinery for transfer between cells via conjugation. These plasmids are also called self-transmissable plasmids. The F plasmid of *E. coli* provides an example. It derives its name from "fertility" factor, because F$^+$ bacteria could transfer the F plasmid to F$^-$ cells. The F plasmid devotes about one-third of its genome to conjugation, illustrating the importance of the process for its persistence.

The Conjugation Pathway (29, 39)

Conjugation begins when the pilus, a long cylindrical structure that extends from the F$^+$ cell, binds the F$^-$ recipient (Fig. 3.6). The pilus then shortens to bring the cell walls of the two bacteria close together. F DNA is then transferred from the F$^+$ to the F$^-$ strain through the mating pore. The F$^+$ cells do not lose their copies of the F plasmid during transfer, implying that the F plasmid, normally present in only 1 or 2 copies per cell, must be replicated during the transfer reaction.

The linked replication/DNA transfer process mediating conjugation is understood in some detail. Transfer begins from a specific sequence on the donor F plasmid, named *oriT* for *ori*gin of *t*ransfer. The process starts by formation of a single-stranded DNA break at *oriT* (Fig. 3.7). The nicking reaction is carried out by an enzyme called "relaxase," which together with other proteins forms the relaxosome complex. Relaxase cleavage at *oriT* results in breakage of the DNA along with covalent linkage of relaxase to an end of the broken DNA. The relaxase-bound DNA end is then unpaired from the complementary DNA strand and transferred to the recipient cell.

Meanwhile, DNA replication initiates using the unbroken DNA strand as template. Synthesis is primed either by the free DNA 3′ end generated by nicking or by specially synthesized primers. Extension of

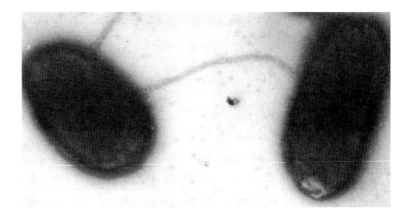

FIGURE 3.6. Electron micrograph of *E. coli* cells undergoing mating. (Courtesy D.P. Allison, Oak Ridge National Laboratory.)

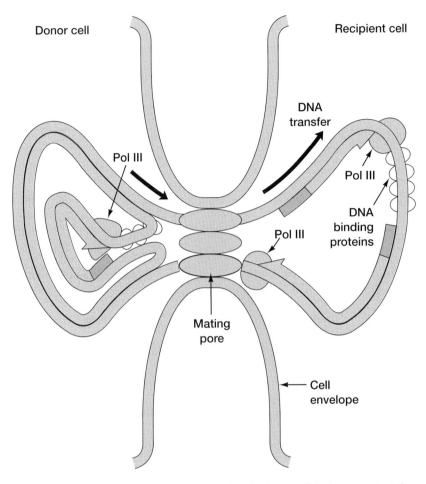

FIGURE 3.7. Action of the "relaxosome" at *oriT*. The donor cell is shown to the left, the recipient to the right. A single strand of DNA threads through the mating pore in the direction indicated by the arrows. DNA polymerase III (DNA Pol III) directs synthesis of a replacement DNA strand in the donor cell and synthesis of the second strand in the recipient cell. Polymerization by Pol III is assisted by single-strand DNA-binding proteins, which coat the template DNA. RNA primers for second-strand synthesis are shown in gray. (Adapted from Neidhardt et al. 1987.[48])

the DNA chain copies the continuous plasmid strand with concomitant displacement of the nicked DNA strand (rolling circle replication). Transfer is actively terminated when a complete single copy of the F plasmid has been transferred. The relaxase again cleaves at *oriT*, this time to free the completed single strand. This reaction is accompanied by circularization of the transferred strand. DNA replication proteins in the recipient cell are then recruited to synthesize the complementary strand, yielding a double-stranded copy of F in the formerly F⁻ bacteria. Replication also occurs in the donor cell to generate a duplex copy of the DNA. After completion of DNA transfer, the two cells separate. The F⁻ cell is now F⁺ phenotypically and can act as the donor cell in subsequent matings.

F-encoded Genes (25, 48)

The F plasmid encodes about 100 genes, including more than 20 *tra* (transfer) genes that are directly involved in DNA transfer. The *traI* and *traY* products comprise the relaxase endonuclease that cleaves at *oriT*. The TraG "coupling" protein binds relaxase and forms a membrane channel, potentially acting as a "DNA pump" that forces single-stranded DNA through the membrane. The structure of a conjugation protein related to TraG, the TrwB protein of plasmid R388, has been solved by X-ray crystallography, revealing that the protein forms a spherical hexamer with a channel down the middle that may allow DNA passage (Plate 2). The TrwB protein also binds and hydrolyzes ATP, providing energy to drive single-stranded DNA through the pore. The TraD and TraM proteins also reside in the inner membrane and bind DNA, potentially helping form the DNA channel for transfer and perhaps coordinating transfer with DNA synthesis.

The main component of the pilus, the pilin protein, is encoded by the *traA* gene. Pilin is synthesized as a precursor, which must be cleaved by the product of the *traQ* gene to be active. Eleven more of the *tra* gene products are required for pilus synthesis, some of which are present in the final structures while others transiently promote assembly. Similar secretion systems mediate transfer of macromolecules among many species of bacteria (Chapter 12; a speculative model for the overall structure is shown in Plate 9).

Other genes encoded on F direct DNA replication, control plasmid copy number, regulate gene expression, and ensure that during cell division both daughter cells inherit a plasmid copy. F also encodes proteins that are displayed on the cell surface and serve to inhibit mating between F⁺ cells. Thus, the F plasmid encodes considerable machinery for transfer and replication, all of which is transported to a new cell by conjugation.

General or homologous recombination is genetic exchange between any pair of DNAs that are similar or identical in sequence. Homologous recombination is important for restarting "collapsed" DNA replication forks, diverse types of DNA repair, and DNA exchange to generate genetic diversity.

Mobilization of Genes by Hitchhiking on Conjugative Plasmids

Genes encoded on a conjugative plasmid can be transferred efficiently even if they do not contribute to the transfer process. Several means are known for incorporating new genes into conjugative plasmids, such as by transposons as described below. Any gene that becomes incorporated into a conjugative plasmid can thereafter be transferred between cells like a normal plasmid gene. More broadly, any plasmids encoding an *oriT* sequence and appropriate relaxosome proteins can be mobilized if a conjugative plasmid present in the same cell can direct pilus formation. Such plasmids are commonly seen in natural settings as "hitchhikers" on conjugative plasmids.

Conjugation can even result in the transfer not only of plasmid genes, but also of genes from the cellular chromosome. On occasion, a conjugative plasmid can become integrated into the host cell chromosome. This can happen by any of several mechanisms. For example, if a sequence on the plasmid has an exact match with a sequence in the host cell DNA, then homologous recombination between the identical sequences can result in incorporation into the cellular DNA (Fig. 3.8).

If conjugation is initiated at *oriT* in an integrated plasmid, DNA transfer can proceed across the plasmid, then out into a cellular chromosome. Transfer can, in principle, proceed indefinitely, transferring

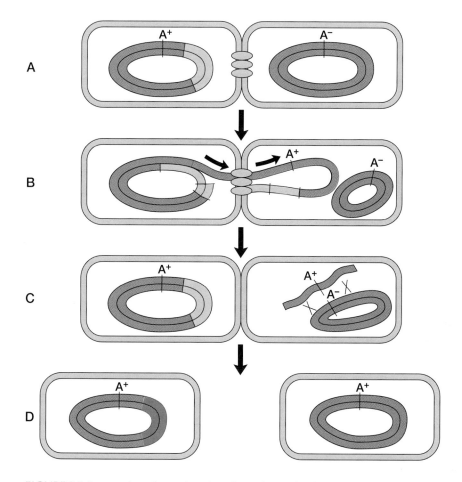

FIGURE 3.8. Integration of a conjugative plasmid into the chromosome and subsequent transfer of chromosomal genes. Integration of a conjugative plasmid (*gray*) into the host chromosome (*green*) is illustrated in one member of a mating pair (*A*). Cleavage at oriT can direct conjugative transfer of the integrated plasmid and also flanking host cell markers, indicated by A^+ in the diagram (*B*). If the recipient bacterium is A^-, then incorporation of A^+ information by recombination (*C*) can convert the recipient from A^- to A^+ (*D*). (Adapted from Salyers 1995.[55])

the entire host cell chromosome to the conjugative recipient cell. In practice, however, transfer often terminates spontaneously in the flanking chromosomal DNA. This results in transfer of cellular genes just adjacent to the integrated plasmid. Such strains with integrated conjugative plasmids have been isolated and named Hfr (high frequency of transfer) strains, since they transfer certain chromosomal genes to conjugative recipient cells with high frequency. Different strains may have conjugative plasmids integrated into different points in the chromosomal DNA, resulting in the transfer of different parts of the bacterial chromosome.

Conjugation can mediate transfer of DNA not only between cells of the same species, but in some cases between quite distantly related bacteria. Analysis of conjugative plasmids has in some cases revealed closely related plasmid sequences in distantly related bacteria, suggesting natural transfer by conjugation. Laboratory tests have directly documented examples of such "wide" transfer. The ability of conjugative plasmids to assimilate diverse DNA sequences and transfer them between distantly related species provides a mechanism for transferring essentially any sequence, including antiobiotic-resistance genes, between bacteria. Conjugation is presently thought to be one of the major means of transferring antibiotic-resistance determinants between bacteria in the environment. Even more extreme examples are presented in Chapter 12, in which a conjugation-like mechanism has mediated transfer between organisms from the different domains of life.

Simple Insertion Elements (7, 60)

We turn to a discussion of one of the most active mechanisms for rearranging DNA sequences, genetic transposition. In this process, segments of DNA present at one location are transferred to a new location in the genome. The DNA segments that move are called "transposable elements" or "transposons." Strictly speaking, most transposons do not direct the transfer of DNA between cells, only the rearrangement of DNA within cells. However, the transposons are major players in the transfer of antibiotic resistance because they can direct the incorporation of resistance genes into mobile elements such as conjugative plasmids, which can then shuttle those genes to new cells. A few types of transposons can even direct lateral transfer on their own. In the following sections, we examine in detail the transposons involved in transferable resistance, starting with the simplest elements.

In bacteria the simple insertion sequences are designated IS followed by a number indicating the particular isolate (Fig. 3.9). The IS elements are distinguished from the large and more complex transposons, designated Tn elements, which are discussed in a later section. The abundance of these elements can be appreciated by the fact that the isolate numbers are now in the thousands.

The transposition donor and target DNAs can be a cellular chromosome, a plasmid, or a viral genome. The particular DNA sites chosen as transposition targets usually include many sequences in the host cell, although transposons with preferences for specific target sites are also known. Often transposition involves DNA replication (replicative transposition), in which a copy of the transposon remains at the starting location as well as appearing as new copies elsewhere in the chromosome.

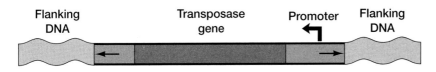

FIGURE 3.9. Simplified genetic map of IS*10*. The element consists of a transposase gene flanked by DNA inverted repeats (*arrows*) that act as transposase-binding sites. (Adapted from Kleckner 1989.[33])

All autonomously active IS elements encode at least one protein, called a transposase, that carries out the initial DNA breaking and joining reactions that mediate transposition. For transposase to act, there must be appropriate transposase-binding sites at each end of the IS element (Fig. 3.9, inverted repeat). Each transposon DNA encodes sites that are specifically recognized and bound by transposase near the DNA ends, thereby positioning the transposase for the chemical steps of the transposition reaction. There are typically multiple transposase-binding sites at each end of the transposon DNA within the inverted repeat. For the transposon to make similar interactions at each end of the element, the binding sites must be in inverted orientation with respect to the DNA sequence, so that when the ends are bound together in a protein–DNA complex, symmetric interactions can be made at each end. These two features, a transposase gene and flanking transposase-binding sites, are all that is required to construct a minimal IS element.

Transposase Proteins (7, 57, 60)

Several functions, common to all prokaryotic transposase proteins, are required for transposition. Transposase proteins bind to the ends of the transposon DNA, thereby directing their action to the transposon sequences and not other sites in the genome. Accordingly, transposase proteins are capable of sequence-specific DNA binding. In some well-studied cases, the transposase determinants for binding have been mapped to subdomains within the larger transposase protein.

Transposase proteins also contain catalytic domains that carry out the chemical steps responsible for breaking and rejoining DNA molecules. This domain must also bind DNA, but at the site of DNA cleavage at the border between the transposon and host DNA. The catalytic domain must also bind the transposition target DNA in order to direct the chemistry of DNA rejoining.

The transposase of IS*50* has been shown bound to transposon DNA by X-ray crystallography (Plate 3). The structure shows a tight-knit interaction between two monomers of the transposase protein and the two transposon DNA ends. The tip of the transposon DNA is bound near three acidic residues (DDE protein sequence motif) that comprise the "active site" where covalent DNA breaking and joining take place. Each transposase monomer binds to both DNA duplexes using different protein domains, thereby knitting the complex together.

Most, but not all, transposases fall into a single protein superfamily, judging by the similarity of the amino acid sequences of their catalytic domains, notably the conserved DDE motif. Where studied, these domains appear to carry out similar or identical chemical trans-

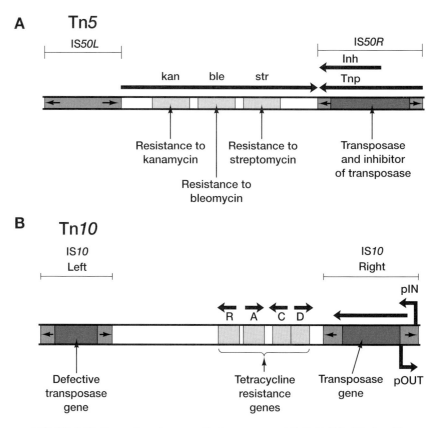

FIGURE 3.10. Examples of composite transposons. (*A*) Tn*5*. (Modified, with permission from Berg 1989.[6]) (*B*) Tn*10*. (Modified, with permission, from Kleckner 1989.[33])

formations of DNA. This class of catalytic domain has also been found in other enzymes involved in DNA breaking and rejoining reactions, such as the retroviral integrase proteins (Chapters 6 and 7) and the RAG1 recombinase (Chapter 11). Evidently this class of catalytic domain represents a widely used and evolutionarily ancient class of enzymes devoted to rearranging nucleic acids.

The transposase protein, once synthesized, is not restricted to acting at the ends of the element from which it was transcribed. If other recognition sites are available in the bacterial genome, the transposase can bind to these sites and carry out DNA cleavage and joining reactions involving these sequences. Thus, sequences that do not encode transposase but do contain two transposon ends in correct orientation can participate in transposition reactions. Such defective elements occur naturally in many transposon systems and can be readily constructed in the laboratory.

A few transposase proteins are actually members of a completely different family of enzymes. As mentioned in the introduction, some bacteriophages are capable of integrating their DNA into host DNA, a mechanism of lateral DNA transfer discussed in Chapter 4. This integration reaction is carried out by the phage integrase enzyme (despite the similarity in names, this is a different class of enzymes from the retroviral integrases). Some of the transposase enzymes of the conjugative transposons are related to the phage integrases rather than the majority of bacterial transposases. As discussed below, the DNA breaking and joining reactions mediating transposition differ for the two transposon classes.

Composite Transposable Elements (32, 34, 70)

Two IS elements located close together in the DNA may form a composite transposon, or Tn element (Fig. 3.10). Tn elements contain four transposon ends that can potentially serve as binding sites for transposases. In those cases where two "outside" ends are recruited by transposase protein, the entire DNA unit between the two can then transpose, carrying along the DNA in between. Tn5 is an example, composed of two IS50 elements flanking genes for resistance to kanamycin, bleomycin, and streptomycin. Tn10 is composed of two IS10 elements flanking a central region containing genes for resistance to tetracycline.

Formation of composite transposons represents a crucial mechanism for mobilizing and disseminating antibiotic-resistance genes. Sequential insertion of IS elements on either side of a resistance gene can allow it to become incorporated into a composite transposon. Such a transposon can then "hop" into some other mobile element, such as a plasmid or a phage, that can promote its dispersal. Tn10 itself, for example, is responsible for widespread dissemination of a gene for resistance to tetracycline, *tetA*.

DNA Breaking and Joining Reactions Mediating Transposition (5, 17, 31, 33, 45, 52, 59, 75)

Transposition reactions presented a bewildering array of DNA products to early investigators. Happily, the picture cleared with the finding that a small number of related pathways could yield the observed products. As is often seen in studies of mobile DNA, a deep understanding of relatively simple DNA breaking and joining reactions explained apparently complicated biological consequences.

One pathway is "cut-and-paste transposition," also called non-replicative transposition, as is seen for Tn10 (Fig. 3.11). The transposition cycle is initiated by the binding of transposase to sites at the ends of the transposon DNA. Protein–protein interactions between transposase proteins and protein–DNA interactions organize the transposon ends into a "nucleoprotein machine" poised to carry out the subsequent reaction steps, as for Tn5 (Plate 3). The transposase protein then cleaves each of the two DNA strands at the ends of the transposon, freeing the transposon DNA from the host cell DNA. The cleavage of the transposon DNA ends actually proceeds in two steps. First, one strand of the DNA is cut, then the free end of the transposon DNA attacks the remaining continuous strand at the transposon–host DNA junction. This frees the transposon while simultaneously forming a DNA hairpin at each end of the transposon. This leaves a break in the transposon "donor" DNA, which must subsequently be repaired by the cell.

The DNA hairpin at each transposon end must be opened and one strand at each end joined to the target DNA. At this stage, the job of the transposase is finished. The partially joined DNA intermediate is

Although some eukaryotic transposons are similar to the prokaryotic transposons, others replicate by a quite different mechanism, involving transcription, reverse transcription, and integration. A few examples of elements employing reverse transcription are also known in bacteria, although they are not known to transfer genes for antibiotic resistance. These elements are discussed in detail in Chapter 8.

Transposon insertion can both activate and inactivate cellular genes. Insertion of a transposon into a coding region or required regulatory region can inactivate gene function. Insertion of a transposon at a suitable location upstream of a gene can also contribute transcription control signals that increase its activity or alter regulation. Many examples have been found of transposons acting as mobile gene control regions in prokaryotes and eukaryotes.

now a substrate for host DNA repair enzymes, which can carry out polymerization and ligation reactions that join the remaining DNA strands (Fig. 3.11). This is typical of many mobile elements, which encode the minimum necessary machinery for moving DNA sequences, recruiting host systems for DNA repair to finish attaching unjoined DNA ends.

The related "replicative" transposition pathway is characterized by the production of a "cointegrate" intermediate, in which the donor (old target) DNA and the new target DNA become fused, with complete transposon copies between the borders of each (Fig. 3.12). An example of replicative transposons is provided by the Tn3 family. Transposase initially binds to the ends of the transposon embedded in continuous target DNA and brings together the ends of the element. Replicative transposition begins with cleavage not of both DNA strands at the ends of the transposon, as in cut-and-paste transposition, but cleavage of only one strand at each end. This exposes a single-stranded DNA end at each border between host and transposon DNA. The complex captures target DNA, and the transposon ends become joined to the target DNA. One strand of the target DNA only becomes joined to the transposon DNA. The points of joining on each of the two strands are close together in the new target DNA, so the paired nicks break the target DNA. Replicative transposition differs from cut-and-paste transposition in that the old flanking host DNA is not removed in this intermediate, but remains attached by one DNA strand at each end of the transposon DNA. Overall, the structure of the DNA resembles that of the Greek letter θ, so these intermediates have been called θ structures. Another name is "Shapiro intermediate," for James Shapiro, who first proposed such a structure.

The θ structure can be transformed into transposition products by either of two pathways. The old flanking DNA can be removed by another DNA cleavage reaction, thereby transferring the transposon to the new target and leaving a DNA gap in the old target. The outcome is the same as for cut-and-paste transposition, although the order of the DNA cleavage and joining reactions differs. Another pathway involves replication across the transposable element DNA. The free target DNA 3´ ends at the junctions between the element and target DNA can serve as primers for a DNA polymerase. Polymerization across the element duplicates the transposon sequences, while leaving both the old and new target DNAs connected. Cleaning up this intermediate by DNA repair enzymes creates a structure in which the old target and new target DNAs are connected in a circle, with a copy of the transposon at each boundary between the two. If both the transposon donor and target are plasmids, the two become fused by this reaction, forming a structure called a cointegrate.

Many replicative transposons encode an enzyme, named resolvase, that separates the two starting replicons fused in a cointegrate (Fig. 3.13). Such transposons also contain a recognition site for resolvase,

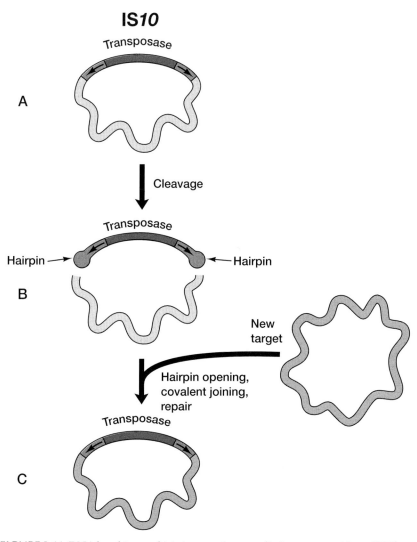

FIGURE 3.11. DNA breaking and joining reactions mediating transposition of IS10. Transposition begins when IS10 present in a replicon (a plasmid or bacterial chromosome) (A) excises from the chromosome (B). First one DNA strand is nicked, then a free 3′ end at the nick attacks the other strand, resulting in formation of a DNA hairpin. In the presence of a target DNA (C), the hairpin is opened and the transposon integrated. The final steps, involving repair to generate the target site duplication, are not shown. (Adapted from Kleckner 1989.[33])

designated *res*. Resolvase catalyzes recombination between pairs of correctly oriented *res* sites. Resolvase recombination converts the cointegrate into two DNA circles matching the starting replicons, except both now contain integrated copies of the transposon.

At this point, one might wonder how cut-and-paste transposons ever increase in number, since the replication mechanism involves only excision from one place and insertion into another. Two pathways probably account for a net increase (Fig. 3.14). In one, the transposon

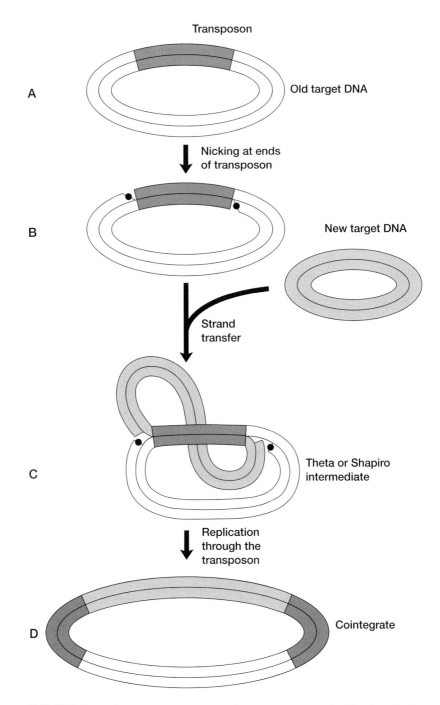

FIGURE 3.12. Replicative transposition to form a cointegrate. A cycle of replicative transposition is initiated when a transposon resident in a replicon (*A*) becomes nicked by transposase to expose 3´ DNA ends in the element DNA (*B*). In the presence of an added target DNA, the 3´ ends of the transposon DNA become joined to protruding 5´ ends in the new target DNA. The old target DNA remains attached by one strand at each end, yielding a θ or Shapiro intermediate (*C*). DNA replication, initiated at the 3´ ends in the new target DNA, extending across the transposon DNA yields the cointegrate structure (*D*). (Adapted from Shapiro 1979.[59])

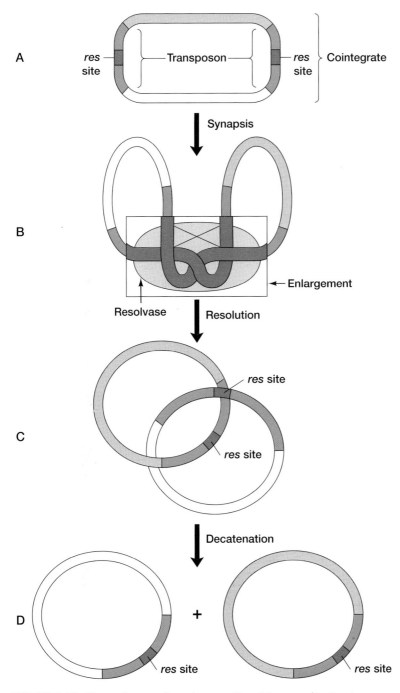

FIGURE 3.13. The pathway of resolvase action. Many replicative transposons encode a resolution system that resolves cointegrates into separate replicons. (*A*) A cointegrate containing two replicons (*gray* and *white*) spaced by transposons (*light green*) containing *res* sites (*dark green*). Binding of resolvase proteins (*B*) wraps the DNA with the indicated configuration, leading to recombination at the indicated cross-over point. (*C*) The product of resolvase action is a pair of DNA catenanes, which become decatenated by cellular enzymes (*D*) to yield two free replicons, each containing a copy of the transposon. (Adapted from Stark et al. 1989.[64])

moves from replicated to unreplicated DNA. Subsequent replication of the new insertion site allows the total number of transposons in a cell to be increased.

Several transposons, including Tn*5* and Tn*10*, have evolved a mechanism to couple transposition to passage of replication fork. In *E. coli*, 5′GATC3′ sequences are substrates for the Dam methylase, which attaches methyl groups to the A residue on each strand. Passage of a replication fork yields a sequence that transiently has a methyl on one strand only, although the newly synthesized strand is quickly methylated as well. The transposon promoters, which direct transposase synthesis, have 5′GATC3′ sites within them. Methylation inhibits transcription, but transient hemimethylation favors transcription, thereby coupling transposase production and transposition to passage of a replication fork.

The second mechanism for multiplication of cut-and-paste transposons involves repair of the empty site from which the transposon was excised using another copy of the transposon. If there are multiple copies of the starting transposon in the host cell, for example a transposon copy on the sister strand after passage of a replication fork, then the double-strand break at the empty site can be repaired by copying

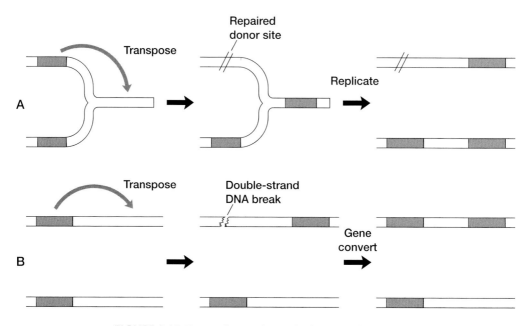

FIGURE 3.14. Two pathways for multiplication of cut-and-paste transposons. (*A*) The transposon can excise after passage of a replication fork and integrate ahead of the fork. Replication of the newly integrated transposon leads to a net increase in transposon number. (*B*) Excision and transposition of a cut-and-paste transposon, followed by repair of the empty donor site by gene conversion using another transposon copy, can also yield an increase in transposon number.

transposon sequence information into the DNA gap. This process, named gene conversion, results in placing an intact copy of the transposon in the empty donor site. In this way, the cut-and-paste pathway can also result in a net increase in transposon number.

There is no reason that only the outside ends of a composite transposon can participate in transposition reactions. As might be expected, other types of events are also known involving transposase action at the other possible sets of ends. For example, action of transposase at the paired inside ends can cause the entire associated genome to be treated as a composite transposon (Fig. 3.15). If such "inside end" transposition uses a plasmid DNA as a target, the net effect is incorporation of the plasmid sequences into the host genome.

Targeted Transposition by Tn*7* (3, 4, 43, 65, 72)

The Tn7 family of transposons displays an added refinement, the ability to transpose to specific target sites that promote the dispersal of the element. A genetic map of Tn7 is shown in Figure 3.16. Tn7 encodes two genes for antibiotic resistance: *dhfr*, which confers resistance to trimethoprim, and *aadA*, which confers resistance to spectinomycin and streptomycin. These antibiotic-resistance determinants have been found in many bacterial strains, sometimes associated with Tn7 and other times not. Unique to Tn7 is a set of five genes, *tnsA–E*, which direct two different transposition pathways (*tns* stands for *transposon seven*).

In one pathway, Tn7 transposes at a high frequency to a unique site in the bacterial chromosome, known as *attTn7*. This site is located just downstream of a gene, named *GlmS*, which encodes a protein important for cell wall biosynthesis. The tns A, B, C, and D proteins are required for transposition into *attTn7*. These proteins form a complex that binds to a DNA sequence just inside the *GlmS* gene and direct transposition into a site located about 20 base pairs away. It turns out that transposition into *attTn7* does not disrupt expression of the nearby genes or otherwise debilitate the host bacteria. If Tn7 were to inactivate some important cellular gene during transposition, it could end up committing suicide by killing its host. The *GlmS* gene seems to be conserved among the bacterial hosts of Tn7, possibly explaining the selection of this gene over others. Several other types of transposons have also developed mechanisms to insert into preselected sites that minimize damage to the host cell. Minimizing the disruption to host-cell genes further promotes the spread of hitchhiking antibiotic-resistance genes.

Tn7 has another class of favored target sites—conjugative plasmids undergoing conjugation. This target site is particularly well suited to promote transfer of Tn7 and associated antibiotic-resistance genes between cells. This pathway requires the TnsA–C proteins and TnsE

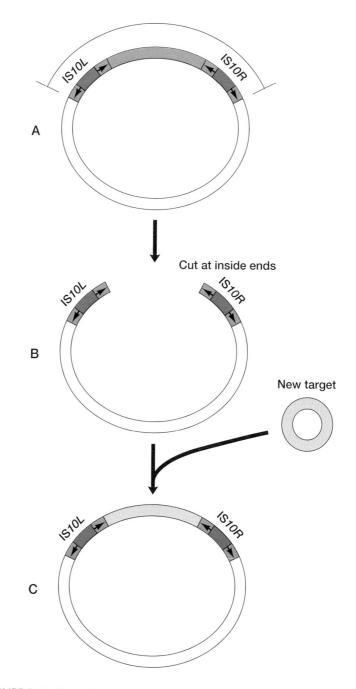

FIGURE 3.15. "Inside" end transposition by a composite transposon, using Tn*10* as an example. (*A*) Tn*10* inserted in a circular replicon, illustrating the IS*10* elements comprising the composite element. (*B*) Cleavage at the inside end of each copy of IS*10* creates a new giant composite transposon comprising the entire replicon. Integration into a circular target replicon (*C*) yields a cointegrate, with the two formally independent replicons joined with copies of the IS elements at the junctions.

Tn7

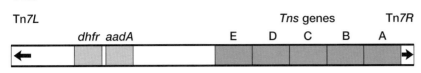

FIGURE 3.16. Genetic map of Tn7. (Modified, with permission, from Craig 1989.[16])

instead of *tnsD* as in the *attTn7* pathway. The sequences selected as transposition target sites in this pathway bear no resemblance to *attTn7* or to each other. When neither *attTn7* nor a conjugating plasmid is available, Tn7 can transpose to diverse locations within the genome, but at 100-fold lower frequency.

The roles of the Tns proteins have been worked out in some detail. TnsA and TnsB are both transposase proteins, and each is dedicated to a specific cleavage in initiating transposition. TnsA cleaves one DNA strand at each end of the element, generating the ends that are first joined to target DNA. TnsB cleaves the other strand. Tn7 displays a sophisticated adaptation in the timing of these cleavages. Only when the target DNA is captured by the transposition complex do the cleavages in the element DNA take place, thereby ensuring that the element only excises itself when a suitable target is at hand.

Normally, Tn7 transposes via a cut-and-paste mechanism. As with other transposons, TnsA and B bind specifically to sites at each end of the element, near the transposon-host DNA boundaries where DNA cleavage takes place. Remarkably, Tn7 derivatives in which TnsB is mutated, and so cannot cut the unjoined strand of the element DNA, can transpose via a replicative pathway (Fig. 3.12). The DNA cleavage and joining reactions in this mutant pathway exactly match those of the replicative pathway discussed above for Tn3. The fact that a single transposon can switch between cut-and-paste and replicative transposition emphasizes the close relationship of the two pathways.

The TnsD and TnsE proteins are devoted to selection of target sites for transposition. TnsD binds specifically to *attTn7*, thereby facilitating insertion at this site. TnsE mediates binding the Tn7 complex to conjugative plasmids. TnsC is a coordinator, responding to positive inputs from TnsD and TnsE, and also negative inputs that help Tn7 avoid inserting into other copies of itself.

Stepping back, we see in Tn7 a greater level of sophistication than in the simpler cut-and-paste transposons. Several proteins encoded by the element act together to favor insertion into either of two targets, both of which have particular advantages to the element. Linkage of antibiotic-resistance genes to Tn7 allows them to be spread with similar efficiency.

Conjugative Transposons (12, 22, 57, 58)

A still more sophisticated class of transposons can not only transpose, but also actually direct, their own conjugative transfer. These "conjugative transposons," found in gram-positive bacteria, can direct efficient transfer between distantly related gram-positive species and, rarely, even to gram-negative species. In gram-negative bacteria, conjugative transposons are capable of transposition although not of further conjugation. Most conjugative transposons contain a gene conferring resistance to tetracycline, and some contain further antibiotic-resistance genes as well. Clinically, the conjugative transposons are a major contributor to failed antibiotic treatments of gram-positive bacterial infections.

One well-studied example is Tn*916*. The DNA breaking and joining reactions mediating Tn*916* conjugative transposition have been worked out in detail and found to be quite surprising (Fig. 3.17). The transposition/conjugation cycle begins with staggered cleavage at each end of the element. The DNA then becomes circularized, with the odd consequence that the protruding ends generated by staggered cleavage form a closed region of non-base-paired DNA (heteroduplex). Nicking at the internal element *oriT* site allows conjugation to be carried out as is normally the case with plasmid conjugation. This leads to transfer of one sequence only from the heteroduplex overlap region. Once the single-stranded DNA copy is transferred to the recipient cell, it is copied by host cell enzymes to generate a fully duplex transposon circle. Expression of the transposase genes can then take place from the unintegrated transposon copy, permitting transposase to carry out the final stages of insertion into the chromosome of the new host cell.

It is also possible for conjugative transposons to transpose to another location within a single cell (Fig. 3.18). In this case, the transposon is excised and circularized as above, but then transposase acts on this intermediate without concomitant transfer, inserting the transposon into host cell DNA in the usual fashion.

The proteins encoded by conjugative transposons differ from their counterparts in other transposons and conjugative plasmids. The transposase genes are related not to the other bacterial transposase families, but rather to the λ integrase enzymes normally involved in bacteriophage DNA integration. These enzymes are treated in detail in Chapter 4. Furthermore, the reactions involving heteroduplex formation differ from phage integration, which requires sequence identity over part of the recombining region. Evidently the conjugative transposons have evolved a unique set of transposases suited to their style of replication.

The genes that direct conjugation by conjugative transposons also appear to differ from their counterparts in conjugative plasmids. Tn*916* does encode an apparent relaxase DNA-nicking enzyme (*orf23*), which may be responsible for initiating DNA transfer by nicking at the

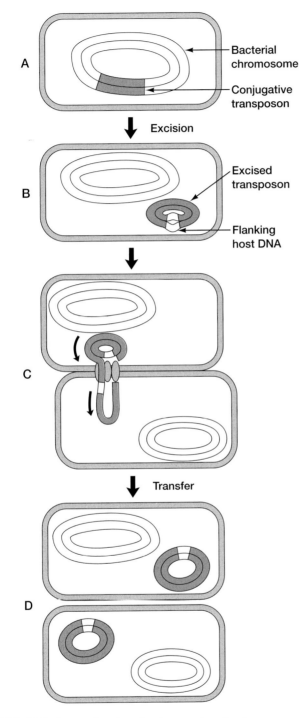

FIGURE 3.17. Pathway of conjugation by a conjugative transposon. (*A*) The transposon (*green*) initially resident in a host cell replicon (*white*) is excised by the action of transposase (*B*), to yield an extrachromosomal circle. Binding of donor and recipient cells, cleavage at oriT, and subsequent conjugative transfer (*C*) install the transposon in the recipient cell. DNA replication yields the circular extrachromosomal form of the transposon (*D*). (Adapted from Scott and Churchward 1995.[57])

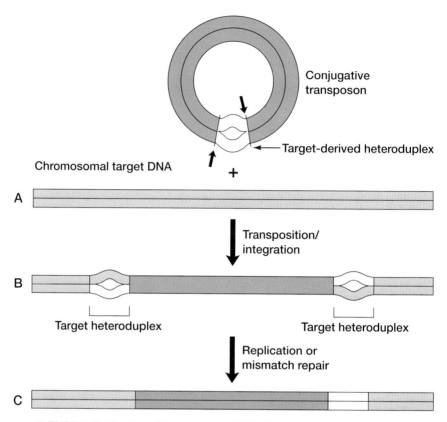

FIGURE 3.18. The fate of heteroduplex DNA during integration of a conjugative transposon. Excision of a conjugative transposon yields a single strand at each end of the transposon derived from the target DNA. Circularization yields a short region where the host-derived sequences do not match in sequence, yielding a short region of heteroduplex (*A*). Transposition into a target replicon results in formation of heteroduplex at each host–virus DNA junction, since the flanking sequences are unlikely to match by chance. The flanking sequences are then homogenized, either by mismatch repair or replication through the heteroduplex. This pathway explains the curious pattern of inheritance of sequences flanking conjugative transposons. (Adapted from Scott and Churchward 1995.[57])

transposon *oriT*. Another gene (*orf14*) encodes an enzyme that may be important for degrading the cell wall to allow penetration and DNA transfer. However, many of the genes important for transfer are of unclear function, and many of the genes expected to be important from studies of conjugative plasmids have not been identified.

In the conjugative transposons, we see another recurrent theme in the mobile DNA field, the formation of new genetic elements that are composites of several independent systems. For Tn*916*, functions for conjugation and transposition have been combined to make a conjugative transposon. Many other such hybrid forms are known. For example, bacteriophage Mu can grow both as a bacterial virus and as a transposon. The antibiotic-resistance elements found in nature are often complex composites built up of many types of mobile DNA elements.

Integrons (14, 15, 41, 44, 53, 54)

Another mechanism that promotes dispersal of antibiotic-resistance genes in gram-negative bacteria involves formation of mobile gene arrays called integrons. The structure of an integron array is shown in Figure 3.19. Many of the antibiotic-resistance genes found in gram-negative bacteria are apparently organized into these clusters.

Integrons are characterized by the following features. An integrase gene is found at one end of the gene array and is apparently responsible for its construction. The integrase gene is a member of the λ integrase family, which includes integrases of bacterial viruses and transposases of conjugative transposons. Upstream of the integrase coding region are two promoters, one (*Pint*) directing expression of the integrase gene, and the other (*Pant*) directing transcription in the opposite direction.

Other genes in the array are located farther upstream of the integrase gene and in the opposite transcriptional orientation, placing them under the control of the *Pant* promoter. This transcription unit can include many genes. Many antibiotic-resistance genes have been found in such arrays, including genes for resistance to rifampin, chloramphenicol, β-lactams, and quaternary ammonium disinfectants. In some strains of *Vibrio cholerae*, the causative agent of cholera, integrons with about 100 gene cassettes have been found.

Integrons are distinguished from simple groups of genes in that each gene in the array is thought to be flanked by a 59-base element that serves as a recombination site recognized by the integrase enzyme. This allows the genes in the integron to be actively rearranged by the action of integrase. The dynamic nature of integron arrays is well documented experimentally. Addition of new genes has been observed, as has the excision of genes as free DNA circles containing a

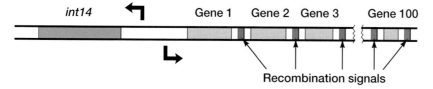

FIGURE 3.19. An integron gene array. The array comprises an integrase gene (*light green*) that acts on recombination sites (*dark green*) between genes in the array (*light gray*). Two promoters lie adjacent to the int gene, one directing expression of integrase and the other directing expression of the genes in the array. Arrays can contain up to 100 genes, many of which have been found to confer resistance to antibiotics. (Adapted from Rowes-Magnus and Mazel 1999.[53])

single 59-base element. However, it is not known whether all the recombination sites between genes are functional.

Much remains to be explained about the nature of integrons. Are integrons derived from conventional mobile elements, such as transposons or phages? Tn7, for example, contains an inactive remnant of a λ integrase gene, indicating that an integron-like mechanism may have incorporated antibiotic-resistance genes into Tn7 DNA. How did the genes in integron arrays come to be flanked by recombination sites? The 59-base elements are actually rather poorly defined sequences, tolerating diverse changes in spacing and base sequence while retaining function. Has the integron recombination system specifically evolved to be sloppy, thereby promoting recruitment of genes with accidental matches to recombination sites? Are 59-base elements actually themselves remnants of transposons, which became inserted randomly and so allowed nearby genes to be recruited into integron arrays? More detailed studies of integrons should provide new insights into the evolution of antibiotic resistance and bacterial genomes generally (a point treated further in Chapter 5).

Resistance Gene Transfer by Transduction or Transformation? (8, 13, 18, 20, 23, 27, 35, 42, 62, 70)

In this section, we briefly review gene transfer by transduction and transformation, mechanisms that are, at most, minor contributors to the dispersal of antibiotic-resistance genes. In the next section, we turn to the composite pathways involving conjugation that are the most prominent transfer mechanisms. Transduction is the focus of Chapter 4, and both transduction and transformation are discussed further in Chapter 5.

Transduction, the transfer of cellular DNA by integrating viruses, is not a major means of transferring antibiotic-resistance genes. One case has been reported by Malke and coworkers, involving transfer of resistance genes among streptococci by phage P13234mo, but reports of phage involvement are rare. It is unclear why transfer of antibiotic-resistance genes by phage is not more common. Transduction is well documented in nature for "pathogenicity islands," sets of genes important for bacterial pathogenesis (see Chapter 4). Possibly size restrictions arise from the need to package enlarged genomes bearing new genes into phage particles. Conjugative plasmids, in contrast, can be very large (see below), so that multiple resistance genes can readily accumulate on the same DNA molecule, perhaps favoring plasmid vehicles over phage.

Transformation is the process by which naked DNA from the extracellular medium is transferred into cells. Most bacteria can be induced to take up DNA from the surrounding medium, as can eukaryotic cells. For example, exposure to an electrical field (electroporation) or $CaCl_2$ (chemical transformation) can induce uptake of DNA. Some bacteria are able to take up DNA in the absence of any special treatment, including *Streptococcus pneumoniae*, *Neisseria gonorrhoeae*, and *Haemophilus influenzae*. These bacteria encode specialized machinery for binding DNA at cell surfaces and internalizing it in an energy-requiring process. For many gram-positive bacteria, the DNA is degraded on one strand, so that a single-stranded linear DNA is introduced into the cell. The single strand can then be incorporated into the host cell chromosome by homologous recombination if appropriate sequence matches are present. For *H. influenzae*, a gram-negative bacterium, transformation takes place by a sequence-specific mechanism. The DNA is bound at the cell surface by a protein that recognizes a specific 9-base-pair sequence in double-stranded DNA, the uptake signal sequence. The DNA in this case is then internalized in double-stranded form and incorporated into the host cell genome, again by homologous recombination.

Transformation and the Discovery That DNA Is the Transforming Principle (2, 28)

Transformation first came to light in the late 1920s in studies of *Pneumococcus pneumoniae*, a serious medical pathogen. Fred Griffith identifed virulent strains (S, for smooth) and avirulent strains (R, for rough) that could be distinguished by their appearance and their pathogenesis in mice. R strains were not lethal when injected into mice, but mixing R strains with heat-killed S strains did yield pathogenic strains. Reisolation of the modified strains revealed that the former R strain had been "transformed" into a pathogenic S strain, despite the fact that no S bacteria had survived the heat treatment.

Griffith gave up this line of study after publishing the discovery of transformation in 1928, but later Avery, McCleod, and McCarty undertook an extensive series of experiments to purify and then identify the chemical nature of the transforming principle. In 1944 they published the discovery that the R to S conversion could be achieved with DNA purified from an S strain, thereby establishing that genetic information could be transmitted with DNA. Within a few years, the transfer of other genes by transformation was also documented, implying that the R to S conversion was not an isolated case. This seminal discovery indicated that DNA was the chemical embodiment of genetic information, leading Watson and Crick to focus on DNA in the work that led to the discovery of the double-helical structure of DNA.

It is not presently clear to what degree transformation mediates transfer of antibiotic-resistance genes. If a conjugative plasmid is found to be transferred from one strain to another, the obvious inference is that conjugation mediated the transfer. However, transformation could potentially have mediated transfer as well. For the sequence-specific transformation systems such as that in *H. influenzae*, it seems likely that transformation will mediate transfer most efficiently between closely related strains, because only close relatives have high densities of the uptake signal sequences in their DNA. The general feeling is that transformation plays a minimal role in antibiotic gene transfer, but this view may change with new data.

A related pathway might also mediate transfer of transposons between cells. William Reznikoff and coworkers have found that com-

plexes of Tn5 DNA bound to Tn5 transposase protein can be introduced into cells intact, after which they can carry out integrative
transposition. It is unknown whether such "transformation" of transposition complexes takes place in nature, but it is efficient enough in
the laboratory to be useful for biotechnological manipulations.
Perhaps future studies will reveal that this pathway mediates gene
transfer in natural settings.

Pathways for Transfer of Antibiotic-resistance Genes (49, 50, 57)

In nature, the systems that transfer antibiotic-resistance genes are often
composites of different types of mobile elements. In this section, we
consider examples of composite gene transfer pathways that result,
and in the next section, some of the very complex hybrid elements
found in antibiotic-resistant organisms.

One means of mobilizing a DNA sequence involves capturing a
chromosomal gene in a composite transposon. Such a mechanism has
been postulated to explain the development of resistance to vancomycin. One set of vancomycin-resistance genes, the *vanB* operon,
appears to have been mobilized by formation of the composite transposon, Tn*1547* (Fig. 3.20). Tn*1547* likely arose by sequential insertions
of IS elements (related to IS*256*) on either side of the *vanB* genes in the
chromosome of the original bacterial host. Although insertion at each
of these locations is likely a low-frequency event, such events nevertheless occur in very large bacterial populations harboring active IS
elements. IS*256* transposase can then act at the "outside" end of each
IS*256* element to mobilize the *vanB* gene cluster. This pathway provides a general mechanism for mobilizing any linked group of genes.

Conjugative plasmids can act as targets for insertion of composite
transposons, allowing the transposon and captured cellular genes to
be transferred between cells just like any other plasmid sequence. In
nature, conjugative plasmids with integrated transposons carrying
resistance genes are common. The *vanB* cluster carried on Tn*1547* was
in fact found on a conjugative plasmid. Many resistance transposons
can accumulate in a single conjugative plasmid.

Another means of incorporating resistance determinants into conjugative plasmids involves formation of plasmid cointegrates. If a
replicative transposon on one plasmid directs formation of a cointegrate, but then fails to resolve, the resulting intermediate is trapped.
This results in the joining of the two plasmids, with copies of the IS element or transposon between the two plasmid sequences. If one plasmid is conjugative, the whole composite can be transferred as a single
unit. Many of the multiple antibiotic-resistance elements found in
nature also contain such trapped cointegrate structures.

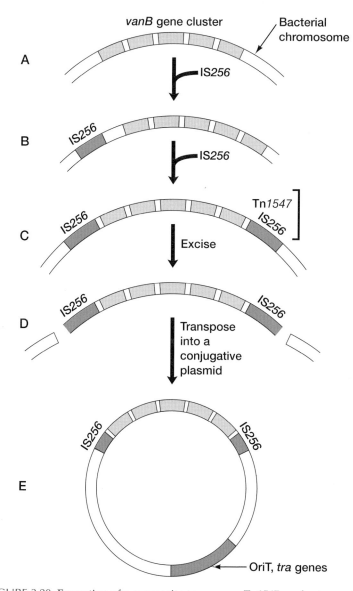

FIGURE 3.20. Formation of a composite transposon Tn*1547* conferring resistance to vancomycin. Genes closely related to the *van* genes have been found to be normal components of some bacterial chromosomes (*A*). Transposition of IS*256*-related elements (named IS*256*-like and IS*16*) to either side of the *van* gene cluster (*B* and *C*) results in formation of a composite transposon. Excision of the transposon (*D*) and integration into a conjugative plasmid (*E*) allows dispersal of the genes by conjugation. (Data from Quintiliani and Courvalin 1996.[50])

Several pathways can mediate mobilization of bacterial chromosomal sequences by conjugation. As discussed above, if a conjugative plasmid becomes integrated into the host cell chromosome, transfer starting at *oriT* can proceed through the plasmid sequences and into flanking cellular sequences, thereby allowing transfer of genes from the host bacteria. Several mechanisms are known for inserting conjugative plasmids into host chromosomal DNA and thereby potentiating transfer. If regions of sequence identity are present in the plasmid and the host chromosome, homologous recombination can incorporate the conjugative plasmid into the cellular chromosome. Transposons can play an important role in this pathway, since they can provide portable regions of sequence homology that can serve as sites for homologous recombination. In a related mechanism, if transposons in both the chromosome and a conjugative plasmid have sites for resolvase (*res* sites), and a source of resolvase is present, resolvase-mediated recombination can incorporate the plasmid into the chromosome. Other transposon-mediated events can also fuse a plasmid and the host chromosome. For example, cointegrate formation involving the chromosome and a plasmid can also incorporate a conjugative plasmid. These pathways involving conjugative plasmids can equally well be used for gene transfer by conjugative transposons.

Naturally Occurring Composite Elements (9, 10, 41, 47, 73)

Naturally occurring antibiotic-resistance elements are often complex composites transporting many resistance genes. Such complexity is illustrated by the composite plasmid R100 (Fig. 3.21), which was isolated in Japan in the late 1950s from the gram-negative bacterium *Shigella flexneri*, which causes dysentery. Many antibiotic-resistance plasmids are designated "R" for resistance, followed by an isolate number.

R100 is a 94.5-kilobase conjugative plasmid encoding typical *tra*, *rep*, and *oriT* sequences. In addition, R100 encodes genes for resistance to tetracycline (*tetA*), chloramphenicol (*catA1*), streptomycin (*aadA1*), spectinomycin (*aadA1*), quaternary ammonium antiseptics (*qacE*), and sulfonamide (*sulI*). Integrated into R100 are several transposons that contributed the resistance determinants. The *tetA* determinant is located on an integrated copy of Tn*10*. The rest of the antibiotic-resistance determinants are carried on Tn*2670*, a complex composite apparently built up from five transposons integrated into one another. Further complicating this array is the presence of an integron in one of the transposon remnants. The Tn*2670*, one of the R100 transposons, was formed by integration of Tn*21* into transposon Tn*9*. Tn*9* contributed the chloramphenicol-resistance determinant. Tn*21* is typically found in association with genes for resistance to mercury (*Mer* operon), and these genes are also found in the Tn*2670* copy. Mercury is highly toxic to bacteria. Selection for resistance to mercury may have occurred in humans during treatment of syphilis with mercury or during leaching of mercury from dental amalgams. The copy of Tn*21* in Tn*2670* contains the integron, which encodes genes for integrase and resistance to antiseptics, sulfonamides, and streptomycin. The integron is apparently a component of yet another transposon, since it is linked to copies

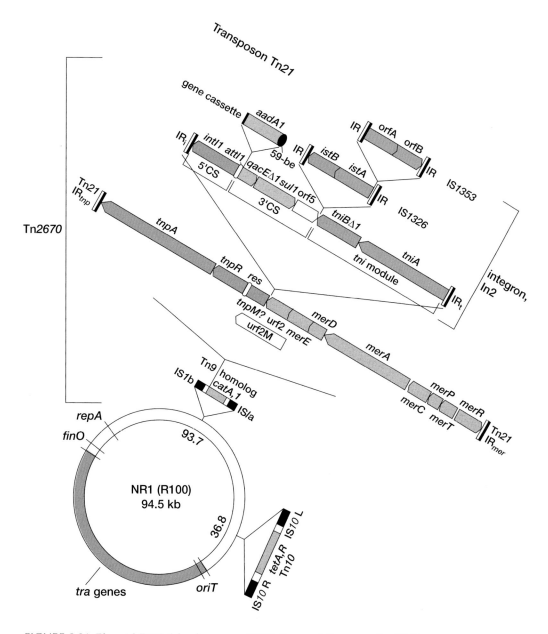

FIGURE 3.21. Plasmid R100 (also known as NR1). Some of the genes for DNA mobilization (transposases and tra functions) are shown in green. Genes for resistance to antibiotics or antiseptics are shown in gray. The thin lines connecting the transposons to NR1 indicate the points of insertion. (Modified, with permission, from Liebert et al. 1999 [© American Society for Microbiology]. [41])

of transposase genes and flanked by characteristic repeats. In this case, however, the transposase genes do not appear to be active and may represent molecular "fossils" that have remained linked to the integron. It may be that the integron can be mobilized by transposase supplied by another element located elsewhere in the genome, since close-

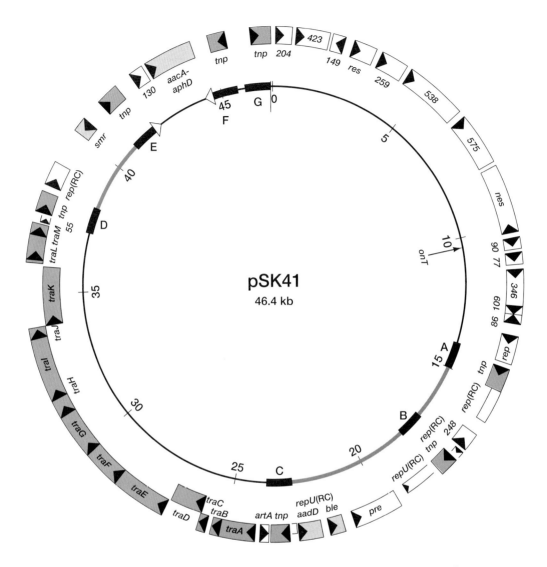

FIGURE 3.22. Plasmid pSK41. Genes for DNA mobilization are shown in green, genes for resistance to antibiotics or antiseptics are shown in gray. Transposon copies are labeled *A–D* on the inner ring. (Modified, with permission, from Berg et al. 1998 [© American Society for Microbiology]. [9])

ly related integrons have been found independently of Tn*21* in other bacteria. An IS*1326* element is inserted into the Tn*2670* integron, and IS*1353* has inserted into IS*1326*, yielding the observed complement of transposons.

The sequence of R100 also contains further probable genes of unknown function. Perhaps some of these genes provide resistance to more antibacterial agents. Some of the additional genes may also encode functions that permit the host bacteria to tolerate the presence of the antibiotic-resistance genes themselves. If the products of resistance genes impose a burden on the bacterial host, additional genes may evolve to reduce the cost. The tet repressor (*tetR*) encoded by Tn*10* is such a gene. The *tetR* gene encodes a DNA-binding protein that regulates transcription of *tetA*. In the absence of tetracycline, tet repressor binds to the *tetA* promoter and inhibits transcription. When tetracycline is present, it binds to the tet repressor and converts it into an inactive form. This increases expression of *tetA* and confers resistance to tetracycline. The expression of *tetA* is known to be disadvantageous to cells in the absence of tetracycline, so it is turned on only when needed.

Another example of consolidated resistance genes is the plasmid pSK41 from *Staphylococcus aureus*. Plasmid pSK41 is a 46.4-kilobase conjugative plasmid that confers resistance to gentamicin, tobramycin, kanamycin (*aacA-aphD*), neomycin (*aadD*), bleomycin (*ble*), and multiple antiseptics (*smr*) (Fig. 3.22). In addition to the resistance genes, and the usual *tra* and *rep* genes, the plasmid contains seven copies of IS*257* and two copies of IS*256*. One set of IS elements and resistance genes comprises Tn*4001*, a transposon also found in other cases of antibiotic resistance. Of particular interest is the incorporation of three small plasmids into the progenitor of pSK41. The genes for resistance to bleomycin, neomycin, and antiseptics are all introduced into pSK41 via these fused plasmids. Each plasmid is flanked by IS elements, indicating a probable role for the transposons in capture of the plasmids. A likely pathway would involve replicative transposition of an IS element present in one plasmid into the fusion partner, yielding a cointegrate. A lack of subsequent resolution would yield the observed structures. Each of the three small plasmids appears to have been subsequently inactivated for replication, making each dependent on the pSK41 origin for growth.

Many other such complex resistance determinants have been documented. As described above, Tn*1547*, the composite transposon responsible for mobilizing the *VanB* cluster, is often found on a conjugative plasmid. These examples of conjugative antibiotic-resistant composites provide only a glimpse of the remarkable diversity of elements found in contemporary antibiotic-resistant strains.

Why are antibiotic-resistance determinants consolidated into composite plasmids? Exposure to antibiotics is of course likely to be involved, but why not maintain populations of many smaller plasmids? It may be that if a bacterium is exposed to one antibiotic, it will likely be exposed to more later. The more resistance genes they possess, the greater the chance of survival for the plasmid and the host bacteria, as long as the resistance genes themselves do not impose too great a burden on the functioning of the bacterial cell. In addition, consolidation of resistance determinants into conjugative plasmids offers the advantage of transferring multiple resistance genes in a single event, to the advantage of both the plasmid and its new host.

Both R100 and pSK41 are members of large families of related plasmids. In some cases, relatives have been isolated from different pathogenic bacteria, indicating likely lateral transfer. The related plasmids in each family differ by insertion of new transposons, deletions adjacent to transposons, and insertion of additional plasmids.

MICROBIAL ECOLOGY AND ANTIBIOTIC RESISTANCE (21, 55, 56, 67, 71)

The human body exists in association with roughly 10 trillion bacteria comprising at least 500–600 species. Twenty to fifty percent of the volume of the human colon is thought to be occupied by bacteria. Many of these bacterial species can become pathogenic to humans, but most of the time they are harmless. Disruptions of normal homeostasis, perhaps by an immunodeficiency in the human host, can allow otherwise harmless strains to multiply and become pathogenic. As discussed in Chapter 4, mobile DNA systems can also incorporate genes for pathogenic determinants into otherwise benign bacteria, leading to disease.

The introduction of antibiotic-resistance genes into any bacteria capable of existing in humans can potentially transfer resistance to a pathogenic strain. The donor bacteria need not even live for very long. For example, selection of tetracycline resistance in ruminates led to the development of resistance in bacteria resident in the animal gut. Such strains do not normally colonize the human gut. Laboratory tests revealed, however, that the tetracycline-resistant determinant could be transferred in the laboratory to *Bacteroides* species found in humans. Studies of later natural isolates of *Bacteroides* from the human gut revealed almost identical tetracycline-resistant determinants to those found in ruminants, supporting a model for transfer into *Bacteroides* by a ruminant strain transiently present in the human gut.

New antibiotic-resistant strains are often detected in hospitals, and many may have evolved there to begin with. Overuse of antibiotics in hospitals, often combined with inadequate hygiene and a centralization of pathogenic strains, arguably provides the most important force for development of resistant bacteria. Weakened patients may not be able to fight off infections as might healthy people, and they also may not tolerate high doses of antibiotics, leading to treatment with ineffective doses. Such patients can serve as incubators for new resistant strains. In cases of incomplete therapy, where bacterial infections are not fully cleared, strains with low-level resistance have an opportunity to expand at the expense of sensitive strains. Further mutation or gene acquisition can then result in increased resistance. Treatments with multiple antibiotics, common in hospital settings, can select for multiply resistant strains, as with the MRSA strains of *S. aureus* discussed at the beginning of the chapter. It has even been suggested that waste water from hospitals may contain high enough concentrations of some antibiotics to select for resistant bacteria in sewage.

Less well known is the effect of antibiotic use in animal husbandry. It is estimated that 50% of antibiotics sold in the developed world are used as growth promoters in agriculture. Excessive use of antibiotics in farm animals has hastened the development of resistance to many

It is estimated that 5% of all patients become infected with bacteria during hospitalization in the United States, so-called nosocomial infection. In 1992, over two million patients suffered nosocomial infections, costing the health-care system an estimated $4.5 billion dollars. Many of these infections were with antibiotic-resistant bacterial strains.

antibiotics used to treat infections in humans. For example, the antibiotic nourseothricin has been used as a growth promoter in animals. Two years after its introduction, resistance was found in *E. coli* from meat of treated animals. Resistance was later found in farmers, their families, and patients suffering from *E. coli* infections. Molecular analysis revealed that the transposon carrying the resistance gene was nearly identical in strains from human and animal sources. Later the resistance determinant was found in *Shigella*, a pathogen found only in humans.

Perhaps most frightening is the rise of vancomycin resistance. Vancomycin is the antibiotic of last resort for treatment of infections by MRSA *S. aureus* and resistant strains of *Enterococcus faecalis*. The effectiveness of vancomycin, however, has been severely eroded by the use of avoparcin in animals. Selection of resistance to avoparcin confers cross-resistance to vancomycin. In Denmark in 1994, for example, 24 kg of vancomycin was sold for use in humans, while 24,000 kg of avoparin was sold for use in animals. From 1992 to 1996, Australia imported 582 kg of vancomycin and 62,642 kg of avoparcin per year. A convincing test of transmission of antibiotic resistance from animals to people is possible in this case, because avoparcin was used extensively in Europe but not in the United States. In Europe, vancomycin-resistant bacteria could be detected in animals and in the guts of European meat-eaters. However, resistant bacteria were not detected in European vegetarians nor in North Americans. Although it has been argued by some that the evidence is inconclusive, it seems highly likely that the use of avoparcin has accelerated the spread of vancomycin resistance to human pathogens.

In another example, in 1996 the fluoroquinoline antibiotics were approved for use in chickens and turkeys, despite strong opposition by the Centers for Disease Control. Resistance quickly spread in *Campylobacter* species in chickens. By 1999, 17.6% of *C. jejuni* and 30% of *C. coli* isolated from humans were resistant. These bacteria are a major source of food poisoning and can cause chronic conditions such as Guillian-Barre syndrome and reactive arthritis, and now a substantial fraction are antibiotic-resistant. Inexplicably, at the time of this writing, a major pharmaceutical company is still pressing to continue use of fluoroquinolines in animals.

The public discussion of antibiotic resistance has often failed to consider the impact of lateral DNA transfer. Mobile DNA systems have a pool of naturally occurring resistance genes to act on. Given the ubiquitous and highly active systems for lateral transfer, it seems inevitable that these resistance genes will appear quickly in medically important pathogens in response to treatment. The threat of losing effective antibiotic treatments warrants much more thoughtful management of antibiotic use, incorporating a realistic appraisal of the high frequency of lateral DNA transfer.

Summary

Antibiotic drugs specifically poison microbes, usually by targeting systems present in microbes but not humans. Unfortunately, the powerful selective pressure exerted by these drugs leads to the development of antibiotic resistance. Much of the resistance to antibiotics arises from the action of dedicated genes that block the function of antibiotics, although some important types of antibiotic resistance also arise from point mutations in genes encoding antibiotic targets. Antibiotic-resistance genes typically encode products that inactivate antibiotics, modify the antibiotic target, or reduce the concentration of antibiotics in cells.

Antibiotic-resistance genes are transferred between cells at alarming rates. The most prominent vehicles for transfer are conjugative plasmids, which themselves encode the machinery for DNA transfer. Among the encoded proteins are those for pili formation, fibrous structures that link cells for conjugative transfer. To carry out transfer, plasmid DNA is nicked and a single DNA strand is passed between cells, again employing plasmid-encoded functions. DNA replication generates the second strand on the plasmid copies in the donor and recipient cells. After conjugation is completed, each cell contains a complete plasmid copy.

Another class of mobile elements, the transposons, is also prominent in transferring antibiotic-resistance genes. Transposons by definition can direct the insertion of element copies at new locations in the host cell. Often the starting copy remains intact, leading to a net increase in transposon number. Bacterial transposons encode a transposase protein that carries out the DNA breaking and joining reactions mediating transposition and DNA sites at the ends of the element that bind transposase and organize the protein and DNA for transposition reactions.

Transposons can capture and transport genes for antibiotic resistance. In composite transposon formation, for example, integration of two IS elements on either side of an antibiotic-resistance gene can lead to transposition of the full unit, including the two transposons and the resistance gene.

Transposition of such units to conjugative plasmids creates elements capable of mobilizing resistance determinants. Combined with the strong selective pressures of multiple drug treatment regimens, such mechanisms can assemble large conjugative elements encoding many antibiotic-resistance genes.

Today the spread of such multiple-resistance determinants threatens the effectiveness of antibiotics. Multidrug-resistant *S. aureus* is a growing threat, particularly coupled with the emergence of resistance to vancomycin, the drug of last resort. After the introduction of a new antibiotic, resistance strains typically emerge after only 2–5 years. Addressing this growing crisis will require a much more realistic appreciation of the widespread lateral transfer of resistance genes.

REFERENCES

1. Atlas R.M. 1984. *Microbiology: Fundamentals and applications.* Macmillan, New York.
2. Avery O.T., MacLeod C.M., and McCarty M. 1944. Studies on the chemical nature of the substance inducing transformation of pneumococcal types. Induction of transformation by a desoxyribonucleic acid fraction isolated from pneumococcus type III. *J. Exp. Med.* **79:** 137–159.
3. Bainton R., Gamas P., and Craig N.L. 1991. Tn7 transposition in vitro proceeds through an excised transposon intermediate generated by staggered breaks in DNA. *Cell* **65:** 805–816.

4. Bainton R.J., Kubo K.M., Feng J., and Craig N.L. 1993. Tn7 transposition: Target DNA recognition is mediated by multiple Tn7-encoded proteins in a purified in vitro system. *Cell* **72:** 931–943.

5. Bender J. and Kleckner N. 1986. Genetic evidence that Tn10 transposes by a nonreplicative mechanism. *Cell* **45:** 801–815.

6. Berg D.E. 1989. Transposon Tn5, In *Mobile DNA* (ed. D.E. Berg and M.M. Howe), pp. 185–210. American Society for Microbiology, Washington, D.C.

7. Berg D.E. and Howe M.M., Eds. 1989. *Mobile DNA.* American Society for Microbiology, Washington, D.C.

8 Berg D.E., Davies J., Allet B., and Rochaix J.-D. 1975. Transposition of R factor genes to bacteriophage lambda. *Proc. Natl. Acad. Sci.* **72:** 3628–3632.

9. Berg T., Firth N., Apisiridej S., Hettiaratchi A., Leelaporn A., and Skurray R.A. 1998. Complete nucleotide sequence of pSK41: Evolution of staphylococcal conjugative multiresistance plasmids. *J. Bacteriol.* **180:** 4350–4359.

10. Brown H.J., Stokes H.W., and Hall R.M. 1996. The integrons In0, In2, and In5 are defective transposon derivatives. *J. Bacteriol.* **178:** 4429–4437.

11. Byrd D.R. and Matson S.W. 1997. Nicking by transesterification: The reaction catalyzed by a relaxase. *Mol. Microbiol.* **25:** 1011–1022.

12. Caparon M.G. and Scott J.R. 1989. Excision and insertion of the conjugative transposon Tn916 involves a novel recombination mechanism. *Cell* **59:** 1027–1034.

13. Chan R.K., Botstein D., Watanabe T., and Ogata Y. 1972. Specialized transduction of tetracycline resistance by phage P22 in *Salmonella typhimurium.* II. Properties of a high-frequency transducing lysate. *Virology* **50:** 883–898.

14. Collis C.M. and Hall R.M. 1992. Gene cassettes from the insert region of integrons are excised as covalently closed circles. *Mol. Microbiol.* **6:** 2875–2885.

15. Collis C.M. and Hall R.M. 1992. Site-specific deletion and rearrangement of integron insert genes catalyzed by the integron DNA integrase. *J. Bacteriol.* **174:** 1574–1585.

16. Craig N.L. 1989. Transposon Tn7. In *Mobile DNA.* (ed. D.E. Berg and M.M. Howe), pp. 211–225. American Society for Microbiology, Washington, D.C.

17. Craigie R. and Mizuuchi K. 1985. Mechanism of transposition of bacteriophage Mu: Structure of a transposition intermediate. *Cell* **41:** 867–876.

18. Danner D.B., Deich R.A., Sisco K.L., and Smith H.D. 1980. An eleven-basepair sequence determines the specificity of DNA uptake in *Haemophilus* transformation. *Gene* **11:** 311–318.

19. Davis C.E. and Anandan J. 1970. The evolution of r factor. A study of a "preantibiotic" community in Borneo. *N. Engl. J. Med.* **282:** 117–122.

20. Davison J. 1999. Genetic exchange between bacteria in the environment. *Plasmid* **42:** 73–91.

21. Falkow S. and Kennedy D. 2001. Antibiotics, animals and people—Again! *Science* **291:** 39.

22. Flannagan, S. E., Zitzow L.A., and Clewell D.B. 1994. Nucleotide sequence of the 18-Kb conjugative transposon Tn916 from *Enterococcus faecalis. Plasmid* **32:** 350–354.

23. Fleischmann R.D., Adams M.D., White O., Clayton R.A., Kirkness E.F., Kerlavage A.R., Bult C.J., Tomb J.F., Dougherty B.A., Merrick J.M., et al. 1995. Whole-genome random sequencing and assembly of *Haemophilus*

influenzae Rd. *Science* **269:** 496–512.

24. Gardner P., Smith D., Beer H., and Moellering Jr., R.C. 1969. Recovery of resistance (R) factors from a drug-free community. *Lancet* **2:** 774–776.

25. Gomis-Ruth F.X., Moncalian G., Perez-Luque R., Gonzalez A., Cabezon E., de la Cruz F., and Coll M. 2001. The bacterial conjugation protein TrwB resembles ring helicases and F1-ATPase. *Nature* **409:** 637–641.

26. Gorbach S.L., Bartlett J.G., and Blacklow N.R. 1998. *Infectious diseases.* W.B. Saunders, Philadelphia, Pennsylvania.

27. Gottesman M.M. and Rosner J.L. 1975. Acquisition of a determinant for chloramphenicol resistance by coliphage lambda. *Proc. Natl. Acad. Sci.* **72:** 5041–5045.

28. Griffith F. 1923. *Bacterial studies: The influence of immune serum on the biological properties of pneumococci.* Ministry of Health, London.

29. Hayes W. 1953. Observations on a transmissible agent determining sexual differentiation in *E. coli. J. Gen. Microbiol.* **8:** 72–88.

30. Hotchkiss R.D. 1951. Transfer of penicillin resistance in pneumococci by the desoxyribonucleate derived from resistant cultures. *Cold Spring Harbor Symp. Quant. Biol.* **16:** 457–461.

31. Kennedy A.K., Guhathakurta A., Kleckner N., and Haniford D.B. 1998. Tn10 transposition via a hairpin intermediate. *Cell* **95:** 125–134.

32. Kleckner N. 1981. Transposable elements in prokaryotes. *Annu. Rev. Genet.* **15:** 341–404.

33. Kleckner N. 1989. Transposon Tn*10*. In *Mobile DNA* (ed. D.E. Berg and M.M. Howe), pp. 227–268. American Society for Microbiology, Washington, D.C.

34. Kleckner N., Chan R.K., Tye B.K., and Botstein D. 1975. Mutagenesis by insertion of a drug-resistance element carrying an inverted repetition. *J. Mol. Biol.* **97:** 561–575.

35. Kornberg A. and Baker T. 1991. *DNA replication.* W.H. Freeman, New York.

36. Lanka E. and Wilkins B.M. 1995. DNA processing reactions in bacterial conjugation. *Annu. Rev. Biochem.* **64:** 141–169.

37. Leclercq R., Derlot E., Duval J., and Courvalin P. 1988. Plasmid-mediated resistance to vancomycin and teicoplanin in *Enterococcus faecium. New Engl. J. Med.* **319:** 157–161.

38. Lederberg J. and Tatum E. 1946. Gene recombination in *E. coli. Nature* **158:** 558.

39. Lederberg J. and Tatum E.L. 1953. Sex in bacteria: Genetic studies 1945–1952. *Science* **118:** 169–175.

40. Levy S.B. and Novick R.P. , eds. 1986. Antibiotic resistance genes: Ecology, transfer, and expression. *Banbury Rep.,* vol. 24. Cold Spring Harbor Laboratory, Cold Spring Harbor, New York.

41. Liebert C.A., Hall R.M., and Summers A.O. 1999. Transposon Tn21, flagship of the floating genome. *Microbiol. Mol. Biol. Rev. 63:* 507–522.

42. Malke H.R., Starke W., Kohler W., Kolesnichenko T.K., and Totolian A.A. 1975. Bacteriophage P12334mo-mediated intra- and intergroup transductiohn of antibiotic resistance among streptococci. *Zentbl. Bakteriol. Mikrobiol. Hyg. I Abt. Orig. A. 233:* 24–34.

43. May E.W. and Craig N.L. 1996. Switching from cut-and-paste to replicative Tn7 transposition. *Science 272:* 401–404.

44. Mazel D., Dychinco B., Webb V.A., and Davies J. 1998. A distinctive class of integron in the *Vibrio cholerae* genome. *Science 280:* 605–608.

45. Murley L.L. and Grindley N.D.F. 1998. Architecture of the γ-δ resolvase synaptosome: Oriented heterodimers identify interactions essential for synapsis and recombination. *Cell* **95:** 553–562.

46. Murray P.R., Rosenthal K.S., Kobayashi G.S., and Pfaller M.A. 1998. *Medical microbiology.* Mosby, St. Louis, Missouri.

47. Nakaya R., Nakamura A., and Murata Y. 1960. Resistance transfer agents in *Shigella. Biochem. Biophys. Res. Commun.* **3:** 654–659.

48. Neidhardt F.C., Ingraham J.L., Low K.B., Magasanik B., Schaechter M., and Umbarger H.E. 1987. Escherichia coli *and* Salmonella typhimurium. American Society for Microbiology, Washington, D.C.

49. Quintiliani R. and Courvalin P. 1994. Conjugal transfer of the vancomycin resistance determinant *vanB* between enterococci involves the movement of large genetic elements from chromosome to chromosome. *FEMS Microbiol. Lett.* **119:** 359–364.

50. Quintiliani R. and Courvalin P. 1996. Characterization of Tn1547, a composite transposon flanked by the IS16 and IS256-like elements, that confers vancomycin resistance in *Enterococcus faecalis* BM4281. *Gene* **172:** 1–8.

51. Rang H.P., Dale M.M., Ritter J.M., and Gardner P. 1995. *Pharmacology.* Churchill Livingston, New York.

52. Reed R.R. 1981. Transposon-mediated site specific recombination: A defined in vitro system. *Cell* **25:** 713–719.

53. Rowe-Magnus D.A. and Mazel D. 1999. Resistance gene capture. *Curr. Opin. Microbiol.* **2:** 483–488.

54. Rowe-Magnus D.A., Guerout A.-M., Ploncard P., Dychinco B., Davies J., and Mazel D. 2001. The evolutionary history of chromosomal super-integrons provides an ancestry for multiresistant integrons. *Proc. Natl. Acad. Sci.* **98:** 652–657.

55. Salyers A.A. 1995. *Antibiotic resistance transfer in the mammalian intestinal tract: Implications for human health, food safety and biotechnology.* Springer-Verlag, Heidelberg and R.G. Landes, Austin, Texas.

56. Schouten M.A., Voss A., and Hoogkamp-Korstanje J.A.A. 1997. VRE and meat. *Lancet* **349:** 1258.

57. Scott J.R. and Churchward G.G. 1995. Conjugative transposition. *Annu. Rev. Microbiol.* **49:** 367–397.

58. Senghas E., Jones J.M., Yamamoto M., Gawron-Burke C., and Clewell D.B. 1988. Genetic organization of the bacterial conjugative transposon Tn916. *J. Bacteriol.* **170:** 245–249.

59. Shapiro J. 1979. A molecular model for the transposition and replication of bacteriophage Mu and other transposable elements. *Proc. Natl. Acad. Sci.* **76:** 1933–1937.

60. Sherratt D.J., ed. 1995. *Mobile genetic elements.* Oxford University Press, Oxford, United Kingdom.

61. Sieradzki K., Roberts R.B., Haber S.W., and Tomasz A. 1999. The development of vancomycin resistance in a patient with methicillin-resistant *Staphylococcus aureus* infection. *New Engl. J. Med.* **340:** 517–523.

62. Smith H.O., Tomb J.F., Dougherty B.A., Fleischmann R.D., and Venter J.C. 1995. Frequency and distribution of DNA uptake signal sequences in the *Haemophilus influenzae* Rd genome. *Science* **269:** 538–540.

63. Sobecky P.A., Mincer T.J., Chang M.C., and Heliniski D.R. 1997. Plasmids isolated from marine sediment microbial communities contain replication and incompatibility regions unrelated to those of known plasmid groups.

Appl. Environ. Microbiol. **63:** 888–895.

64. Stark W.M., Boocock M.R., and Sherratt D.J. 1989. Site-specific recombination by Tn3 resolvase. *Trends Genet.* **5:** 304–309.

65. Stellwagen A.E. and Craig N.L. 1998. Mobile DNA elements: Controlling transposition with ATP-dependent molecular switches. *Trends Biochem. Sci.* **23:** 487–490.

66. Tenover F.C., Biddle J.W., and Lancaster M.V. 2001. Increasing resistance to vancomycin and other glycopeptides in *Staphylococcus aureus. Emerg. Infect. Dis.* **7:** 327–332.

67. Tschäpe H. 1994. The spread of plasmids as a function of bacterial adaptability. *FEMS Microbiol. Ecol.* **15:** 23–32.

68. Walsh C.T., Fisher S.L., Park I.-S., Prahalad M., and Wu Z. 1996. Bacterial resistance to vancomycin: Five genes and one missing hydrogen bond tell the story. *Chem. Biol.* **3:** 21–28.

69. Watanabe T. 1963. Infective heredity of multiple drug resistance in bacteria. *Bacteriol. Rev.* **27:** 87–115.

70. Watanabe T., Ogata Y., Chan R.K., and Botstein D. 1972. Specialized transduction of tetracycline resistance by phage P22 in *Salmonella typhimurium.* I. Transduction of R factor 222 by P22. *Virology* **50:** 874–882.

71. Witte W. 1998. Medical consequences of antibiotic use in agriculture. *Science* **279:** 996–997.

72. Wolkow C.A., DeBoy R.T., and Craig N.L. 1996. Conjugating plasmids are preferred targets for Tn7. *Genes Dev.* **10:** 2145–2157.

73. Womble D.D. and Rownd R.H. 1988. Genetic and physical map of plasmid NR1: Comparison with other IncFII antibiotic resistance plasmids. *Microbiol. Rev.* **52:** 433–451.

74. Wu S., Piscitelli C., de Lencastre H., and Tomasz A. 1996. Tracking the evolutionary origin of the methicillin resistance gene: Cloning and sequencing of a homologue of mecA from a methicillin susceptible strain of *Staphylococcus sciuri. Microb. Drug Resist.* **2:** 435–441.

75. Yang W. and Steitz T.A. 1995. Crystal structure of the site-specific recombinase γ-δ resolvase complexed with a 34 bp cleavage site. *Cell* **82:** 193–207.

Phage Transduction and Bacterial Pathogenesis

G ENE TRANSDUCTION, OR THE TRANSFER OF A DNA sequence from one cell to another by a virus, takes place in the environment at a remarkably high rate. Transduction of genes between marine bacteria may take place in the oceans at the rate of 20 million billion times per

Bacteriophages are termed virulent or temperate according to their style of growth. Virulent phages always lyse the infected cell, whereas temperate phages have another alternative, establishing a quiescent relationship with the host, termed lysogeny.

second (Chapter 1). In recent years, phage transduction has also emerged as a central player in human disease.

In this chapter, we review mechanisms and consequences of gene transfer by bacteriophages. We begin with descriptions of phages and their life cycles, with a focus on the integrating or temperate phages, since these are often involved in stable transfer of genes for bacterial pathogenesis. Beautiful mechanistic studies have been carried out on temperate phages for decades, making them some of the best understood gene transfer systems. The mechanisms of DNA transduction are then examined in more detail. The chapter concludes with a survey of phages transporting blocks of genes that increase the pathogenicity of bacterial invaders, the "pathogenicity islands."

TRANSDUCTION AND BACTERIAL PATHOGENESIS (32, 51, 62)

Many bacterial species can be either benign passengers in the human body or malignant invaders, depending on whether they possess specific genes for pathogenic determinants. These genes are often clustered together in the DNA, leading to the name pathogenicity islands for such gene sets. Pathogenicity islands are discussed in detail in the later sections of this chapter, but as an initial example, diphtheria toxin is a well-studied case. Diphtheria is caused by *Corynebacterium diphtheriae*, a gram-positive bacterium. The disease begins with colonization of the throat by *Corynebacterium* that harbor the diphtheria pathogenicity island. The first symptoms include fever, sore throat, and loss of appetite, and then a grayish "pseudomembrane" forms in the throat, composed of bacteria, immune system cells, and other components. In the most serious cases, diphtheria toxin is produced at sites of infection and spreads systemically, causing irregular heartbeat, difficulty in swallowing, coma, and sometimes death. Today in the developed world diphtheria is largely controlled, thanks to an effective vaccine, but in the developing world diphtheria is still a major threat to children.

Pathogenic and harmless strains of *C. diphtheriae* differ only by the presence of a bacteriophage, either corynephage β or corynephage ω. These phages encode a secreted toxin that leads to the destruction of host cells. Diphtheria toxin is taken up by cells in the throat and other tissues, where it directs ribosylation of the host cell EF2 protein. This modification disrupts translational elongation and is ultimately fatal to the cell. The benefit of this to the bacteria is probably at least in part that it allows iron to be obtained from the lysed host cell. Iron, it turns out, is often a limiting resource for pathogenic bacteria, and many of them encode iron-scavenging systems. Sickening the host organism is counterproductive, since this can destroy the comfortable niche occupied by the pathogenic bacterium, but obtaining iron is a high priority. Many pathogenicity genes can be understood as promoting the survival of the bacterium in which they reside, with pathogenesis in the host organism a secondary consequence.

Early Studies of Bacteriophage (4, 5)

Bacteriophages have been the subject of intense study since their discovery independently by Frederick W. Twort (1915) and Felix d'Herelle (1917). Early workers, much impressed by the ability of lytic phages to kill bacteria, attempted at length to develop phages for therapy of bacterial diseases. Although ultimately unsuccessful, work during this period (1915 to around 1930) made possible later studies of bacteriophage replication.

In the late 1930s, a talented group of physicists became interested in phage model systems. Phages were clearly small, because they passed through filters that retained bacteria, and relatively simple, because they were able to multiply only in the presence of their bacterial host. As early as 1922, Hermann Muller suggested that the study of phages might lead to insights into the nature of the gene. The "Phage Group," led by Max Delbrück, focused tightly on phages as a model that captured much of the problem of biological replication. The Phage Group carried out an influential series of experiments on phage replication that comprises much of the underpinnings of modern molecular biology today. These early studies of phage have been the topic of several excellent books, notably *Phage and the Origins of Molecular Biology*, edited by John Cairns, Gunther S. Stent, and James D. Watson, and *Felix d'Herelle and the Origins of Molecular Biology* by William Summers.

Early studies of phage growth revealed an unexpected twist. Many phages simply multiply lytically when mixed with bacteria, killing the host and liberating new phage particles in a straightforward way. Others, however, were found to adopt an unexpectedly benign relationship with their hosts. Some pure cultures of bacteria seemed to produce phages spontaneously for no obvious reason. Study of this odd observation and others eventually revealed that some phages could stably coexist with bacteria, a style of growth termed lysogeny. Delbrück initially did not believe in lysogeny and was not convinced to the contrary until André Lwoff developed methods for experimentally inducing lysogens, ultimately allowing the observation of induction in single bacterial cells. Today it is clear that lysogeny in most cases is mediated by covalent attachment of phage DNA to bacterial chromosomal DNA. Delbrück was initially impatient with the complicated models for lysogeny and encouraged work on the T phages (T1–T7), which only grow lytically (although, less well known, Delbrück also encouraged Weigle, Meselson, and others in studies of lysogeny during this period). Delbrück called the T phages "Snow White and the Seven Dwarfs."

In the next sections, we review the most thoroughly studied examples of integrating phages, starting with phage λ. Later sections survey phages more broadly (including the T phages favored by Delbrück), and then return to the mobilization of pathogenicity islands.

Lysogeny is the hereditary ability to produce phage; it is a genetic property.

PHAGE λ REPLICATION (7, 26, 58)

Early studies by Lwoff and colleagues revealed that certain *Escherichia coli* strains displayed a surprising property. If irradiated with ultraviolet light, these strains would spontaneously lyse and spew phages into the growth medium. If reinoculated onto naive bacteria, the phages, named lambda (λ), could multiply and lyse the new host. However,

The temperate phages were named as such by Elie Wollman in 1952 in a reference to J.S. Bach's musical composition The Well-Tempered Clavier. *Phage λ was named in 1952 by the Lederbergs, Norton Zinder, and Lively; the name λ was chosen to follow the letter κ designation for the κ particles in* Paramecium *because λ was initially thought to be a cytoplasmic symbiont.*

some of the bacteria in the new culture would survive. When these bacteria were repurified and grown out, they too would lyse and release phage after exposure to UV. These and other experiments revealed that λ was capable of two modes of growth. It could grow lytically, producing around 100 new phages and bursting the host cell, or it could grow lysogenically, forming a stable association with the host *E. coli* cell.

Figure 4.1A shows an electron micrograph of a λ particle. Figure 4.1B shows a diagram of the phage particle, indicating the head and tail. The head consists of a proteinaceous shell surrounding the λ DNA genome. The 48,500 base pairs of DNA that comprise the λ chromosome are tightly packaged inside the phage head, probably wound up like a spool of thread. Attached to the phage head is a proteinaceous tail, which is responsible for binding to the λ receptor protein on cells and injecting the phage DNA.

A simplified diagram of the phage chromosome is shown in Figure 4.2. A notable feature of phage genomes generally is that the genes for different functions tend to be clustered. Continuous blocks encode the genes for head proteins, tail proteins, replication proteins, recombination enzymes, and regulatory functions. In many cases the sites of action are also located in the same segments of the genome. For example, the phage origin of DNA replication is embedded within the genes for replication proteins. This modular organization has probably

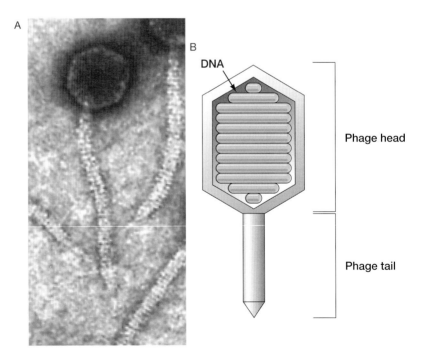

FIGURE 4.1. (*A*) A λ phage particle visualized by electron microscopy. (Reprinted, with permission, from Hendrix et al. 1983. [26]) (*B*) Diagram of a phage λ particle.

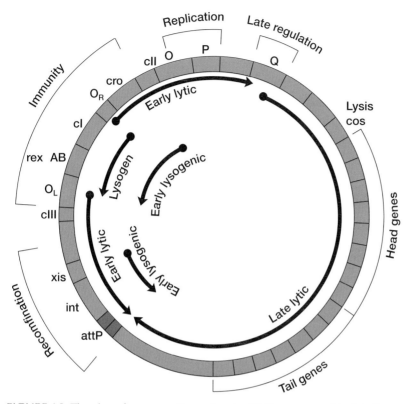

FIGURE 4.2. The phage λ genome. Green denotes DNA sequences, black indicates the major transcripts. (Modified, with permission, from Hendrix et al. 1983. [26])

evolved to facilitate exchanges of gene blocks between phages (or to prevent destruction upon exchange).

A simplified outline of the λ life cycle is presented in Figure 4.3. Following binding of λ to a sensitive cell, the phage genome is injected into the bacterial cytoplasm. There the linear DNA genome is circularized by ligation at the phage cohesive ends (*cos* sites). The subsequent steps differ for the lytic and lysogenic modes of growth.

One of the key determinants of the pathway chosen is the abundance of the transcriptional activator protein cII. The cII protein activates genes for lysogenic growth, whereas lack of cII leads to lytic growth. The steady-state levels of cII are determined by host cell factors, such as proteases and their regulators, thereby providing a sensor that couples the phage lysis/lysogeny decision to cellular inputs that are sensitive to the state of the *E. coli* host.

In the lytic pathway, the λ DNA does not become integrated but is replicated independently of the bacterial chromosome. λ DNA replication takes place in two phases. Initially, the chromosome becomes circularized and replication initiates at the single origin, yielding θ-shaped DNA structures. Later, replication proceeds via a rolling circle

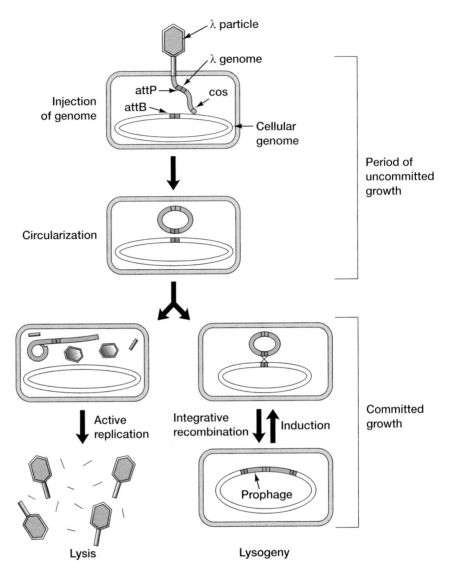

FIGURE 4.3. The life cycle of bacteriophage λ. The λ genome is injected into a cell (*top*), after which it becomes circularized by ligation of the cohesive ends (*cos* sites; *second from top*). Up to this point the phage can grow either lytically or lysogenically. In the lytic pathway (*lower left*), the genome replicates by a rolling circle mechanism, late genes encoding structural proteins are expressed, and new phages are assembled and are released by lysis of the host cell. In the lysogenic pathway (*lower right*) the circular phage genome becomes integrated by site-specific recombination between *att*P and *att*B, yielding the quiescent prophage. Induction, for example in response to DNA damage signals in the host cell, proceeds with excision of the phage genome and a resumption of lytic growth. (Modified, with permission, from Hendrix et al. 1983. [26])

mechanism, yielding many copies of the phage DNA in a long concatemer. Meanwhile, the phage late genes are transcribed and translated to yield the protein components of the heads and tails. The head and tail substructures assemble independently in the bacterial cytoplasm. The DNA is then spooled into the new phage heads. Once a headful of DNA has been packaged, the packaging machinery cuts the λ DNA at *cos*, thereby liberating a complete phage chromosome from the concatemeric precursor. The DNA-containing heads are then joined to the preformed tails. Further phage-encoded proteins then burst the cell and liberate the progeny phages into the medium.

In the lysogenic pathway, the circular phage genome undergoes site-specific recombination to become attached to the host cell chromosome. The cII protein activates transcription of the gene for integrase, the phage enzyme responsible for integration. A site in the phage chromosome, called *attP* for "attachment site of the phage," binds the phage integrase protein and additional cofactors. This complex, called the intasome, then captures a site in the bacterial chromosome, called *attB* (attachment site of the bacteria). Phage and chromosomal DNAs become joined by an ordered series of DNA breaking and joining reactions. The mechanism of phage λ integration, one of the best understood of any mobile DNA system, is considered in detail in the next section.

A prophage is the integrated viral DNA form present in a lysogen.

As a result of integrative recombination, the phage chromosome becomes attached to the bacterial chromosome, afterward resembling a normal cellular gene. The cII protein also activates transcription of λ repressor, a regulatory protein that turns off transcription of genes required for lytic growth. Subsequent replication of the bacterial chromosome results in concomitant replication of the prophage. Only a few regulatory phage genes, notably the gene for λ repressor, are expressed in a lysogen.

However, phage λ can return to lytic growth when a suitable inducing signal is present. UV light, for example, can damage the host cell DNA, signaling to λ that it is time to leave the present host cell in favor of new opportunities. The accumulation of DNA damage leads to the inactivation of repressor, which in turn derepresses the gene for λ integrase and the excision cofactor Xis, allowing excision of the phage genome from the host cell chromosome. Functions for replication and synthesis of heads and tails are also transcribed in the absence of repressor, allowing the phage to return to lytic growth, replicate DNA, assemble heads and tails, and burst the host cell.

Early during lytic growth, genes for replication functions and recombination enzymes are expressed. Later, the genes for lytic proteins and phage heads and tails are produced. During lysogenic growth, these banks of genes are not transcribed. The only active transcription in the lysogen takes place in the regulatory region, leading to production of phage λ repressor and two other regulatory genes.

Beautiful studies of the regulatory cascade dictating the choices between lysis, lysogeny, and induction are only summarized here briefly. A detailed overview can be found in *A Genetic Switch* by Mark Ptashne. Briefly, λ repressor protein is produced in a lysogen and is responsible for maintaining the quiescent lysogenic state, holding replication in check. λ phage with inactivating mutations in the gene for repressor are unable to form lysogens, and can only grow lytically. λ repressor binds to the promoters for early lytic genes, thereby blocking their activity. Proteins synthesized from the early lytic transcripts are required for expression of the late genes, such as those encoding head and tail proteins. Consequently, all lytic transcription is turned off by repressor. Genetic studies of phage λ repressor, together with studies of *E. coli lac* repressor, gave us the first look at gene control, work for which François Jacob and Jacques Monod, along with André Lwoff, were awarded the Nobel Prize in 1965.

Today it is clear that λ repressor is the central signal transducer that mediates induction. Work by Ptashne and coworkers demonstrated that λ repressor is a sequence-specific DNA-binding protein. Repressor acts by binding to the phage promoters active during lytic growth and thereby blocks their activity. Repressor also controls the level of its own expression by activating transcription of its own gene *cI* and repressing *cI* in the presence of too much repressor.

Damage to cellular DNA, as with UV irradiation or other means, induces the SOS emergency repair system and leads to inactivation of the repressor protein. The SOS system is activated by DNA damage, causing activation of the cellular RecA protein. The RecA protein then directs proteolytic cleavage of a repressor of DNA repair functions, LexA, and also the λ repressor protein. The decline in active repressor levels permits derepression of lytic genes, expression of integrase, excision, and the resumption of lytic growth.

Another phage-encoded repressor, Cro, is also transcribed from the promoters for lytic growth. Cro binds the same sites as repressor but binds most tightly to the site that suppresses transcription of the repressor gene. Thus, Cro serves to sharpen the response of the induction switch; once enough Cro has accumulated, any further transcription of repressor is blocked and the prophage is committed to induction.

Repressor protein is also responsible for excluding further λ phages from infecting the same cell. If new λ phage genomes are introduced into the cell, λ repressor binds and blocks transcription. The incoming phages are unable to execute either the lytic or lysogenic pathways of growth, and so are blocked from further replication. Thus, the action of repressor allows λ to optimize its replication strategy and exclude competing phages.

λ SITE-SPECIFIC RECOMBINATION (20, 23, 26, 39, 53, 60, 64)

The phage λ integration system provides one of our best views of a protein–DNA machine for site-specific recombination. λ integrase was the first described member of a large class of recombination enzymes found in myriad phages and even including the variant transposase proteins of the conjugative transposons described in Chapter 3. Many phages that are known or suspected to carry out lateral gene transfer use λ integrase-type proteins to direct DNA incorporation in a newly infected cell. In this section, we examine the λ integration mechanism in detail as a model for these systems.

The DNA sites involved in λ integrative recombination are shown in Figure 4.4. The *attP* site is 250 bases in length, containing two types of binding sites for λ integrase, arm-type sites (labeled P) and core-type sites (labeled C and C´). *attB*, in contrast, is only 21 base pairs long and contains only the core-type integrase-binding sites (labeled B and B´). A short overlap sequence (labeled O) of 7 base pairs matches exactly between *attP* and *attB*. As discussed below, DNA cleavages that initiate strand exchange take place on either side of this overlap region. Integration by phage λ is greatly favored at *attB* over other chromosomal sites. If *attB* is inactivated by mutation, λ integration can still take place, but at reduced frequency, at secondary sites with sequences similar to *attB*.

The λ integrase protein contains separate binding domains for the arm-type and core-type sites (Fig. 4.5). The amino-terminal part of the protein contains the binding site for the P-type arm-binding sites, whereas the carboxy-terminal domain contains the binding site for the C-type core-binding sites. The carboxy-terminal domain also contains the catalytic site that carries out the DNA breaking and joining reactions mediating recombination.

Studies of phage λ integration have been greatly advanced by the establishment of a test tube (in vitro) system by Howard Nash and coworkers that reproduced integration. One discovery from studies of the reaction was the requirement for several host factors in addition to the phage-encoded integrase. One host protein, named integration host factor (IHF), was found to be a small DNA-binding protein encoded by the *E. coli* host. IHF is composed of two small subunits that bind together tightly to form a DNA-binding surface. IHF binding drastically alters the conformation of the DNA site, introducing a remarkable U-shaped bend (Plate 4). Steven Goodman and Howard Nash showed that the DNA bending itself activates recombination, because they could replace IHF and the IHF site with a static DNA bend and demonstrate efficient recombination.

A speculative model of the intasome, the nucleoprotein machine responsible for integration, is shown in Figure 4.6. Integrase is shown by the "bowling pins," organized in a symmetric tetramer. IHF is shown by the oval shapes. The core-binding sites are at the thick end of the integrase proteins and the arm-binding sites are at the thin end. Binding of IHF bends the DNA into the curved path required to permit integrase to make the DNA contacts. The *attB* site comes into the intasome "naked" and binds to the free core sites in the integrase tetramer.

The curved path of the DNA through the intasome has an interesting consequence. The DNA double helix inside bacterial cells is itself

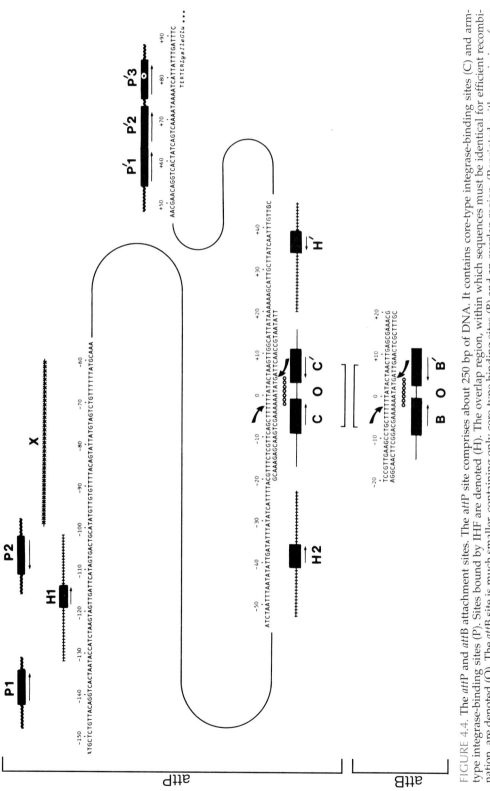

FIGURE 4.4. The *att*P and *att*B attachment sites. The *att*P site comprises about 250 bp of DNA. It contains core-type integrase-binding sites (C) and arm-type integrase-binding sites (P). Sites bound by IHF are denoted (H). The overlap region, within which sequences must be identical for efficient recombination, are denoted (O). The *att*B site is much smaller, containing only core-type binding sites (B) and an overlap region. (Reprinted, with permission, from Hendrix et al. 1983. [26])

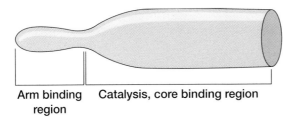

Arm binding   Catalysis, core binding region
region

FIGURE 4.5. Schematic diagram of λ integrase, illustrating the DNA-binding domains that bind the arm sites and core sites in the DNA. (Redrawn, with permission, from Nash 1996. [53])

coiled (supercoiled); that is, wound up like a twisted telephone cord, where the long axis of the DNA helix tends to become wrapped around itself. The direction of superhelical wrapping found in *E. coli* DNA favors the DNA wrapping needed to form the intasome. In test tube reactions, this results in another requirement for a host factor, one that introduces supercoiling in the *attP* DNA substrate. Historically,

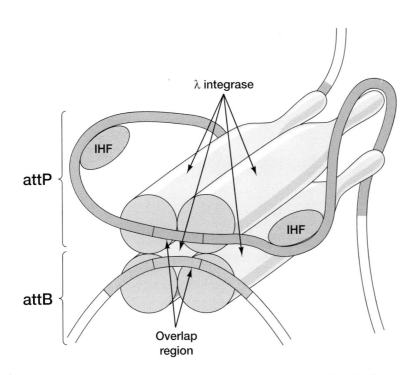

FIGURE 4.6. Schematic diagram of the intasome. The four "bowling pins" represent a tetramer of λ integrase. The *att*P DNA is wrapped in the intasome (*green*) so as to make three contacts to the arm-binding sites and two contacts in the core site. Two of the three IHF dimers are shown, the third is behind the integrase tetramer. The *att*B site makes contacts only at the core-type binding sites. (Redrawn, with permission, from Nash 1996. [53])

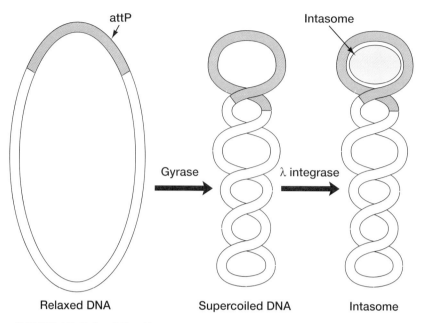

FIGURE 4.7. Role of *E. coli* gyrase and DNA supercoiling in intasome assembly. Relaxed DNA (*left*) when underwound (*middle*) adopts a right-handed superhelical structure. DNA gyrase introduces such negative supercoils into DNA. Intasome formation is favored by negative supercoiling (*right*). Gyrase was initially discovered for its role in promoting intasome formation in vitro. (Adapted from Mizuuchi et al. 1980. [50])

the discovery of this supercoiling requirement clarified a series of puzzling observations with the concomitant discovery of an *E. coli* enzyme that actually adds supercoils to circular DNAs, named DNA gyrase (Fig. 4.7). Gyrase, in fact, came up in Chapter 3 as the target of the quinolone antibiotics and is discussed again below.

The DNA breaking and joining reactions that mediate phage λ integration are shown in Figure 4.8. A λ integrase monomer binds to each side of the overlap region. Initially, one pair of monomers attacks and breaks each of the two DNA strands. The λ integrase mechanism involves the formation of a covalent bond between the cleaved DNA and the enzyme. One DNA cleavage takes place in *attP* and one in *attB* (Fig. 4.8A,B). The free DNA ends created by cleavage then attack across the complex, exchanging the integrase–DNA bond for a new DNA–DNA bond (Fig. 4.8C). This accomplishes the transfer of one pair of DNA strands between recombination partners. The resulting four-armed DNA structure is called a "Holliday junction," named for Robin Holliday, who first recognized the importance of such branched DNAs. A second similar pair of steps, attack by an integrase side chain to form a covalent intermediate, followed by a cross-complex attack of the free

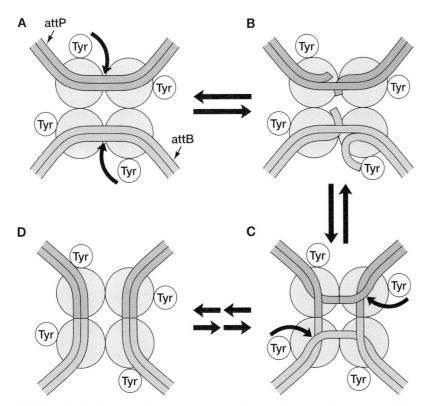

FIGURE 4.8. Mechanism of integrative recombination. Integrase is represented by green circles, which correspond to the bottoms of the bowling pins in Fig. 4.6. In the first step (*A*), a tyrosine residue on each of two integrase monomers attacks the bound DNA to form a covalent phosphotyrosine bond (*B*). This breaks one strand in *attP* and one in *attB*. The binding partners are then exchanged by attack of the free DNA ends on the phosphotyrosine linkages across the complex (*C*). The resulting DNA is a Holliday intermediate, in which four DNA arms are connected by base-pairing. The second pair of strands is then attacked by enzyme tyrosine residues, followed by joining to free DNA ends, yielding the integration product (*D*). (Adapted from Yang and Mizuuchi 1997. [75])

strand, completes the recombination reaction (Fig. 4.8D). Integration of phage λ DNA has the net effect of converting *attP* and *attB* into the recombination products flanking the integration prophage, *attL* and *attR*.

Aberrant versions of λ excision from the bacterial host chromosome can result in capture of flanking host genes. Subsequent infection and integration in a new cell can result in stable transfer of these genes, a process called "specialized transduction." The role of lambdoid phages and specialized transduction is discussed in detail below, but first we compare the life-styles of several more phage that are important in gene transfer.

Controlling λ Integration (26, 53)

The use of covalent intermediates in the integration reaction conserves the energy in the chemical bonds of the DNA phosphate backbone. No high-energy cofactor such as ATP is required for DNA rejoining. A DNA phosphate bond is swapped for a DNA–integrase bond, and then swapped back again to make a DNA phosphate bond, but this time with new DNA connectivity.

How, then, can the reaction ever have a direction? If each chemical step of DNA breaking and joining is energetically equivalent, why doesn't the integration system flicker back and forth between substrates and products? Obviously, this would be bad for the phage, since having made a life-style choice, it would need to execute the required steps as efficiently as possible. This issue comes up for many mobile DNA systems, and the answer appears to be general, at least in the broadest terms. Although the chemical steps of integration are energetically equivalent, the overall reaction is favorable because the protein–DNA complexes at early steps in the pathway are less stable than the products, thus allowing the reaction to run "down hill" energetically.

This leads to another puzzle. How can the phage genome ever excise itself from the bacterial chromosome? After all, it was just argued that the forward integration reaction is energetically favorable. Therefore, how can it ever run backward? The answer is that the excision reaction is not a simple reverse of integration but different in key characteristics. λ encodes a protein, named Xis, that is expressed only when a lysogen has committed to excision. The presence of Xis alters the nature of the complexes formed on *attL* and *attR*, thereby favoring the formation of the excision products *attP* and *attB*.

PHAGE MU (8, 9, 29, 42, 45, 48, 49, 69–71)

Phage Mu is a DNA phage of *E. coli* with a typical "lunar lander" morphology (Fig. 4.9), but Mu replicates by a radically different mechanism. Early studies revealed that infection with Mu induced a high level of mutation in *E. coli*. Thanks to previous work on mutator elements in corn by Barbara McClintock (discussed in Chapter 9), the idea was already in place that mobile genetic elements could cause mutations in their hosts. As a result, the mutagenic capacity of Mu was appreciated early by Larry Taylor, leading to the name Mu for mutator. Further studies revealed that Mu was inserted into the host chromosome so as to preserve the linear order of Mu genes, but, unlike λ, almost any point on the chromosome could serve as an integration acceptor site. Thus, it seemed that Mu was disrupting host genes by integrating into them, implying a life-style very different from that of λ.

The modern view of the Mu life cycle is shown schematically in Figure 4.10. The first steps mimic those of λ, involving binding of the Mu particle to a susceptible cell followed by injection of the linear viral genome. Subsequent steps differ between λ and Mu. The Mu genome does not become circularized, but rather the linear genome becomes integrated into the bacterial genome. A further difference is that Mu replication always proceeds through this integration step, whereas λ can grow lytically in an entirely extrachromosomal fashion.

After integration, Mu can adopt the quiescent prophage life-style, dependent on a lysogenic repressor as with λ. Upon induction, Mu repli-

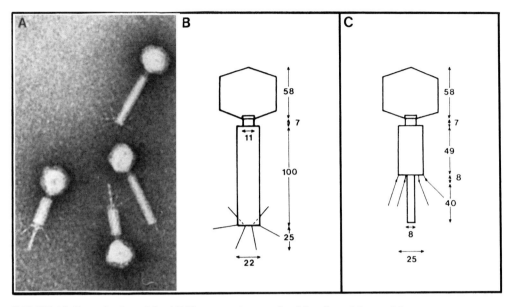

FIGURE 4.9. Bacteriophage Mu. (*A*) Electron micrograph of the phage Mu particle. (*B*) Diagram of a phage Mu particle. (*C*) Diagram of a particle after contraction and DNA injection. The numbers are in nanometers. (Reprinted, with permission, from Symonds et al. 1987. [70])

cates to produce new virions and lyse the cell. Remarkably, new Mu genomes are formed by replicative transposition, as described in Chapter 3 for Tn3. Following induction, the integrated Mu genome is cleaved on one DNA strand at each end of the element by the action of the Mu transposase protein, MuA. The free DNA ends then attack new locations in the host DNA, to form the θ or Shapiro intermediate (see Fig. 3.12). Replication through the element then yields two copies, one at the old donor location and the other at the new target site. Repeated cycles of replicative transposition yield new copies of the Mu genome distributed at diverse locations around the circular bacterial chromosome.

The phage chromosomes are then packaged into Mu phage heads, again by a mechanism different from that of λ (Fig. 4.11). The Mu DNA packaging machinery binds to a site near the left end of the genome, named the *pac* site. The packaging machinery, probably comprising a nearly completed phage head, cleaves the DNA to be packaged, but at a site about 100–200 bp to the left of the *pac* site, within host cell DNA. The DNA then threads into the phage head, packaging the flanking host DNA and the Mu genome. The second cleavage, however, does not happen until some of the host DNA on the other side of the Mu lysogen has also been packaged. The signal for cleavage is derived from the complete filling of the head and not from a specific DNA sequence (consequently, this is termed a "headful" mechanism of packaging). An unresolved question in Mu biology is how the transition from replicative transposition to DNA packaging is coordinated to maximize phage production.

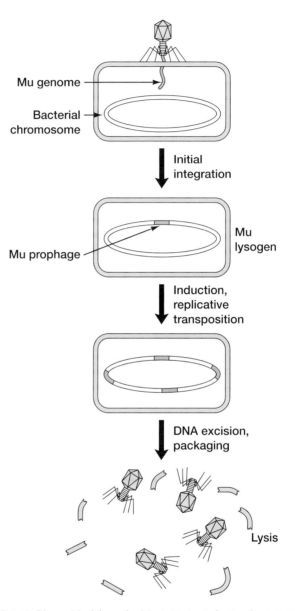

FIGURE 4.10. Phage Mu life cycle. Mu initiates infection by injecting its genome into the host cell (*top*) and integration of the Mu DNA (*second from the top*). The Mu phage can then persist as a prophage. Upon induction, Mu genomes increase in number by replicative transposition (*lower middle*). Mu genomes are then packaged by excision from the host cell chromosome; subsequent lysis releases the newly formed phage particles (*bottom*).

Following encapsidation of the genome, the filled head is assembled with tails. Phage-encoded proteins for lysis then break open the host bacterium, yielding a burst typically containing several hundred Mu phage.

The mechanism of Mu transposition has been elucidated in detail, thanks in large measure to studies of the in vitro transposition system established by Kiyoshi Mizuuchi. Two Mu proteins, called MuA and

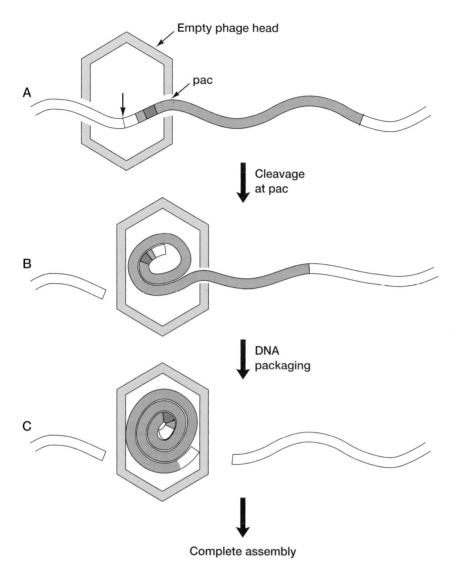

FIGURE 4.11. Phage Mu DNA packaging. An empty Mu phage head binds at the pac (packaging) site and directs cleavage in the nearby flanking host DNA. DNA is then spooled into the phage head until the head is filled (headful packaging mechanism). Mu DNA is shown as light green, pac as dark green, and flanking cellular DNA as white. Normally, packaging extends across the phage genome and into host cell DNA on the right of the genome. A second cleavage then completes packaging. The filled head is then assembled with tail proteins and released from the cell by lysis.

MuB, are required for transposition. MuA is the transposase enzyme, a member of the large class of bacterial transposases similar in amino acid sequence to each other and to the retroviral integrase enzymes (discussed in Chapters 3, 6, and 7). MuA contains multiple DNA-binding domains, one of which is responsible for binding to the phage DNA ends and another for binding internal DNA sites.

The second phage-encoded protein required for transposition, MuB, is a cofactor that activates and targets MuA to appropriate trans-

The Mu Genome

The Mu genome is shown in Figure 4.12. The chromosome is 37,000 base pairs long. Genes for regulation, integration, and replication are present to the left of the chromosome. The middle and right segments of the genome encode mainly structural proteins for heads and tails (D–W). The Mu chromosome contains several structural features distinct from λ. Each end of the Mu chromosome as packaged in viral particles contains short sequences at each end derived from the host DNA, a consequence of the headful mechanism of packaging. Thus, Mu replication necessarily results in the transfer of host cell DNA between bacteria infected by Mu, providing a mechanism for lateral DNA transfer.

The Mu chromosome also contains a DNA segment that can actually invert relative to the rest of the phage chromosome. Because this segment lies near the G gene, it was called the G-loop, and the recombinase responsible was called gin. It turns out that each of the two orientations of the G-loop results in production of different tail fiber proteins, which in turn confer different host ranges on the Mu particles. Phages with the G⁺ orientation can infect *E. coli* K-12 and *Salmonella arizonae*, whereas G⁻ phages can infect *Citrobacter freundii*, *Shigella sonnei*, and *E. coli* C. This type of inversion system has been found widely in prokaryotes, affording a simple means for switching gene expression between two states. These fascinating systems come up again in Chapter 11, in our discussion of the related *hin* system, which is used by *Salmonella* to vary its surface proteins to evade the host immune system.

position target sites. MuB binds to both MuA and the target DNA. MuB activity helps ensure that MuA does not direct transposition into the Mu phage chromosome itself, a reaction that would result in a suicidal rearrangement of the Mu chromosome. MuB provides "target site immunity" to sequences containing MuA-binding sites, the apparent signal that a Mu chromosome is present. The mechanism of this communication between MuB and MuA is not fully clarified, although it is clear that MuB uses chemical energy in the form of ATP hydrolysis to drive the signaling cycle.

Extensive studies using the in vitro system have provided a detailed picture of the reaction pathway (Fig. 4.13). MuA binds to three sites at each end of the Mu chromosome. In test tube reactions, productive binding requires supercoiling of the Mu end DNA sequences, implying wrapping of the DNA in the MuA complex. Initial assembly of MuA at the phage DNA ends is promoted by another DNA sequence, the

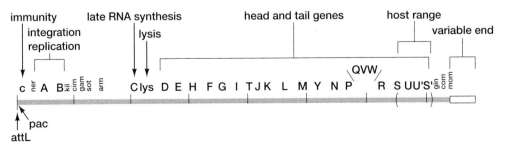

FIGURE 4.12. Phage Mu genome. The variable end at the right indicates the longer segment of host cell DNA found attached to the Mu genome. (Redrawn, with permission, from Howe 1998 [© Springer-Verlag]. [29])

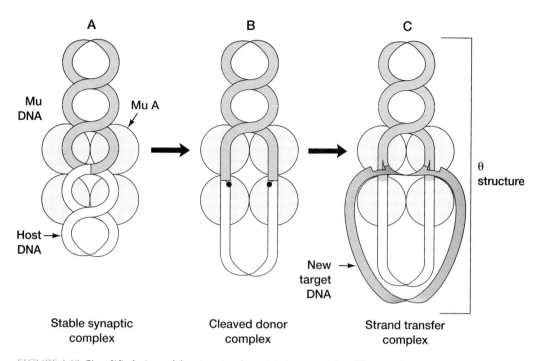

FIGURE 4.13. Simplified view of the steps in phage Mu transposition. The reaction begins with a Mu genome (*green*) carried on a circular replicon. A tetramer of MuA becomes assembled at the junction between the Mu DNA and host DNA (*left*). Formation of the active tetramer requires the participation of additional DNA sites and proteins (not shown). MuA protein then cleaves the host–virus DNA junctions on one DNA strand, leading to relaxation of the flanking host DNA. The MuA DNA complex then captures target DNA (shown in *gray*). This reaction requires the action of MuB protein (not shown). Joining of the free 3′ ends at the ends of the Mu DNA yields a θ structure (also known as the Shapiro intermediate, shown at the right). The Shapiro intermediate can then be processed as with bacterial transposons (Chapter 3). Cleavage of the second strand yields a simple insert, whereas replication through the Mu DNA yields a cointegrate. (Modified, with permission, from Mizuuchi et al. 1995 [© Elsevier]. [49])

recombinational "enhancer." This sequence overlaps the binding sites for the Mu repressor, helping couple regulation of transcription and recombination. The enhancer also binds the architectural DNA-binding protein IHF, encountered above as a cofactor for phage λ integration. The binding of MuA protomers to the enhancer and the phage DNA ends leads to the assembly of a tetramer of MuA proteins at the ends of the phage DNA.

Another architectural DNA-binding protein, HU, is required for formation of the active tetrameric complexes at the ends of the Mu DNA. HU resembles IHF in structure, but differs in binding DNA without sequence specificity. HU is required for recombination at the phage left end, which is somewhat less active for recombination than the phage right end. This is likely because the *pac* site required for DNA packaging is interdigitated among the MuA-binding sites, thereby slightly impairing function. HU overcomes this deficit by binding to a specific site at the left end in the stable synaptic complex. Although HU normally binds DNA nonspecifically, a specific HU-binding site is apparently formed by nearby MuA proteins and the changes in DNA conformation resulting from complex formation.

Following formation of the stable synaptic complex, the DNA junctions between Mu and host sequences are then cleaved on one strand. The free Mu DNA end then attacks the new target DNA, forming the θ or Shapiro intermediate. In fact, the hypothesis that this intermediate existed was first demonstrated conclusively by studies with the Mu in vitro system.

Replication through the Mu DNA sequences then completes transposition. However, this step can only be completed after the removal of MuA from the newly formed DNA junctions. As discussed above, the chemical steps of many DNA exchange reactions are energetically neutral, with the same number of chemical bonds present in the DNA at the end as at the beginning. The reaction is given a direction by the production of increasingly stable protein–DNA complexes at each reaction step. Consequently, by the time the Mu strand transfer step is completed, the MuA–DNA complexes are quite stable. MuA is then actively removed from strand transfer intermediates by the action of the ClpX protein, which uses ATP energy to dissociate MuA. In fact, MuA has evolved a specific docking site to recruit ClpX for this task. Removal of MuA then allows a DNA polymerase to load at the junction between Mu and target DNA, thereby permitting DNA synthesis to duplicate the Mu DNA. The completion of the reaction has not been clarified, but presumably involves action of *E. coli*-encoded nucleases and ligases to trim and seal the remaining DNA strands.

A Bestiary of Bacteriophage (6, 11, 13)

Bacterial viruses comprise a remarkably diverse group, ranging from very large viruses with DNA inside their capsids to tiny viruses with RNA genomes. Some viruses are long, thread-like structures, whereas others display the "lunar lander" morphology of λ and Mu. One phage even possesses a capsid composed of proteins covalently linked like molecular chain mail. Phages can have protein shells, or membrane layers around their capsids, or even protein shells around a membrane. Below we review a few of the most prominent families in addition to the lambdoid and Mu-like phages, emphasizing aspects of phage biology important in gene transduction. Some families of bacteriophages are summarized in Table 4.1.

The T Phages (33)

Phages T1 through T7 are examples of the large DNA phages that package double-stranded linear DNA chromosomes. The T phages were initially chosen by Delbrück from a large collection gathered by Milislav Demerec. According to lore, Delbrück decided on studying the T phages because they formed clear plaques, thus avoiding complications due to lysogeny! The T phages are thus "lytic" phages, because they always lyse their hosts following infection. It is now known that the "T-even" phages form a related group (T2, T4, and T6). T3 and

T7 are related to each other, and T1 and T5 each form their own groups. Electron micrographs of some T4 are shown in Figures 4.14, 4.15, and 4.16.

The large sizes of the T phages have allowed the process of DNA injection to be seen in some detail. Injection by phage T4 is shown schematically in Figure 4.15A, and electron micrographs are shown in Figure 4.15B. First, the phage lands on the sensitive bacterium, making contact using only the tail fibers. Subsequently, the full tail assembly of the phage binds the host cell (pinning), followed by tail contraction. This results in the penetration of the bacterial cell wall, allowing subsequent penetration into the bacterium. Last, the DNA is injected into the cytoplasm, where transcription initiates lytic infection.

The chromosomes of the T phages are quite large, fully 166 kilobases for T4. The DNA is greatly compressed in T4 heads, since the contour length of the phage DNA is about 50 micrometers, whereas the diameter of the phage head is close to 500 times smaller! The dramatic compaction of phage DNA is emphasized by the electron micrograph in Figure 4.16, which shows myriad loops of DNA protruding from a burst phage head. The T4 chromosome contains around 160 genes, although the exact number is uncertain due to the difficulty of identifying genes in the primary DNA sequence. Genes are divided into immediate-early, delayed-early, quasi-late, and late sets, depending on the timing of transcription.

Following injection, the T4 replication cycle begins when early genes are transcribed by host RNA polymerase and phage mRNAs translated by host ribosomes (Fig. 4.17). The immediate-early phage proteins carry out diverse functions, including blocking host DNA synthesis, degrading the host chromosome, and directing synthesis of cytosine, which is modified by hydroxymethlyation, and which the phage incorporates instead of dCTP in its DNA and then further modifies by attachment of a sugar group. This latter reaction probably helps protect the phage DNA from attack by host defenses in the form of restriction enzymes (discussed in Chapter 13). Replication proceeds with intermediate steps of transcription, DNA synthesis, and the formation of precursors for heads and tails. Phage genomes are formed as monomer units, which then are thought to prime further replication using the end of one chromosome as primer on another, forming a complex DNA network. Packaging of the resulting DNA concatemer takes place by a headful mechanism, explaining how T4 can also occasionally package chromosomal DNA and carry out generalized transduction (discussed below). Finally, late transcription directs synthesis of proteins important for completing assembly, the phage chromosomes become packaged, and the host cell is lysed to liberate the newly synthesized phage. Remarkably, this entire program requires only 20–30 minutes to complete.

The Filamentous Phages (34, 35)

The filamentous phages f1, fd, and M13 adopt a much different style of growth. These closely related phages contain single-stranded DNA genomes of about 6407 bases. The genome encodes 11 proteins, the functions of which are understood in some detail. The capsid, composed of gene products pIII, pVI, pVII, pVIII, and pIX, adopts a long, thread-like structure (Fig. 4.18). The remaining genes encode functions for capsid assembly (pI, pIV, and pXI), DNA replication (pII and pX), and DNA packaging (pV).

To replicate, the filamentous phages bind initially to the F mating pilus of conjugation-competent gram-negative cells (Fig. 4.19). The phage genome then traverses the outer and inner membranes facilitated by the pilus structure, a mechanism that is not fully clarified. Once the genome enters the cytoplasm, the complementary DNA strand is synthesized by the action of host DNA synthetic enzymes. This DNA then serves as a template for transcription of phage genes, and as a template for synthesis of new plus-strand genomes. Meanwhile, the phage pV single-stranded DNA-binding protein accumulates. Single-stranded genomes bound to pV protein do not participate in replication, but migrate to the site of phage assembly at the inner membrane, where the newly synthesized capsid proteins also accumulate. Three phage proteins, pI, pIV, and pXI, together with the host protein thiore-doxin, bind to the inner membrane to form the assembly site. Here the phage capsid proteins assemble with genomes bound to pV and become extruded across the inner and outer membranes. In contrast to the lytic phage, the production of filamentous phage is relatively benign for the infected cell. Despite producing some 1000 phage in the first generation after infection, the growth rate of bacterial cells is only slowed 50%. An unusual relative of the filamentous phages found in *Vibrio cholerae*, CTXφ, is capable of integrating and, in fact, transduces the genes for cholera toxin.

Small RNA Phages (18)

Some of the simplest viruses known are the small RNA phages of *E. coli* such as MS2 and Qβ. The RNA genomes can be as short as 3569 bases, as is the case for MS2, and encode a mere four genes. A priori, one might think the simplest RNA virus might encode just two genes: an RNA-dependent RNA polymerase, because bacteria do not encode such enzymes normally, and a coat protein. However, in MS2 and Qβ, we see two further genes, a protein for binding and adsorption (the A protein) and a protein for lysing host cells.

The RNA phages are not known to be able to transduce host cell genes, although hypothetical scenarios can be constructed. Some bacterial cells do contain a reverse transcriptase protein capable of copying RNA into DNA, so transduction of RNA followed by reverse transcription and integration is a formal possibility. As described in Chapters 6 and 7, in eukaryotic cells this means of gene transduction by the retroviruses is one of the most prominent mechanisms for exchange of DNA sequences, although no examples are known in prokaryotes.

Marine Phages–The Most Abundant Viruses in the World? (19, 61, 67)

More types of phages are known than can be described in detail here, but before leaving the topic of phage diversity, we return briefly to the marine bacteriophages, probably the most abundant group of viruses on earth. Tests of ocean waters worldwide confirm that bacteriophages are incredibly numerous and typically are present at concentrations of 10^7 phages or more per milliliter. This puts the abundance of phages well above that of their hosts, the marine bacteria. Little is known about this gigantic population, although credible estimates suggest that phage-mediated lysis of marine bacteria may limit the rate of capture of carbon from the atmosphere (carbon fixation) worldwide. As discussed in Chapter 1, two studies with populations of marine phages and marine bacteria document the ability of marine phages to carry out generalized transduction of genes between populations of marine bacteria.

To date, only one marine bacteriophage has been fully sequenced, Roseophage SIO1, which infects the marine bacterium *Roseobacter SIO67*. Comparison of the sequence to the public database reveals Roseophage SIO1 to be a relative of coliphages T7 and T3, at least for parts of its genome. Although studies are just beginning, Roseophage SIO1 has been found to be quite abundant in seawater samples from different seasons and different locations, making it a candidate for the most abundant virus on earth.

How many different kinds of phage are there? The number is unknown, but studies of marine systems hint that it is very high. A study of 900 culturable marine bacteria revealed that at least one-third were susceptible to infection by one or more lytic phages. A single marine bacterium, *Vibrio parahaemolyticus*, was tested against many phage isolates and found to be sensitive to six distinct types. None of the isolates was found to be able to infect closely related *Vibrio* species. Similarly, *Synechococcus* species have been found to be sensitive to diverse cyanophages that differ in their specificity and morphology. Extrapolating from this admittedly small sample, if each species of marine bacteria is host to several unique phage types—and there are thought to be millions of species of marine bacteria—then there must be several million types of phages.

TABLE 4.1. Some Bacteriophages

Phage family or type	Genome	Structure	Sample hosts
Lambdoid phages (λ, 434, P22, φ80, HK022, ES18, HK97)	dsDNA, 48 kb linear	head/tail	*E. coli*
Mu-like phages (Mu, D108)	dsDNA, about 39 kb linear (and flanking host DNA)	head/tail	*E. coli*, many others
Filamentous phages (M13, f1, fd)	ssDNA, 6.4 kb circular	filamentous	*E. coli*
T1	dsDNA, 47 kb linear	head/tail	*E. coli*
T2, T4, T6	dsDNA, 166 kb linear	head/tail	*E. coli*
T3, T7	dsDNA, 40 kb linear	head/tail	*E. coli*
T5	dsDNA, 113 kb linear	head/tail	*E. coli*
T12	dsDNA, 35.0 kb linear		*S. pyogenes*
Roseophage SIO1	dsDNA, 39.9 kb linear	head/tail	*Roseobacter SIO67*
Small RNA phages (Qβ, MS2)	ssRNA, 3.6 kb linear	spherical	*E. coli*
φX174, S13, G4	ssDNA, 5.4 kb circular	icosahedron	*E. coli*
Corynephages (β, γ)	dsDNA, 35 kb linear	head, tail	*C. diphtheriae*
CTXφ	ssDNA, 9.9 kb linear (including 2 RS segments)	filamentous	*V. cholerae*
VPIφ	ssDNA 40 kb linear	filamentous?	*V. cholerae*
80α	dsDNA		*S. aureus*

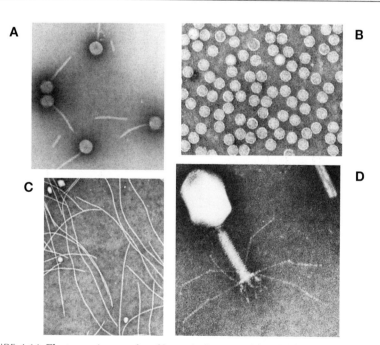

FIGURE 4.14. Electron micrographs of bacteriophage particles. (*A*) λ. (*B*) MS2. (*C*) Fd. (*D*) T4. (Courtesy Robley Williams, University of California, Berkeley; see Atlas 1984. [1])

A

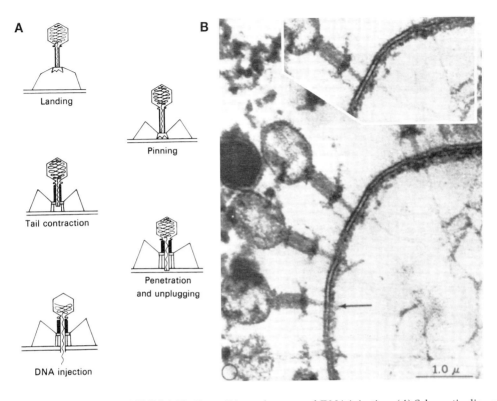

Landing

Pinning

Tail contraction

Penetration
and unplugging

DNA injection

B

1.0 μ

FIGURE 4.15. Phage T4 attachment and DNA injection. (*A*) Schematic diagram of the steps of attachment, penetration, and injection. (*B*) Thin-section electron micrograph illustrating attachment and injection. (Reprinted, with permission, from Fraenkel-Conrat et al. 1988 [© Kluwer Academic/Plenum]. [18])

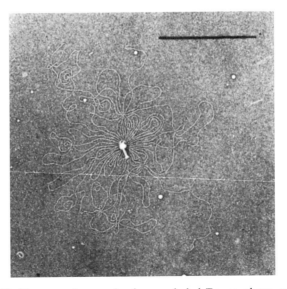

FIGURE 4.16. Electron micrograph of an exploded T-even phage, revealing the remarkable length of DNA packaged in the phage head. Bar, 1 μm. (Reprinted, with permission, from Fraenkel-Conrat et al. 1988 [©Pearson Education, Inc., Upper Saddle River, New Jersey]. [18])

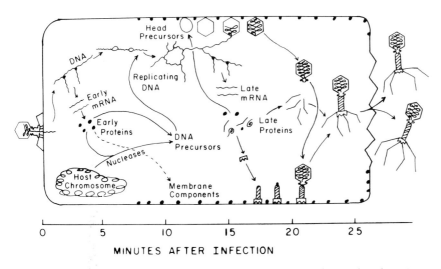

FIGURE 4.17. Diagram of the T4 life cycle, proceeding from left to right. The axis below the diagram illustrates the time after infection in minutes. Destruction of the host cell chromosome provides some of the precursors for phage DNA synthesis. Phage assembly takes place linked to the host cell membrane. The full process takes only 25 minutes. (Reprinted, with permission, from Fraenkel-Conrat et al. 1988 [© Kluwer Academic/Plenum]. [18])

EXCHANGE OF DNA SEQUENCES AMONG PHAGE (3, 10, 27, 31, 73, 74)

Analysis of phage chromosomes illustrates frequent gene swapping among these viruses. Pair-wise comparison of the chromosomes within phage families often yields patchy matches between sequences. Some blocks of genes are closely related, whereas others are divergent or even unrelated. Phage genomes seem to comprise relatively independent gene cassettes arrayed together in the phage chromosomes. In this section, we review swapping of phage cassettes in some detail, because this provides a rich record of lateral DNA transfer. Many of the conclusions reached for swapping of phage genes find parallels in Chapter 5, which discusses the swapping of bacterial operons. In subsequent sections, we begin our survey of phage gene transduction in bacterial pathogenesis.

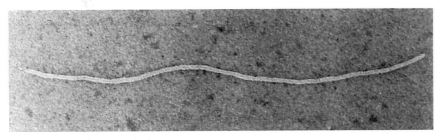

FIGURE 4.18. Electron micrograph of a filamentous phage particle. (Reprinted, with permission, from Kay et al. 1996 [©Academic Press]. [35])

98 ◆ *Chapter 4*

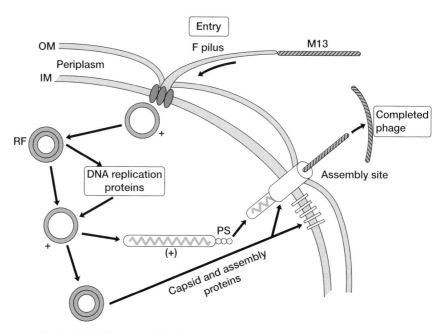

FIGURE 4.19. Diagram of the filamentous phage life cycle. Initially, the phage binds the mating pilus to enter the cell. The single-stranded plus-strand DNA is then converted to a double-stranded form, which serves as the template for synthesis of new plus strands. The double-stranded form also serves as a transcription template for synthesis of mRNAs. Phage structural proteins assemble at a membrane site and are extruded across the membrane to complete the phage life cycle. Release of phage is not associated with lysis of the host cell. (OM) Outer membrane; (IM) inner membrane; (RF) replicative form; (PS) packaging signal. (Modified, with permission, from Kay et al. 1996 [©Academic Press]. [35])

Figure 4.20 shows the genetic map of lambdoid phage HK022 and regions of similarity with three other family members: HK97, λ, and P22 (dark bars under the figure). The regions of homology are clearly patchy, with abrupt transitions from similar to unmatched sequences. These boundaries are interpreted as fossils of the recombination events between family members that assembled the observed DNA sequences. The recombination boundaries are not random, however, but show a strong bias in favor of gene boundaries, consistent with a need to preserve function. The few recombination events that violate this rule often have recombination breakpoints between segments encoding domains within a single protein, again a placement likely to preserve function. These nonrandom arrangements almost certainly arose not by some specialized site-specific recombination mechanism that favored these breakpoints, but rather by random recombination with subsequent selection for functional phage.

The cassette organization has apparently evolved to allow efficient reassortment of gene blocks. If a block of genes is nonfunctional upon introduction into a new host, it will likely be quickly lost. If a gene cassette can be immediately active, however, it has a much greater chance of persisting, leading to clustering of sites of action with the DNA

Gene cassettes in the lambdoid phages can also encode determinants of pathogenesis. The Stx gene, found in lambdoid phages from certain clinical isolates of E. coli, encodes Shiga toxin, a potent poison of protein synthesis.

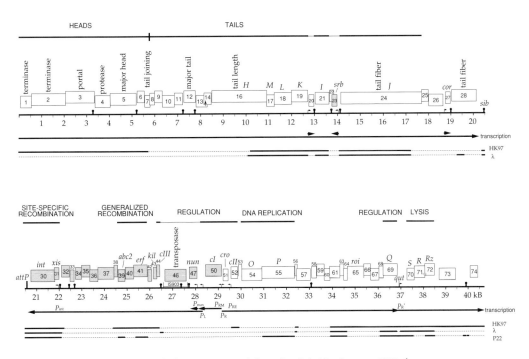

FIGURE 4.20. Comparison of the genomes of four lambdoid phages—P22, λ, HK97, and HK022—illustrating the cassette organization of phage genomes. The dark lines along the bottom of the genome indicate regions of similarity, dotted lines indicate regions of difference. (Reprinted from Weisberg et al. 1999. [73])

A Comparison of Lambdoid Phage Genomes

Starting from the left end of the lambdoid phage genome, we see a block of genes encoding head proteins that is closely similar in HK022 and HK97 but different from λ and P22. Right of that is the b-region, which contains several genes of unknown function in λ, but is nearly absent in HK022. Right of that are the genes for integrase and excisionase, and their site of action, *attP*. These regions are similar for all the phages, although other more divergent cassettes are known within the λ family. Rightward of the integrase cassette is a block of genes and gene fragments of unknown function that are highly divergent among family members. Next comes the immunity region, which is important for regulation of the transcriptional choices that specify lysogeny or lytic growth. Diverse cassettes specify immunity regions, operating by related but distinct mechanisms. The repressor and Cro regulatory proteins, for example, have different DNA-binding specificities in λ, HK022, and P22. HK022 has an IS*903* element inserted into its genome in this region, presumably an evolutionarily recent and transitory event. To the right of the immunity region lie the origin of replication and the coding regions for the replication proteins O and P (genes *54* and *55* in HK022). The binding sites for O protein that comprise part of the origin actually lie within the coding region for O protein. Although all lambdoid P proteins serve to recruit a DnaB-like protein, the mechanisms differ between phages. λ P protein binds the host DnaB, whereas HK022 P resembles a full DnaB protein and fulfills the role itself. Between the replication genes and the Q regulatory protein lies a set of about 10 genes, found sporadically among lambdoid phages. Few of these genes have known functions, leading to the suggestion that they may help fine-tune growth but are not strictly required. To the right of this set is the conserved Q antiterminator gene and conserved R and S genes promoting lysis.

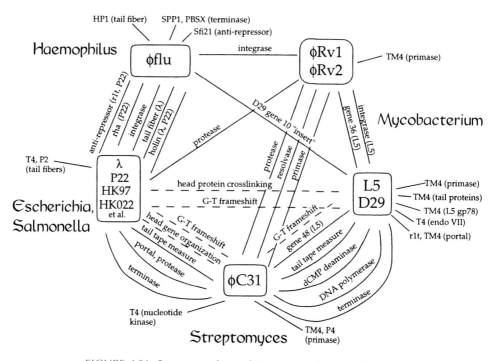

FIGURE 4.21. Sequence relationships among phages and prophages. Detectable similarities between genomes are shown by the solid lines; similarities in gene organization are shown by the dashed lines. Groups of related phages are boxed; bacterial hosts are shown in large letters. (Reprinted, with permission, from Hendrix et al. 1999 [©National Academy of Sciences]. [27])

encoding the proteins that act on those sites. The lambdoid family in nature can be thought of as all the combinatorial possibilities resulting from reassortment of all the functional cassettes.

Even phages that replicate in evolutionarily distant hosts can exchange gene cassettes. Roger Hendrix and coworkers compared the lambdoid family to phages inhabiting distantly related bacteria, specifically phages of *Mycobacteria* (φRv1, φRv2, L5, and D29), *Streptomyces* (φC31), and *Haemophilus* (φflu). Many genes and blocks of genes were found to be common to two or more phage families (Fig. 4.21). Of these, many spanned the great evolutionary distances separating the gram-negative bacteria (*E. coli, Salmonella, Haemophilus*) from the gram-positive bacteria (*Streptomyces*), and both of these from the Mycobacteria. However, the greater the evolutionary distance between hosts, the less the similarity between gene blocks, suggesting that the cassettes acquired many changes during the long journeys required to reach their divergent hosts.

Many phages have multiple hosts, so that infection of a single host by two different phages provides an opportunity for genetic exchange between them. Hendrix and coworkers speculate that phage host ranges may be sufficiently overlapping that genes can be transferred between species with no common phage by infection of intermediate bacterial species. It even seems likely that all the genes present in the isometric tailed DNA phages of bacteria may be accessible to any phage, although to reach a given phage, some sequences might need to pass through many intermediate hosts. As Hendrix and coworkers observe, if this is true, then "All the world's a phage!"

GENERALIZED TRANSDUCTION (14, 15, 28, 54, 79)

Phages such as Mu that package DNA by a headful mechanism can occasionally misfire and package a headful of host cell DNA instead. Upon infection of a new host cell, the DNA can sometimes become incorporated into the genome of the new host. This process, known as generalized transduction, is a prominent means of gene transfer between bacteria. Generalized transduction is so named because essentially any cellular gene—bacterial DNA from any location in its genome—can be transferred between cells by this mechanism. Phages such as λ, which package DNA by recognizing and cleaving at specialized att sites, do not typically carry out generalized transduction but do carry out specialized transduction (discussed in the next section).

Phage Mu can actually direct generalized transduction by several different mechanisms. The most prominent occurs when the DNA packaging machinery binds to a DNA sequence within the host chromosome rather than to *pac* in the Mu DNA, resulting in the packaging of a headful of host DNA. Infection of a population of new cells with a Mu stock results in the occasional injection of purely host DNA into the infected cell. This DNA can on occasion become incorporated in the chromosome of the new host. In one pathway, DNA from the transduction donor cell is incorporated into the target cell by homologous recombination, as with DNA assimilation after conjugation (Chapter 3). One prominent system employs the host RecA enzyme, which together with its cofactors carries out recombination between matching sequences in the incoming DNA and the chromosome of the new host. Transduction of a given *E. coli* gene occurs about once in 10^7 to 10^8 Mu infections.

Another mechanism for generalized transduction arises because the length of DNA packaged in a Mu head is a little longer than the Mu chromosome. Because each Mu genome is excised from continuous cellular DNA, a segment of flanking host DNA from either side of the genome is also packaged to fill the head to capacity, about 100 bp of host DNA at the left end and 2000 bp at the right end. These short DNA fragments are not generally recovered in the new host cell, but under special circumstances transduction of sequences by this mechanism can be detected. If the Mu genome is shortened, the total length of DNA packaged remains unchanged, resulting in packaging of

In generalized transduction, the phage can be viewed as a nonspecific carrier of the bacterial host genes. When this DNA is inserted into a new host cell, either it can undergo homologous recombination or it can just remain in the bacterial cell cytoplasm, a state termed "episomal." However, an episome must be capable of replication to persist during cell division.

longer pieces of flanking host DNA and more efficient transduction.

Mu can also carry out generalized transduction by yet another means, this time under the action of the Mu transposition system. This mechanism only works with abnormally small Mu genomes, and so has been given the colorful name "mini-Mu-duction!" In cases where two mini-Mu genomes integrate into the host chromosome in nearby locations, the two genomes can be excised from the host chromosome using the "outside" ends of each, resulting in packaging of the two genomes and the intervening host DNA. Upon infection of a new cell, the composite can be integrated just as with a normal Mu chromosome. This integration pathway does not require the presence of the RecA homologous recombination system in the target cell, because integration is directed by Mu enzymes. In many respects, this mechanism resembles the formation of composite transposons, but with the added twist that the product can then be packaged in phage particles and transferred between cells.

Another mechanism of DNA transfer by phages involves incorporation of new DNA by transposition into a prophage genome. Many studies confirm that prophage DNA can be a target for transposon insertion just like other cellular DNA. If such a prophage is induced and infects a new cell, the transposon DNA will be transferred as well. Transposition of a composite transposon into a phage, followed by induction and infection, allows sequences transported on the composite transposon to appear in the newly infected cell.

A phage chromosome bearing a transposon can have several fates after injection into a new cell. The genome can be incorporated into the host cell genome by phage integration enzymes, but this is only one of several pathways. If the phage chromosome is unable to integrate, the transposon may still excise and insert itself into the new host. If the phage chromosome contains a composite transposon, and if the phage chromosome circularizes upon entry into the cell, then "inside-end" transposition can insert the phage into the host chromosome (Chapter 3). Stepping back, it is clear that insertion of a transposon into a phage genome provides a rich set of opportunities for transfer of DNA between cells.

Phages Capable of Generalized Transduction

Generalized transduction was first discovered by Norton Zinder and Joshua Lederberg, who found that heritable traits could be transferred between *Salmonella typhimurium* strains by a filterable agent. Further studies revealed that transduction was mediated by phage P22, and that about 2% of the P22 particles in a typical stock contained cellular DNA. Later, phage P1 was also reported to support generalized transduction, and because P1 has a larger genome than P22, larger pieces of host DNA were found to be transduced. Further phages known to be capable of generalized transduction include T1, T4, KB1, and ES18. Rates of gene transfer vary with the phage and host gene studied, but generally range from once in 10^5 to once in 10^8 per plaque-forming unit.

SPECIALIZED TRANSDUCTION (21, 38, 63, 65, 66)

Specialized transduction, in contrast to generalized transduction, results in the transfer of only those host cell genes flanking the attachment site of an integrated prophage. Phage λ provides the most thoroughly studied example. The region surrounding *attB*, the integration site for the λ chromosome, is shown in Figure 4.22. To the left of *attB* lie the *gal* genes, which are required for metabolism of the sugar galactose. To the right are the *bio* genes, which direct the synthesis of the vitamin biotin. Induction of a λ lysogen, followed by infection of a new cell, can occasionally result in transfer of *gal* or *bio* DNA. Transfer can be detected if the recipient cell is mutant in *gal* or *bio* genes, so that transduction restores *gal* or *bio* gene function.

Specialized transduction results from a misfiring of the normal excision mechanism. On rare occasions, excisive recombination involves not *attL* and *attR*, but instead a flanking site in *E. coli* DNA and an internal site in λ. Such chromosomes are named λ*dgal* (for deleted but containing *gal*) or λ*dbio* (deleted but containing *bio*). Packaging of this chromosome, followed by normal infection, can result in the introduction of the λ*dgal* or λ*dbio* chromosome into a new cell.

The newly introduced DNA can be incorporated by either of two means. If the *dgal* or *dbio* chromosome enters a cell along with a wild-type phage chromosome, the wild type can supply λ integrase enzyme. Both chromosomes then can integrate, yielding a double lysogen, in which the two λ genomes are integrated side by side. Double lysogen formation is necessary because the λ*dgal* or λ*dbio* carries either *attL* or *attR*, derived from the prophage chromosome, and not the normal *attP*, and recombination of *attL* or *attR* with *attB* is inefficient. However, prior integration of a wild-type phage *attP* sequence with chromosomal *attB* yields *attL* and *attR*, which can recombine with *attL* or *attR* on the deleted λ chromosome more readily.

The second means of incorporating λ*dgal* or λ*dbio* sequences involves homologous recombination. If a localized mutation is present in the recipient cell, for example in a *gal* gene, then homologous recombination with wild-type sequences on a λ*dgal* can transfer *gal*+ sequences.

Specialized transduction is probably not a major contributor to gene transfer in the environment. Because the types of sequences that can be transferred are so restricted, it seems likely that specialized transduction is a minor contributor compared to generalized transduction. A few examples of phages capable of specialized transduction in addition to λ have been reported.

There are, however, several means for broadening the repertoire of genes that can be transported by specialized transduction. If chromosomal genes adjacent to *attB* are rearranged, it is possible for λ to pack-

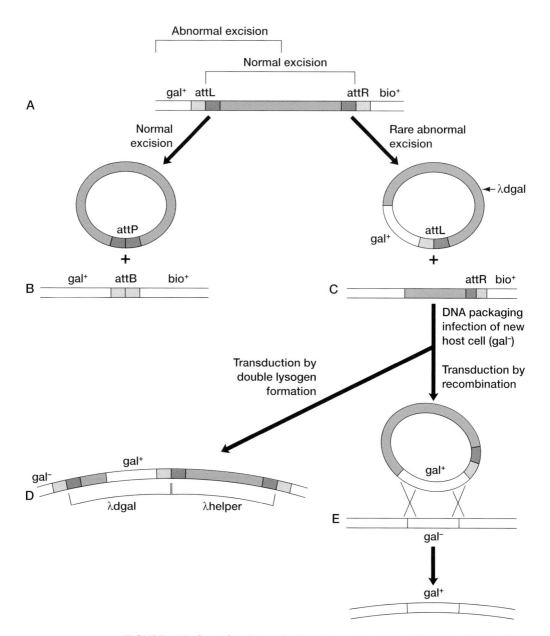

FIGURE 4.22. Specialized transduction. Normal excision (*A*) during λ induction excises the phage genome from the host chromosome by recombination between attL and attR, yielding the circular phage genome (*B*). Rare abnormal excision events take place by recombination between one site within λ and another in flanking DNA, yielding a circular genome containing part of the phage chromosome and part of the flanking DNA (*C*). In the example shown, cellular DNA from the *gal* operon is captured as a λ*dgal* phage. Upon infection of a new cell mutant in the *gal* genes, *gal*+ information can be stably incorporated into the host cell chromosome by either of two means (*D* and *E*). The λ*dgal* defective phage can become integrated by coinfection with an intact phage, yielding a double lysogen (*D*). Alternatively, recombination between *gal* sequences on the phage and in the chromosome can install *gal*+ sequences in the newly infected cell (*E*). (Adapted from Griffiths et al. 1993. [22])

age the newly apposed sequences instead of *gal* or *bio*. Another way to change the flanking sequences is to delete *attB*. λ then integrates into diverse secondary sites, allowing specialized transduction of the new flanking genes. In a third mechanism, if a transposon is present in both the phage genome and the host chromosome, homologous recombination can join the two DNAs. Sequence flanking the host transposon integration site can then be transferred by specialized transduction. To what extent such mechanisms operate in the environment is unknown.

Specialized Transduction and Genetic Engineering

Specialized transduction is exceptionally useful in the laboratory for genetic engineering. DNA sequences of interest, for example, human DNA, can be spliced into phage chromosomes, thereby allowing the purification and manipulation of genes. Phage λ in particular has been exploited as a vector in gene cloning. As long as the phage vector contains the minimal *cis* sequences required for packaging, DNA replication, and regulation, many of the other phage genes can be removed and replaced with the DNA of interest. Because phage stocks can be grown to extremely high titers, and it is possible to make DNA "libraries" in phage vectors in which each phage contains a different segment of human DNA, these libraries can be large enough to contain every sequence in the human genome in a modest volume of highly concentrated phage particles. This remarkable technical achievement laid the groundwork for isolating human genes and ultimately sequencing the human genome.

TRANSPORTING GENES FOR BACTERIAL PATHOGENESIS BY PHAGE INFECTION (17, 25, 32, 51, 62)

Bacteria that cause disease are often only slightly different from harmless relatives, differing only by the presence or absence of a few blocks of genes encoding pathogenic determinants. Epidemiological and laboratory studies have established that these gene clusters are often transferred between bacteria, commonly (although not exclusively) by bacteriophages. In the next sections, we review the nature of pathogenic determinants and their mobilization by phages.

Pathogenic Determinants

Two general properties distinguish pathogenic from nonpathogenic bacteria. One is the presence or absence of factors that directly damage tissue. Examples include toxins that poison specific host cell functions, proteases that damage tissues, and bacterial products that cause damaging hyperactivity of the immune system. Two prominent groups are termed "exotoxins" and "endotoxins" (the names are historical and not very descriptive). Exotoxins are proteins synthesized and secreted by pathogenic bacteria. Often these molecules are highly potent, and only very small amounts are required to damage host tissue. This is in contrast to another bacterial molecule, endotoxin, which is the lipopolysaccharide (LPS) component of the gram-negative bacterial cell wall. LPS is active only at higher concentrations, acting in part by provoking an excessive and inappropriate host immune response. Genes for synthesis of endotoxins are not typically found in pathogenicity islands and so are not discussed further here. Pathogenic strains of *C. diphtheria, Staphylococcus aureus*, and *Vibrio cholerae* all contain mobile genes encoding exotoxins.

The second class of pathogenic determinants are protein factors that permit the pathogenic bacteria to persist in an infected patient. Examples include adhesins that mediate adherence of bacteria to tissue and invasins that permit penetration of bacteria into cells. Uropathogenic *E. coli*, for example, cause urinary tract infections due in part to the presence of a bacterial adhesin protein that binds the bacteria to tissues of the urinary tract. *E. coli* lacking the adhesin do not cause infections because the flux of fluid through the urethra washes the bacteria out. Invasin proteins promote the entry of certain pathogenic bacteria into cells, where they hijack cellular systems to promote their own multiplication.

Another group of genes for pathogenesis promote persistence by helping to elude the host immune response. Normally, invaders are identified as foreign by the host immune system and cleared by the immune response (the subject of Chapter 11). Many bacteria dodge this recognition by cloaking themselves in a dense network of sugar chains. Molecules of the immune system have a hard time penetrating this polysaccharide capsule, so the bacteria become difficult to detect. Other cloaks involve specific proteins, named protein A and protein G, that bind host antibody proteins on the cell surface. Antibodies bound in this fashion are unable to do their normal job of recruiting immune functions for destruction of the invaders, and the surface of the bacterium is thus masked by the bound host antibodies. Still other proteins inactivate the killing mechanisms used by immune system cells. Immune cells called macrophages engulf and destroy bacteria, in part by producing activated oxygen species. Some bacteria encode enzymes such as catalase or superoxide dismutase that inactivate these toxic oxygen species, thereby permitting the bacteria to survive inside the macrophage.

Another prominent set of pathogenic genes mediates iron uptake. In the human body the concentration of free iron is low, because it is bound up by carriers such as lactoferrin, transferrin, ferritin, and hemin. Bacterial growth is often limited by the availability of iron. Bacterial genes for proteins that capture iron, called siderophores, promote survival of bacteria in the infected host and so are considered to be virulence factors. Even some of the genes for exotoxins that kill host cells may serve this role, because free iron from dead host cells can be used by the bacteria in some cases. Consistent with this idea, some of the genes for exotoxins are known to be turned on in the presence of low concentrations of iron.

The genes for pathogenesis can be difficult to define precisely, because they grade continuously into genes that are required for bacterial persistence. Functions that allow bacteria to survive a little longer in the human body may lead to greater damage to the host, although the changes may be in mundane steps of bacterial metabolism. In the following sections, we emphasize the pathogenic genes with known functions, while recognizing that this is only part of the picture.

Pathogenicity Islands (17, 24, 25, 32)

The term pathogenicity island was first used to describe large segments of chromosomal DNA found in pathogenic *E. coli* but not in nonpathogenic strains. Subsequent studies revealed that many pathogenic bacteria, both gram-positive and gram-negative, contained pathogenicity islands, while closely related avirulent strains lacked them. These gene blocks can be quite large, up to at least 200 kb, and can code for many virulence genes. The LEE pathogenicity island of enterohemorrhagic *E. coli* encodes 41 known genes. Pathogenicity islands contain diverse genes important for pathogenesis, including adhesins, invasins, and exotoxins. Other genes encode regulators of virulence factors and integrases that are likely involved in mobilization. In some cases, the regulatory cascades have been worked out, revealing that genes for pathogenesis are often activated under conditions where their expression is most beneficial to the host bacteria. Many more of the genes in pathogenicity islands are of unknown function, although for some the conservation in diverse bacterial isolates testifies to their importance. Some pathogenic strains have multiple pathogenicity islands. Uropathogenic *E. coli*, for example, has eight.

How do we know that pathogenicity islands are mobile? The observation that they are present only in pathogenic strains and not in closely related nonpathogenic strains is consistent with recent acquisition as mobile DNA but also consistent with loss of an island present in a progenitor strain. In favorable cases, the pathogenic bacteria can be cultivated in the laboratory, and transmission of the pathogenic determinant to new strains can be demonstrated directly. Commonly, however, mobility cannot be easily established. In these cases, sequence analysis often provides indirect evidence. Bacterial genomes have distinctive ratios of A/T to G/C, with some more rich in A/T and others more rich in G/C. In general, the base composition skew found in any one part of the genome tends to be similar throughout the genome. It is presumed that base composition is optimized for each organism, so that newly introduced sequences tend to mutate over time to match more closely the base composition of the new host. Analysis of microbial sequences, however, occasionally reveals blocks of sequence, such as the genes for pathogenicity islands, that contain base compositions much different from the host. Such sequences are inferred to be newly introduced from another organism (although supporting data are helpful to strengthen this inference). Another indication of mobility is the observation that genes for pathogenicity are often intermingled with phage or transposon sequences that may have contributed the machinery involved in mobilization. Similarly, genes for tRNAs have been found to lie near *att* sites for many phages, and pathogenicity islands are often found to integrate near tRNA genes. Why phages favor integration near tRNA genes is not entirely clear, although it may be that tRNA genes mark benign locations for insertion, thus avoiding damaging the host genome. Several eukaryotic transposons also integrate near tRNA genes, probably because this represents a safe haven in the genome.

In some cases, genes for pathogenesis can be shown to be unstable. During prolonged growth, the genes for pathogenesis can sometimes disappear, commonly associated with the "curing" of a strain of a resident prophage. This too provides evidence for mobility of pathogenic determinants. In the next several sections, we review examples of pathogenicity islands, with particular emphasis on those transported by phage.

PATHOGENIC *E. COLI* AND THE DISCOVERY OF PATHOGENICITY ISLANDS (17, 32, 57)

Pathogenicity islands were first discovered in *E. coli* strains associated with urinary infections, so-called uropathogenic *E. coli* or UPEC. *E. coli* was long thought to be a benign passenger in the human gut, but in recent years the bacterium has been identified as the causative agent of a variety of serious infections. Some 80% of urinary tract infections are caused by UPEC. In severe cases, the infecting bacteria can enter the bloodstream, causing life-threatening septicemia. UPEC colonizing the urinary tract of pregnant women can also infect the developing fetus, causing newborn meningitis. Other pathogenic strains of *E. coli* cause diseases of the gastrointestinal tract and extraintestinal tissues.

At the molecular level, the different types of pathogenic *E. coli* are distinguished by their complement of pathogenicity islands. Figure 4.23 shows a map of the UPEC chromosome, indicating the pathogenic islands characteristic of this strain. Natural isolates of *E. coli* can differ by as much as a megabase, in part due to acquisition of diverse pathogenicity islands (Chapter 5). For UPEC, the pathogenicity islands range in size from 25 kb to 190 kb. Each is integrated near a tRNA, and four have identifiable integrase genes associated, indicative of mobilization by phage or a phage-like mechanism.

The virulence factors encoded in the pathogenicity islands have been studied in some detail, revealing those functions that convert a commensal bacterium into a pathogen. Six genes for hemolysin pro-

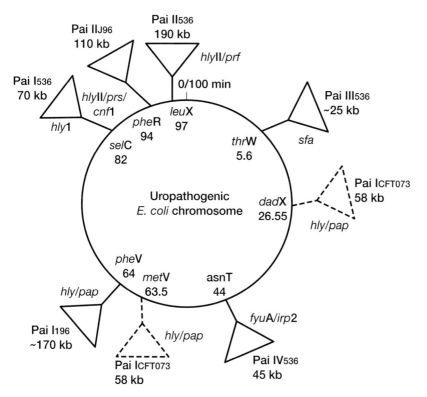

FIGURE 4.23. Pathogenicity islands in uropathogenic *E. coli* (UPEC). The indicated sites of insertion correspond to eight pathogenicity islands characteristic of UPEC. The map positions refer to the *E. coli* K12 chromosome. The dotted lines indicate uncertainty in the map position of tPAI$_{ICFT073}$. (Redrawn, with permission, from Kaper and Hacker 1999. [32])

teins are present, proteins that direct the lysis of host red blood cells by forming pores in the membrane. Lysis of host cells releases iron that can be assimilated by the UPEC, and lysis of immune cells may have the further advantage of inhibiting the host immune response. The hemolysin proteins are matched with iron uptake systems (enterbactin, aerobactin, and yersiniabactin) that scavenge the released metal.

At least six adhesin proteins are also encoded in the pathogenic islands. These proteins bind the bacteria to the mucosal wall of the bladder and urethra, allowing the bacteria to persist despite the washing action of the urine. The mucosa is composed of several types of sugar polymers that cloak the surfaces of the underlying cells that are bound by specific adhesins. For example, type 1 fimbrial adhesins bind to mannose oligosaccharides, and P fimbrial adhesins bind to Gal α (1–4)Gal chains. Experiments with animal models establish that both the hemolysin proteins and adhesins contribute to uropathogenesis.

The UPEC pathogenicity islands are associated with a rich array of mobile DNA sequences (Table 4.2). Present are diverse phage-like inte-

TABLE 4.2. Main Features of PAIs of UPEC Strains

Strain	Designation of PAI	Insertion site (min)	Target tRNA	Size (kb)	Virulence factors encoded	Integrase	Boundary[a]
536	PAI I$_{536}$	82	selC	70	α-hemolysin	φR73	16-bp DR
536	PAI II$_{536}$	97	leuX	190	α-hemolysin, P fimbriae (Prf)	P4	18-bp DR
536	PAI III$_{536}$	5.6	thrW	~25	S fimbriae	DLP12	?
536	PAI IV$_{536}$	44	asnT	~40	Yersiniabactin, iron-repressible proteins P4	?	
J96	PAI I$_{J96}$	64	pheV	170	α-hemolysin, P fimbriae (Pap)	?	?
J96	PAI II$_{J96}$	94	pheR	110	α-hemolysin, P fimbriae (Prs), cytotoxic necrotizing factor 1	?	135-bp IR
CFT073	PAI I$_{CFT073}$	63.5?	near metV	58	α-hemolysin, P fimbriae (Pap)	?	9-bp DR

Reprinted, with permission, from Kaper and Hacker 1999. (32)
[a]DR, Direct repeat; IR, indirect repeat.

grase genes, transposases, and other sequences matching open reading frames in the transposons and phage. The inference is that these sequences mediated the insertion of the pathogenicity islands into the chromosome. The integrases are particularly implicated, since the pathogenicity islands are integrated near tRNA genes, sites favored by phage integration systems. The borders of many pathogenicity islands are flanked by inverted repeated sequences resembling classic phage *att* sites. Reconstructing the mechanism of transfer is complicated by the fact that important sequences may have been lost by deletion after the transfer event. In some of the cases described below, the mechanisms of transfer have been amenable to direct analysis in the laboratory, providing further support for the involvement of phage-mediated transduction.

There are now two completely sequenced *E. coli* genomes, one the laboratory K12 strain and the other the enterohaemorrhagic O157:H7 strain, which is responsible for outbreaks of severe and sometimes fatal food poisoning. Numerous differences are found between the two genomes, including 1387 new genes in O157:H7 in 177 strain-specific clusters. Thus, a more complete look reveals a very large number of pathogenicity islands. A detailed comparison of the two genomes is presented in Chapter 5 in the context of the complete prokaryotic sequences.

SCARLET FEVER, TOXIC SHOCK-LIKE SYNDROME, AND *STREPTOCOCCUS PYOGENES* PHAGE T12 (12, 46, 77)

The gram-positive bacterium *Streptococcus pyogenes* (group A streptococcus) causes scarlet fever and, more recently, the lethal toxic shock-like syndrome (TSLS). Both diseases result in part from the expression of an exotoxin, streptococcal pyrogenic exotoxin (SpeA), encoded by bacteriophage T12. Scarlet fever was a devastating disease of 19th-century Europe and North America, characterized by a diffuse rash with fever and, in severe cases, progressive organ failure and death. Children were especially susceptible to this disease until the 1950s, when penicillin treatment of the sore throats caused by *Streptococcus* infections also killed *S. pyogenes*. *S. pyogenes* also causes rheumatic fever, impetigo, and necrotizing fasciitis.

The decline in scarlet fever, however, has been matched by the rise of a new streptococcal disease, TSLS. The symptoms resemble toxic shock and scarlet fever, involving rash, pain and redness at the site of infection, fever, shock, and loss of skin from the palms and soles of the feet. The infection spreads very rapidly with hospitalization required only 24–48 hours after the symptoms begin. In fatal cases, death results from shock and organ failure. The death rates of TSLS in some hospitals are as high as 30–60%. Unlike toxic shock (discussed below),

Enterotoxigenic E. coli *provides one of the rare examples of mixed gene blocks containing not only pathogenicity genes, but also genes for antibiotic resistance. The genes for the enterotoxin, causing gastrointestinal poisoning, are found on plasmids also encoding resistance to tetracycline, streptomycin, and sulfonamides. These composite plasmids are obviously a still greater threat to human health than either antibiotic-resistant plasmids or toxigenic plasmids alone.*

Scarlet fever was the scourge of 17th century Italy, Spain, and northern Europe. In 1736, about 4000 people were killed by scarlet fever in the new American colonies. By the late 1880s, 25–30% of children died from scarlet fever in New York City, Chicago, and Norway, but by the 1900s, this rate of death dropped to just 2% in these same locations, perhaps due to the reduced expression of streptococcus virulence factors. Some studies have attributed the cycles of scarlet fever occurrence to drifts in antigenic or virulence properties. Following a 50- to 60-year period of relatively benign disease caused by group A streptococci, there has been a recent increase in severe infections.

Streptococcus pyogenes is transmitted by direct contact with infected nasal or throat secretions and open sores of the infected patient. People with chronic illnesses are at increased risk of infection. S. pyogenes infection is treated with penicillin.

the TSLS death rate is roughly the same in men and women. Some fatalities have occurred in people with underlying conditions, such as diabetes or AIDS, but many have occurred in otherwise healthy adults.

The disease can strike very suddenly. TSLS is commonly caused by infection of wounds with *S. pyogenes*, followed by bacterial invasion of the bloodstream, but even this is not always clear. In an outbreak in Ontario between 1987 and 1991, a sizable minority of patients were unable to recall any injury that could have led to the invasive *S. pyogenes* infection. Patients had few symptoms until massive progression of the infection, complicating treatment. In this outbreak, fatalities reached nearly 50% of cases.

The SpeA toxin, short for streptococcal pyogenic exotoxin A, mediates much of the toxicity of *S. pyogenes* infection, but its mode of action is not fully clarified. SpeA protein is secreted by pathogenic *S. pyogenes*, leading to hyperactivation of the immune system of the infected person. SpeA is thought to act at least in part as a "superantigen," a molecule that bridges between immune system cells, thereby hyperactivating the bound cells and causing them to produce excitatory immune signals in toxic quantities (Fig. 4.24).

SpeA may have further activities as well, including sensitizing the immune system to activation by other bacterial antigens. The overactivated immune system can be toxic to many organ systems, in extreme cases leading to shock and death.

The speA protein is encoded by the *S. pyogenes* phage T12 (Fig. 4.25). The genome is 35,066 bp long and contains 49 probable genes. A

Normally, foreign antigens are recognized by the host cellular immune response in the context of a presenting molecule called MHC (major histocompatibility complex). A host T cell with a T-cell receptor specific for the displayed antigen then binds, initiating a signaling cascade that initiates an immune response (the molecular basis of the immune response is discussed in detail in Chapter 11). Superantigens also stabilize the interaction between a T cell and antigen-presenting cell, but by binding to conserved regions on the T-cell receptor and MHC molecules. This activates a much larger number of T cells than is typically activated by antigen, leading to a hyperactivation of the immune response. Cytokines and other immune signaling molecules are produced in abnormally high amounts, in advanced cases leading to toxicity to multiple organ systems. Hyperactivation by superantigens also disrupts normal function of the immune system, assisting infecting pathogens in evading the host immune system.

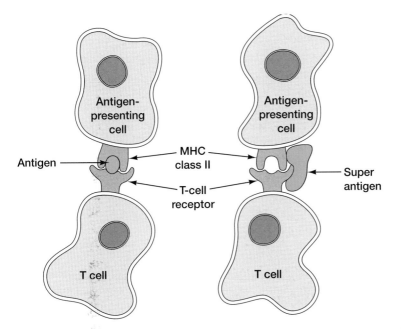

FIGURE 4.24. Superantigen mechanism. Normal antigen presentation (*left*) involves contact between MHC molecules bound to antigen on the antigen-presenting cell (APC) and the T-cell receptor (TCR) on a T cell, leading to T-cell activation. Superantigens bypass the need for antigen by binding to conserved regions of both the TCR and MHC, thereby stimulating large numbers of T cells, leading to toxic hyperactivation of the immune response. (Adapted from Salyers and Whitt 1994. [62])

few of the genes are recognizable, such as those for head proteins, tail proteins, speA, and a λ-family integrase. The T12 genome differs from those of any of the phages described here so far, but is related to phage r1t of the bacterium *Lactococcus lactis*.

T12 actually integrates within a tRNAser gene, but because the phage itself encodes a segment of a tRNAser gene, the sequence is pre-

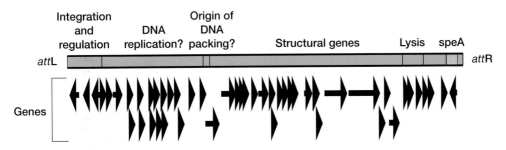

FIGURE 4.25. Phage T12 genome. Arrows indicate particular genes, with the direction of the arrows indicating the direction of transcription. The *Spe* gene is encoded at the extreme right of the genome. (Redrawn, with permission, from McShan and Ferretti 1997 [© Kluwer Academic/Plenum]. [46])

served after integration. Several other phages are known to integrate into this site, and the tRNAser gene is conserved among several streptococcal species, indicating that this may be a frequently used *att* site in this group of bacteria.

The *speA* gene seems to be mobile in populations of *S. pyogenes* phages. The *speA* gene has been detected in phages φ270 and φ49, which otherwise have no known similarity to T12. Reciprocally, not all T12 phage isolates carry *speA*, although those from pathogenic sources do. It seems likely that the presence of the *speA* gene confers some benefit on the phage, the host, or both, but at present the benefit of SpeA is unclear.

PHAGES TRANSPORTING PATHOGENICITY ISLANDS: *VIBRIO CHOLERAE* (34, 47, 55, 56, 62, 72)

Epidemic cholera has killed millions of people worldwide, often taking its worst toll on infants and children. The disease is caused by *Vibrio cholerae*, an aquatic gram-negative bacterium. Cholera is spread by ingestion of contaminated water, often fouled by the presence of human feces. The *V. cholerae* bacterium can exist in a benign form or in the pathogenic form, depending on the presence or absence of pathogenicity islands. A recently emerged strain of *V. cholerae*, designated O139, has proven to be a particularly devastating form of the disease, probably as a result of acquiring a new pathogenicity island.

Cholera begins when pathogenic bacteria ingested from contaminated water bind to the mucosal surface of the small intestine. Normally, mucosal cells maintain a constant intracellular concentration of dissolved ions such as Na^+, Cl^-, K^+, and HCO_3^- by actively pumping them into cells. Water flows freely in and out of cells, but is maintained at a constant intracellular concentration by the need to match the ionic strength inside and out. Pathogenic strains of *V. cholerae* synthesize a secreted toxin, cholera exotoxin, that alters this balance.

Cholera exotoxin is composed of two chains, A and B (Fig. 4.26). The B chains mediate docking of toxin on mucosal cells by binding to cell-surface sugars, called G_{M1} gangliosides, which are covalently attached to the lipid membrane. The A subunit becomes introduced into cells, where it covalently modifies a cellular regulatory protein, Gs, by attaching ADP-ribose groups. Normally, Gs regulates the activity of adenylate cyclase, an enzyme that synthesizes the signaling molecule cyclic AMP (cAMP). ADP-ribosylation of Gs alters the regulatory cycle, causing adenylate cycles to be locked in the "on" position, resulting in the overproduction of cAMP.

This signal strongly affects the systems responsible for maintaining the ionic balance in cells. Ions flow out of cells into the intestinal lumen, and since ion concentrations control water fluxes, water follows. Consequently, vast quantities of water and ions enter the intestines and are lost as violent diarrhea. At the most intense stage of disease, infect-

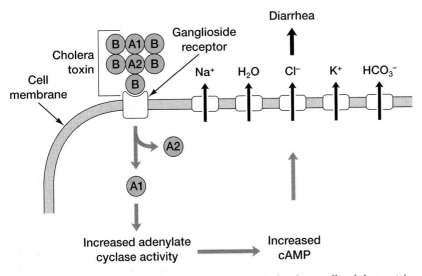

FIGURE 4.26. Action of cholera toxin. The A-B toxin binds to cells of the gastric epithelium via the B subunit, allowing entry of the A1 and A2 components. The A1 component activates adenylate cyclase, leading to an increase in the intracellular concentrations of the signaling molecule cyclic AMP. As a result, the ionic balance in cells changes so that water and dissolved ions flow outward, giving rise to violent diarrhea. (Modified, with permission, from Murray et al. 1998. [51])

ed people can lose 20 liters of water per day. Untreated, the consequences are often fatal, but if water and ions are replaced, the disease resolves without further intervention, in part because the massive diarrhea helps wash out the infecting bacteria.

Two phages transport pathogenicity islands responsible for epidemic cholera, and in so doing interact in a unique fashion. The genes for cholera toxin *ctxA* and *ctxB* are transported on phage CTXφ (Fig. 4.27). The boundaries of the CTXφ chromosome were first determined by comparing pathogenic and nonpathogenic strains, revealing a colinear bank of genes including the toxin genes. Isolation of phage particles permitted characterization of the genomes in particles. CTXφ turns out to be an unusual member of the filamentous phages (Fig. 4.28), the genome of which comprises a 4.5-kb core sequence flanked

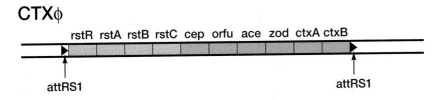

FIGURE 4.27. CTXφ phage/pathogenicity island. Cholera toxin is encoded by the *ctxA* and *ctxB* genes at the right of the genome. (Portion modified, with permission, from Waldor and Mekalanos 1996 [©American Association for the Advancement of Science]. [72])

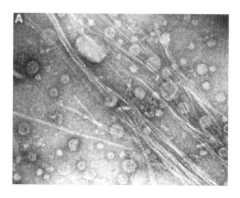

FIGURE 4.28. Electron micrograph of CTXφ particles, illustrating their filamentous nature. (Reprinted, with permission, from Waldor and Mekalanos 1996 [©American Association for the Advancement of Science]. [72])

by repeated 2.7-kb sequence (RS) elements. The core sequence encodes four genes in addition to *ctxA* and *ctxB* (*cep, orfU, ace,* and *zot*) that are similar in sequence and probable function to genes VIII, III, VI, and I of M13 and f1. These CTXφ genes almost certainly encode capsid and assembly proteins although, oddly, they have also been implicated in pathogenesis. The repeated RS elements themselves terminate with a repeated sequence matching the *attRS* site in *V. cholerae*, thus constituting a typical phage attachment site. The RS genes *rstABCR* encode a site-specific recombination system that acts on *attRS* sites.

Remarkably, the cellular receptor for CTXφ, TcpA, is actually encoded by a second phage, VPIφ (Fig. 4.29). VPI was first studied as a pathogenicity island important in cholera pathogenesis, hence the name *V. cholerae* *p*athogenicity *i*sland, and only later found to be a bacteriophage. VPI has a much different base composition from its host. VPI is 35% G/C, whereas *V. cholerae* is 47–49% G/C. As with CTXφ, VPIφ has been proposed to be an integrating filamentous phage. At the right end of the genome, a λ integrase family gene is present, likely the recombinase responsible for integration into the host *att* site. Other genes of VPIφ are related to genes found in viruses (e.g., *tagE, orfZ,* and *orfV*), but the functions of these are less clear than in the case of CTXφ.

TcpA is a remarkable multifunctional protein. TcpA is the major structural gene for a pilus known to be important for pathogenesis by

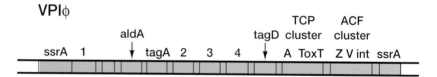

FIGURE 4.29. VPIφ phage/pathogenicity island. (Adapted from Karaolis et al. 1999. [34])

V. cholerae. The name derives from "toxin-coregulated pilus," since *tcpA, ctxA,* and *ctxB* are all regulated by the same host cell transcription factors. As mentioned above, TcpA is also the receptor for CTXϕ. Thus, the receptor for one phage is transported by the other, so that infection by CTXϕ requires prior infection by VPIϕ to provide the receptor for entry, a unique interdependence between viruses. But this strange story does not end here. TcpA is also the capsid protein for VPIϕ. TcpA actually has three roles: (1) as a pilus important for pathogenesis, (2) as a receptor for CTXϕ, and (3) as a coat protein for VPIϕ.

CTXϕ and VPIϕ, once thought to be two blocks of cellular genes captured by lateral transfer, turned out to be two phages. This finding challenges the distinction between bacterial genes and mobile elements. Like normal cellular genes, the phages probably promote the welfare of the bacterium, in this case by allowing persistence in a new niche, a tissue of the human body. On the other hand, CTXϕ and VPIϕ are viruses by all the usual criteria. As shown in the next chapter, the issue of what are cellular sequences and what are mobile elements is even more blurred by studies of the complete microbial genome sequences.

STAPHYLOCOCCUS AUREUS AND TOXIC SHOCK: A PATHOGENICITY ISLAND AS A SATELLITE PHAGE (36, 37, 43, 51, 62)

In 1980 a new infectious disease emerged targeting mainly women between 20 and 40 years of age. The disease resulted in fever, diffuse rash, and loss of skin on the extremities, a condition resembling septic shock. About 3% of cases proved fatal, and in the absence of effective treatment, the disease recurred in 65% of survivors.

The condition, called *toxic shock syndrome* (TSS), was traced to newly introduced superabsorbent tampons released in 1979 and composed of compressed beads of carboxymethyl cellulose and polyester. Because these tampons did not need to be removed as often as conventional tampons, they provided a new niche for growth of *S. aureus.* Normally, *S. aureus* is a minor component of the vaginal microflora and relatively harmless, in part because the vaginal tract is practically anaerobic and *S. aureus* produces TSS toxin only under aerobic conditions. However, the superabsorbent tampons trapped microscopic air pockets, providing the aerobic conditions needed for toxin production. The realization of this mechanism of toxicity led to the removal of the tampons from the market and a reduction in cases of toxic shock. However, the story was a little more complicated than this, and extensive immunological and molecular studies were required to determine the true nature of TSS. It turned out that the new superabsorbent tam-

Cases of TSS still occur, today primarily associated with wound, respiratory, and skeletal infections; postoperatively; and as a post-influenza syndrome.

pons were actually a passive factor in development of the disease. The real culprit was a new form of *S. aureus* that contained a set of genes coding for pyogenic exotoxin A and that probably evolved in the 1970s by acquiring genes for making this toxin and for withstanding penicillin antibiotics. The environmental conditions provided by the new tampons allowed it to flourish.

This toxic shock toxin protein, called toxic shock syndrome toxin-1, like the TSLS toxin of *Streptococcus pyogenes*, acts as a superantigen. The TSS toxin appears to bind to pairs of immune system cells, bridging the V_β region of the T-cell receptor and MHC class II, thereby activating immune cell signaling as in the immune response. However, the degree of immune activation is excessive, resulting in an overproduction of lymphokines and cytokines, leading to multiple organ toxicity and toxic shock.

Intensive study of the DNA determinant for TSS toxin led to the realization that it is in fact a mobile element. The gene for TST toxin (*tst*) was found clinically on either of two different DNAs, each of which appears to fit the definition of a pathogenicity island. SaPI1, the most thoroughly characterized, is about 15.2 kb in length and encodes 16 predicted genes, 3 of which have the potential to encode pathogenicity factors. These include *tst*, another superantigen (*ent*), and a homolog of the *vapE* gene found in the *vap* pathogenicity island of *Dichelobacter nodosu*. Epidemiological studies show that about 20% of natural *S. aureus* isolates encode a *tst*-containing pathogenicity island, the first pathogenicity islands described in gram-positive bacteria.

The S. aureus strain that produces TSST-1 grows much faster than normal strains of S. aureus, and many isolates are insensitive to penicillin-type antibiotics.

The sequence of *SaPI1* also reveals a λ integrase family member and 17-bp direct repeats forming the borders of the element (Fig. 4.30). Strains lacking *SaPI1* contain only a single copy of the sequence, suggesting that it functions as a standard λ-like *att* site. The *SaPI1* system is unique among pathogenicity islands, however, in that it has been shown to be mobilizable *in trans* by infection of a cell with an *S. aureus* phage. Infection of an *S. aureus* strain containing *SaPI1* by phage 80α, followed by inoculation of the progeny phage on a strain lacking *SaPI1*, resulted in transfer of *SaPI1* about as often as the 80α chromosome itself. All insertions were found at a single site, consistent with a λ-like integration mechanism.

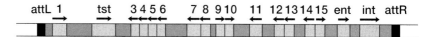

FIGURE 4.30. The toxic shock syndrome pathogenicity island. The toxic shock toxin is encoded by the *tst* gene at the right of the genome. Other genes encode another potential toxin (*ent*), an integrase (*int*), or genes of unknown function (numbered). (Adapted from Lindsay et al. 1998. [43])

Although SaPI1 contains an integrase gene, superinfection by phage 80α was necessary for excision, replication, and transfer of SaPI1 DNA. However, the SaPI1 integrase was sufficient by itself for integration, even in the absence of 80α, indicating that the SaPI1 genome does contribute at least the integrase activity used for this step. Evidently 80α supplies the integrase for excision only.

The demonstration that phage 80α mobilizes *SaPI1* raises a difficulty in nomenclature. *SaPI1* clearly satisfies the definition of a pathogenicity island, since it contains multiple pathogenic determinants and is demonstrably mobile. However, there is essentially no distinction between *SaPI1* and a satellite phage—a defective virus that can multiply only in the presence of a helper virus. *SaPI1* may fit more closely with the pathogenicity island definition, because it seems to contribute only the integrase to its own replication, but there is really a continuum extending between pathogenicity islands and satellite phages. The different names obscure the larger commonalities among these parasites of parasites.

In fact, pathogenicity islands may be transmitted as satellite phages quite commonly. Many islands encode both virulence determinants and integrase enzymes flanked by apparent *att* sites, as with the UPEC islands, potentially paralleling the SaPI1 case. An alternative to the satellite phage model holds that the islands were once complete phages, but most of the phage genes were deleted after integration. The deletion model may hold in some cases, but the frequency of the structure with *att* sites flanking integrase and pathogenicity genes makes the satellite phage model quite attractive.

OTHER MEANS OF MOBILIZING PATHOGENIC DETERMINANTS (16, 17, 24, 32, 41, 51, 78)

In this chapter, we have emphasized transduction by phage because this is the most prominent mechanism of pathogenicity gene transfer. However, there are also examples of transmission by other mobile DNA mechanisms, such as conjugation and transformation, and many more cases for which the mechanism of transduction has not been fully clarified.

There are several examples of virulence factors that are maintained and transported on plasmids. Anthrax toxin, synthesized by the bacterium *Bacillus anthracis*, is an example, as are the enterotoxins of *E. coli* and tetanus toxin of *Clostridium tetani*. The finding that a pathogenic determinant can be transported on a plasmid, however, does not rule out the involvement of a phage. Phage DNA may become integrated into an extrachromosomal plasmid, forming a composite element, as may be the case for some pathogenicity islands of *Streptomyces*.

Transposons can also apparently mobilize pathogenicity islands. In a few cases, the pathogenicity islands are still associated with active transposons. Examples include the plague bacteria (*Yersinia pestis*) pathogenicity island *HPI*, which is flanked by IS*100* elements, and the *E. coli aerobactin* gene, which is flanked by IS*1* elements. In other cases, pathogenicity determinants are inferred to have been transferred by transposons from sequence data. Evidence comes from the presence of inverted repeats flanking the pathogenic determinant, from target site duplications in flanking DNA, and from the presence of sequences homologous to transposases.

The common occurrence of inactive genes for transposases or phage integrases associated with pathogenic determinants raises the question of whether mutational inactivation may be a part of the gene capture process. Perhaps evolutionarily useful pathogenic sequences associated with active mobile elements are lost at a high rate. If so, inactivating the transposase or integrase gene might confer a selective advantage. However, although mutant genes are commonly seen, it is not clear whether the rate of inactivating integrase and transposase genes is higher than the rate of inactivating any other dispensable cellular gene.

In some cases, apparently mobile pathogenicity islands are not associated with any obvious mobile DNA sequences, phage or otherwise. For example, the complete sequence of *Neisseria meningitidis* contains three apparent pathogenicity islands, none of which is associated with mobile DNA elements. In these cases, however, phage transduction or conjugative transfer could still have been involved—pathogenicity islands could have been incorporated by homologous recombination, which does not leave any signature in the final DNA product, or phage or transposase genes could have been lost by deletion after integration.

PATHOGENICITY ISLANDS AND PATHOGEN EVOLUTION

It may seem odd that phage would shuffle about genes for making people sick, but the clinical manifestations are probably secondary consequences of systems that exist for other reasons. Phage-encoded toxins that promote lysis of host cells liberate iron, which can then be used for growth by the lysogenic bacteria. In many other cases, the benefits of the pathogenicity islands to the phage and host are unclear, and for most islands at least some of the genes are of unknown function. Although the process is incompletely understood, it is likely that we are seeing the product of a sophisticated coevolution of host and phage, in which lateral transfer has been exploited to allow access to new biological niches.

The pathogenicity islands and mobile antibiotic-resistance genes provide specific examples of the role of lateral DNA transfer in microbial evolution. In each case, the action of conjugative plasmids, transposons, phage, and transformation systems provides microbes with the benefit of sequence diversity far greater than that present in any single genome. Capture of appropriate pathogenicity islands allows bacteria to tailor their metabolism to the diverse environments of the mammalian host. Treatment with antibiotics only selects for those microbes fortunate enough to have acquired the appropriate resistance determinants. The combination of mobile resistance determinants and mobile pathogenicity islands ensures that new virulent strains which are increasingly resistant to our most effective therapies will arise regularly.

As shown in the next chapter, the central role of lateral transfer in microbial evolution seen in pathogenesis and antibiotic resistance holds for many other types of genes as well.

GENE TRANSFER AGENTS (2, 40, 76)

Before leaving the topic of phage-mediated transduction, we review a mysterious group of lateral transfer elements, the gene transfer agents (GTAs). Studies of several bacterial and archaeal species revealed spontaneous transfer of genetic markers between strains (see Table 4.3). Studies have shown that gene transfer takes place by an agent that can pass through filters that retain bacteria, ruling out conjugation as the transfer mechanism. Simple alternatives would include transformation or phage transduction, but neither of these seems to explain the phenomenon. The GTA was insensitive to digestion with the DNA-degrading enzyme DNase, suggesting that normal transformation by naked DNA did not mediate gene transfer. GTA filtrates did not con-

TABLE 4.3. Gene Transfer Agents

Microbe	Gene transfer agent	Reference
Bacteria		
Rhodobacter capsulatus	GTA	Lang and Beatty 2000 (40)
		Marrs 1974 (44)
		Yen et al. 1979 (76)
Desulfovibrio desulfuricans		Rapp and Wall 1987 (59)
Serpulina hyodysenteriae	VSH-1	Humphrey et al. 1997 (30)
Myxococcus xanthus		Starich et al. 1985 (68)
Archaea		
Methanococcus voltae	VTA	Bertani 1999 (2)

tain detectable phage as measured by plaque-forming activity or the presence of phage genomes. How, then, were genes transferred between these cells?

It turns out that culture supernatants with GTA activity do contain particles resembling tailed phages, suggesting a connection to normal transduction. Isolation of genes required for GTA activity in the purple bacterium *Rhodobacter capsulatus* revealed a set of 19 coding regions. Four of these genes were found to be similar in sequence to head and tail genes of known bacteriophages. Unlike typical phage, however, analysis of the DNA contained within these particles shows an extremely diverse pattern, consistent with packaging of random pieces of the host DNA rather than phage DNA. Typical generalized transducing phage, such as P22 or P1, package cellular DNA in only 0.3–6% of particles, the rest packaging normal phage genomes. GTA agents, in contrast, appear to package host DNA exclusively, or nearly so.

The GTA thus appear to be defective prophages that have lost the ability to package their own DNA but retained the ability to package and transduce cellular DNA. The fragments of packaged host DNA are rather small, 4–5 kb, about the size expected for the small phage particles observed. The GTAs, where tested, were able to transfer a very wide range of host cell markers, indicating that most, and perhaps all, host sequences could be packaged and transduced.

The GTA do not seem to be rare in the species in which they are found. On the contrary, for *R. capsulatus*, strains isolated from distant geographical locations were found to harbor GTA. Similarly, for the Archaeon *Methanococcus voltae*, several strains have shown evidence of GTA-like DNA elements. Whether GTA can transfer genes between different bacterial species is unknown.

These observations raise intriguing questions on the evolution and function of GTAs. It seems likely that GTAs were formed by integration of phage genomes, followed by accumulation of mutations that block normal DNA packaging. Thus, GTAs may be just another example of defective prophages fixed in the host chromosome after loss of phage genes for replication. However, the wide distribution of GTA elements within species, and the recurrence of GTAs in diverse species, raises the question of whether these elements may be beneficial to the host microbe. Might it be advantageous for a bacterium to exchange genes with its nearest neighbors, which in a bacterial population are likely closely related? In this view, GTAs may resemble the natural transformation systems present in many bacteria. Another question is whether related phages are involved in formation of GTAs. The phages of *Rhodobacter* and *Methanococcus* are both quite small, reinforcing the possibility that they are related. If so, might there be more to the story? For example, might a phage have evolved a lysogenic lifestyle in which it contributes generalized transduction functions in the prophage state? It should be possible to address these questions by further characterization of the genetics of GTAs.

Summary

Bacteriophages provide one of the most efficient vehicles for moving DNA sequences between bacterial cells. Among the consequences of transduction is the dissemination of sequences causing bacteria to become more pathogenic.

Phage λ, the classic model temperate phage, can follow either of two growth programs. λ can grow lytically, bursting the cell and liberating hundreds of new phage particles, or lysogenically, adopting a quiescent state in which its chromosome becomes integrated into the host chromosome. Integration takes place at a specific point on the host chromosome, *attB*, and a site on the phage chromosome, *attP*, by means of an ordered series of reactions involving covalent protein–DNA intermediates. Today hundreds of λ integrase family enzymes have been found associated with mobile DNA systems.

Phage Mu replicates by a different strategy involving replicative transposition. Mu integration and replicative transposition are catalyzed by MuA transposase and its cofactor MuB. Following replicative transposition, Mu genomes are excised from each site in the chromosome and packaged. Intensive studies of the phage Mu transposition reaction have provided a detailed picture of its mechanism and regulation, including the experimental demonstration that transposition involves the θ or Shapiro intermediate.

Occasionally host DNA sequences can be transported in particles of λ, Mu, or other phages. Phage λ engages in specialized transduction, the transfer of specific host sequences flanking the *attB* site. Induction of a λ lysogen occasionally goes awry, with abnormal cleavages resulting in packaging and transduction of a segment of the λ chromosome joined to flanking host sequences. Mu carries out generalized transduction, the transfer of any chromosomal sequence between cells, by occasionally packaging a headful of chromosomal DNA. Upon entry into a new cell, the chromosomal DNA can become incorporated by homologous recombination or other means. Other phage that package by a headful mechanism are also known to carry out generalized transduction.

Bacteriophages are extremely abundant in many natural settings, and more than a million different types may exist on earth. The genomic structure of many phages is modular, with genes of related functions clustered in the genome with their sites of action. This allows the cassettes to be exchanged between related phage by coinfection of host cells and recombination between phage genomes. Cassettes appear to move over great phylogenetic distances by transfer through intermediate hosts, resulting in the formation of new phage types from these genetic composites.

Pathogenicity islands, blocks of genes converting bacteria from being benign to pathogenic, are also often transported by phage. In some cases, phage-mediated transduction can be demonstrated directly in laboratory experiments. In many other cases, phage-mediated transfer of pathogenicity islands is inferred by (1) codon utilization in the pathogenicity island different from that of the host, (2) association with phage sequences, and (3) proximity to tRNA genes, known favored sites of phage integration. Genes in pathogenicity islands may encode toxins, such as the toxic shock syndrome toxin (*tst*) of *S. aureus*, and products that promote persistence in the host, such as the adhesins of uropathogenic *E. coli*. The *S. aureus* pathogenicity island containing the *tst* gene was found to be mobilizable *in trans* by infection with phage 80α. A similar pathway may mediate mobilization of many islands containing integrase genes but not capable of independent replication.

REFERENCES

1. Atlas R.M. 1984. *Microbiology: Fundamentals and applications.* Macmillan, Inc., New York.
2. Bertani G. 1999. Transduction-like gene transfer in the methanogen *Methanococcus voltae. J. Bacteriol.* **181:** 2992–3002.

3. Botstein D. and Herskowitz I. 1974. Properties of hybrids between *Salmonella* phage P22 and coliphage lambda. *Nature* **251:** 584–589.

4. Brock T.D. 1990. *The emergence of bacterial genetics*. Cold Spring Harbor Laboratory Press, Cold Spring Harbor, New York.

5. Cairns J., Stent G.S., and Watson J.D. 1966. *Phage and the origins of molecular biology*. Cold Spring Harbor Laboratory, Cold Spring Harbor, New York.

6. Calendar R.E. 1988. *The bacteriophages*. Plenum Press, New York.

7. Cerritelli M.E., Cheng N., Rosenberg A.H., McPherson C.E., Booy F.P., and Steven A.C. 1997. Encapsidated conformation of bacteriophage T7 DNA. *Cell* **91:** 271–280.

8. Chaconas G. 1999. Studies on a "jumping gene machine": Higher-order nucleoprotein complexes in Mu DNA transposition. *Biochem. Cell Biol.* **77:** 487–491.

9. Craigie R. and Mizuuchi K. 1985. Mechanism of transposition of bacteriophage Mu: Structure of a transposition intermediate. *Cell* **41:** 867–876.

10. Davis R.W. and Davidson N. 1968. Electron microscope visualization of deletion mutants. *Proc. Natl. Acad. Sci.* **60:** 243–250.

11. Delbrück M. and Luria S.E. 1942. Interference between bacterial viruses. I. Interference between two bacterial viruses acting on the same host, and the mechanism of virus growth. *Arch. Biochem.* **1:** 111–141.

12. Demers B., Simor A.E., Vellend H., Schievert P.M., Byrne S., Jamieson F., Walmsley S., and Low D.E. 1993. Severe invasive group A streptococcal infections in Ontario, Canada: 1987–1991. *Clin. Infect. Dis.* **16:** 792–800.

13. Duda R.L. 1998. Protein chainmail: Catenated protein in viral capsids. *Cell* **94:** 55–60.

14. Ebel-Tsipis J., Botstein D., and Fox M.S. 1972. Generalized transduction by phage P22 in *Salmonella typhimurium*. I. Molecular origin of transducing DNA. *J. Mol. Biol.* **71:** 433–448.

15. Faelen M., Toussaint A., and Resibois A. 1979. Mini-muduction: A new mode of gene transfer mediated by mini-Mu. *Mol. Gen. Genet.* **176:** 191–197.

16. Fetherston J.D. and Perry R.D. 1994. The pigmentation locus of *Yersinia pestis* KIM6+ is flanked by an insertion sequence and includes the structural genes for pesticin sensitivity and HMWP2. *Mol. Microbiol.* **13:** 697–708.

17. Finlay B.B. and Falkow S. 1997. Common themes in microbial pathogenicity revisited. *Microbiol. Mol. Biol. Rev.* **61:** 136–169.

18. Fraenkel-Conrat F., Kimball P.C., and Levy J.A. 1988. *Virology*. Prentice Hall, Englewood Cliffs, New Jersey.

19. Fuhrman J.A. 1999. Marine viruses and their biogeochemical and ecological effects. *Nature* **399:** 541–548.

20. Goodman S.D. and Nash H.A. 1989. Functional replacement of a protein-induced bend in a DNA recombination site. *Nature* **341:** 251–254.

21. Gottesman S. and Beckwith J. 1969. Directed transposition of the arabinose operon: A technique for the isolation of specialized transducing bacteriophage for an *Escherichia coli* gene. *J. Mol. Biol.* **44:** 117–129.

22. Griffiths A.J.F., Miller J.H., Suzuki D.T., Lewontin R.C., and Gelbart W.M. 1993. *An introduction to genetic analysis*. W.H. Freeman, New York.

23. Guo F., Gopaul D.N., and Van Duyne G.D. 1997. Structure of Cre recombinase complexed with DNA in a site-specific recombination synapse. *Nature* **398:** 40–46.

24. Hacker J., Blum-Oehler G., Muhldorfer I., and Tschape H. 1997. Pathogenicity islands on virulent bacteria: Structure, function and impact on microbial evolution. *Mol. Microbiol.* **23:** 1089–1097.

25. Hacker J., Bender L., Ott M., Wingender J., Lund B., Marre R., and Goebel W. 1990. Deletions of chromosomal regions coding for fimbriae and hemolysis occur in vitro and in vivo in various extraintestinal *Escherichia coli* isolates. *Microb. Pathog.* **8:** 213–225.

26. Hendrix R.W., Roberts J.W., Stahl F.W., and Weisberg R.A. 1983. *Lambda II.* Cold Spring Harbor Laboratory, Cold Spring Harbor, New York.

27. Hendrix R.W., Smith M.C.M., Burns R.N., Ford M.E., and Hatfull G.F. 1999. Evolutionary relationships among diverse bacteriophages and prophages: All the world's a phage. *Proc. Natl. Acad. Sci.* **96:** 2192–2197.

28. Howe M.M. 1973. Transduction by bacteriophage Mu-1. *Virology* **55:** 103–117.

29. Howe M.M. 1998. Bacteriophage Mu. *NATO ASI Ser. Ser. H Cell Biol.* **103:** 65–80.

30. Humphrey S.B., Stanton T.B., Jensen N.S., and Zuerner R.L. 1997. Purification and characterization of VSI I-1, a generalized transducing bacteriophage of *Serpulina hyodysenteriae. J. Bacteriol.* **179:** 323–329.

31. Kaiser A.D. and Jacob F. 1957. Recombination between related temperate bacteriophages and the genetic control of immunity and prophage localization. *Virology* **4:** 509–521.

32. Kaper J.B. and Hacker J. 1999. *Pathogenicity islands and other mobile virulence elements.* American Society for Microbiology, Washington, D.C.

33. Karam J.D., Drake J.W., and Kreuzer K.N. 1994. *Molecular biology of bacteriophage T4.* American Society for Microbiology, Washington, D.C.

34. Karaolis D.K.R., Somara S., Maneval D.R.J., Johnson J.A., and Kaper J.B. 1999. A bacteriophage encoding a pathogenicity island, a type-IV pilus and a phage receptor in cholera bacteria. *Nature* **399:** 375–379.

35. Kay B.K., Winter J., and McCafferty J. 1996. *Phage display of peptides and proteins.* Academic Press, San Diego, California.

36. Kreiswirth B.N., Projan S.J., Schievert P.M., and Novick R.P. 1989. Toxic shock syndrome toxin-1 is encoded by a variable genetic element. *Rev. Infect. Dis.* **11:** S75–S82.

37. Kreiswirth B., Lofdahl S., Betley M., O'Reilly M., Schievert P., Bergdoll M., and Novick R.P. 1983. The toxic shock syndrome exotoxin structural gene is not detectably transmitted by a prophage. *Nature* **305:** 709–712.

38. Kurihara T. and Nakamura Y. 1983. Cloning of the nusA gene of *Escherichia coli. Mol. Gen. Genet.* **190:** 189–195.

39. Kwon H.J., Tirumalai R., Landy A., and Ellenberger T. 1997. Flexibility in DNA recombination: Structure of the lambda integrase catalytic core. *Science* **276:** 126–131.

40. Lang A.S. and Beatty J.T. 2000. Genetic analysis of a bacterial genetic exchange element: The gene transfer agent of *Rhodobacter capsulatus. Proc. Natl. Acad. Sci.* **97:** 859–864.

41. Leblond P. and Decaris B. 1994. New insights into the genetic instability of Streptomyces. *FEMS Microbiol. Lett.* **123:** 225–232.

42. Levchenko I., Luo L., and Baker T.A. 1995. Disassembly of the Mu transposase tetramer by the ClpX chaperone. *Genes Dev.* **9:** 2399–2408.

43. Lindsay J.A., Ruzin A., Ross H.F., Kurepina N., and Novick R.P. 1998. The gene for toxic shock toxin is carried by a family of mobile pathogenicity

islands of *Staphylococcus aureus. Mol. Microbiol.* **29:** 527–543.

44. Marrs B.L. 1974. Genetic recombination in *Rhodopseudomonas capsulata. Proc. Natl. Acad. Sci.* **71:** 971–973.

45. Martuscelli J., Taylor A.L., Cummings D.J., Chapman V.A., DeLong S.S., and Canedo L. 1971. Electron microscopic evidence for linear insertion of bacteriophage Mu-1 in lysogenic bacteria. *J. Virol.* **8:** 551–563.

46. McShan W.M. and Ferretti J.J. 1997. Genetic studies of erythrogenic toxin carrying temperate bacteriophages of *Streptococcus pyogenes. Adv. Exp. Med. Biol.* **418:** 971–973.

47. Mekalanos J.J. 1983. Duplication and amplification of toxin genes in *Vibrio cholerae. Cell* **35:** 253–263.

48. Mizuuchi K. 1983. In vitro transposition of bacteriophage Mu: A biochemical approach to a novel replication reaction. *Cell* **35:** 785–794.

49. Mizuuchi M., Baker T.A., and Mizuuchi K. 1995. Assembly of phage Mu transpososomes: Cooperative transitions assisted by protein and DNA scaffolds. *Cell* **83:** 375–385.

50. Mizuuchi K., Gellert M., Weisberg R.A., and Nash H.A. 1980. Catenation and supercoiling in the products of bacteriophage lambda integrative recombination in vitro. *J. Mol. Biol.* **141:** 484–494.

51. Murray P.R., Rosenthal K.S., Kobayashi G.S., and Pfaller M.A. 1998. *Medical microbiology.* Mosby, St. Louis, Missouri.

52. Nash H.A. 1975. Integrative recombination of bacteriophage lambda DNA in vitro. *Proc. Natl. Acad. Sci.* **72:** 1072–1076.

53. Nash H. 1996. The HU and IHF proteins. In *Regulation of gene expression in* Escherichia coli (ed. E.C.C. Lin and A.S. Lynch), pp. 149–179. R.G. Landes, Austin, Texas.

54. Neidhardt F.C., Ingraham J.L., Low K.B., Magasanik B., Schaechter M., and Umbarger H.E., Eds. 1987. Escherichia coli *and* Salmonella typhimurium: *Cellular and molecular biology.* American Society for Microbiology, Washington, D.C.

55. Pearson G.D.N. 1989. "The cholera toxin genetic element: A site-specific transposon." Ph.D. thesis, Harvard University School of Medicine, Cambridge, Massachusetts.

56. Pearson G.D.N., Woods A., Chiang S.L., and Mekalanos J.J. 1993. CTX genetic element encodes a site-specific recombination system and an intestinal colonization factor. *Proc. Natl. Acad. Sci.* **90:** 3750–3754.

57. Perna N.T., Plunkett III., G., Burland V. , Mau B., Glasner J.D., Rose D.J., Mayhew G.F., Evans P.S., Gregor J., Kirkpatrick H.A., Posfai G., Hackett J., Klink S., Boutin A., Shao Y., Miller L., Grotbeck E.J., Davis N.W., Lim A., Dimalanta E.T., Potamousis K.D., Apodaca J., Anantharaman T.S., Lin J., Yen G., Schwartz D.C., Welch R.A., and Blattner F.R. 2001. Genome sequence of enterohaemorrhagic *Escherichia coli* 0157:H7. *Nature* **409:** 529–533.

58. Ptashne M. 1992. *A genetic switch*, 2nd edition: *Phage λ and higher organisms.* Cell Press and Blackwell Scientific, Cambridge, Massachusetts.

59. Rapp B.J. and Wall J.D. 1987. Genetic transfer in *Desulfobacterium desulfuricans. Proc. Natl. Acad. Sci.* **84:** 9128–9130.

60. Richet E., Abcarian P., and Nash H.A. 1988. Synapsis of attachment sites during lambda integrative recombination involves capture of a naked DNA by a protein-DNA complex. *Cell* **52:** 9–17.

61. Rohwer F., Segall A., Steward G., Seguritan V., Breitbart M., Wolven F., and Azam F. 2000. The complete genomic sequence of the marine phage Roseo-

phage SIO1 shares homology with nonmarine phages. *Limnol. Oceanogr.* **45:** 408–418.

62. Salyers A.A. and Whitt D.D. 1994. *Bacterial pathogenesis: A molecular approach.* American Society for Microbiology, Washington, D.C.

63. Schrenk W.J. and Weisberg R.A. 1975. A simple method for making new transducing lines of coliphage lambda. *Mol. Gen. Genet.* **137:** 101–107.

64. Segall A.M., Goodman S.D., and Nash H.A. 1994. Architectural elements in nucleoprotein complexes: Interchangeability of specific and non-specific DNA binding proteins. *EMBO J.* **19:** 4536–4548.

65. Shimada K., Weisberg R.A., and Gottesman M.E. 1972. Prophage lambda at unusual chromosomal locations. I. Location of the secondary attachment sites and the properties of the lysogens. *J. Mol. Biol.* **63:** 483–503.

66. Shimada K., Weisberg R.A., and Gottesman M.E. 1973. Prophage lambda at unusual chromosomal locations. II. Mutations induced by bacteriophage lambda in *Escherichia coli* K12. *J. Mol. Biol.* **80:** 297–314.

67. Siebert P.D., Chenchik A., Kellog D.E., Lukyanov K.A., and Lukyanov S.A. 1995. An improved PCR method for walking in uncloned genomic DNA. *Nucleic Acids Res.* **23:** 1087–1088.

68. Starich T., Cordes P., and Zissler J. 1985. Transposon tagging to detect a latent virus in *Myxococcus xanthus. Science* **230:** 541–543.

69. Surette M.G., Buch S.J., and Chaconas G. 1987. Transpososomes: Stable protein-DNA complexes involved in the in vitro transposition of bacteriophage Mu DNA. *Cell* **49:** 235–262.

70. Symonds N., Toussaint A., van de Putte P., and Howe M.M. 1987. *Phage Mu,* p. 70. Cold Spring Harbor Laboratory, Cold Spring Harbor, New York.

71. Taylor A.L. 1963. Bacteriophage-induced mutation in *E. coli. Proc. Natl. Acad. Sci.* **50:** 1043–1051.

72. Waldor M.K. and Mekalanos J.J. 1996. Lysogenic conversion by a filamentous phage encoding cholera toxin. *Science* **272:** 1910–1914.

73. Weisberg R.A., Gottesman M.E., Hendrix R.W., and Little J.W. 1999. Family values in the age of genomics: Comparative analyses of temperate bacteriophage HK022. *Annu. Rev. Genet.* **33:** 565–602.

74. Westmoreland B.C., Szybalski W., and Ris H. 1969. Mapping of deletions and substitutions in heteroduplex DNA molecules of bacteriophage lambda by electron microscopy. *Science* **163:** 1343–1348.

75. Yang W. and Mizuuchi K. 1997. Site-specific recombination in plane view. *Structure* **15:** 1401–1406.

76. Yen H.C., Hu N.T., and Marrs B.L. 1979. Characterization of the gene transfer agent made by an overproducer mutant of *Rhodopseudomonas capsulata. J. Mol. Biol.* **131:** 157–168.

77. Yu C.-E. and Ferretti J.J. 1991. Molecular characterization of new group A streptococcal bacteriophages containing the gene for streptococcal erythrogenic toxin A (speA). *Mol. Gen. Genet.* **231:** 161–168.

78. Zagaglia C., Casalino M., Colonna B., Conti C., Calconi A., and Nicoletti M. 1991. Virulence plasmids of enteroinvasive *Escherichia coli* and *Shigella flexneri* integrate into a specific site on the host chromosome: Integration greatly reduces expression of plasmid-carried virulence genes. *Infect. Immun.* **59:** 792–799.

79. Zinder N.D. and Lederberg J. 1952. Genetic exchange in *Salmonella. J. Bacteriol.* **64:** 679–699.

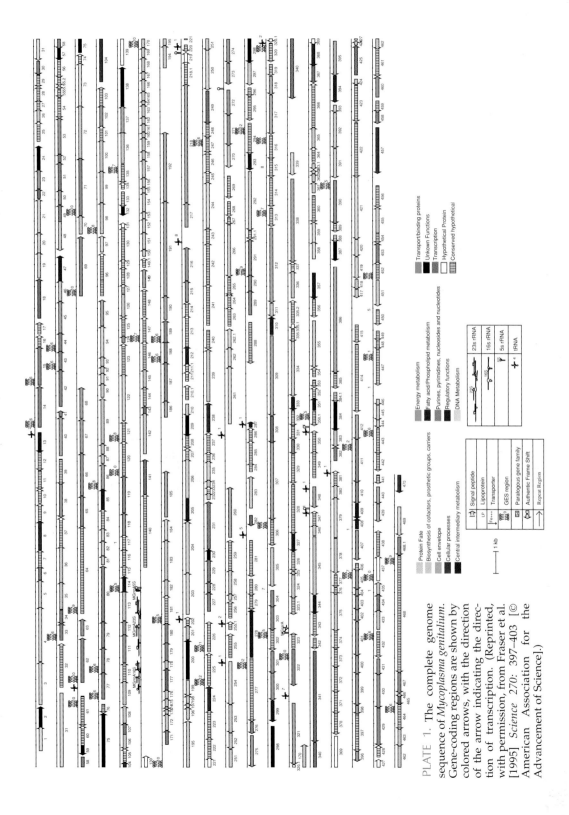

PLATE 1. The complete genome sequence of *Mycoplasma genitalium*. Gene-coding regions are shown by colored arrows, with the direction of the arrow indicating the direction of transcription. (Reprinted, with permission, from Fraser et al. [1995] *Science* 270: 397–403 [© American Association for the Advancement of Science].)

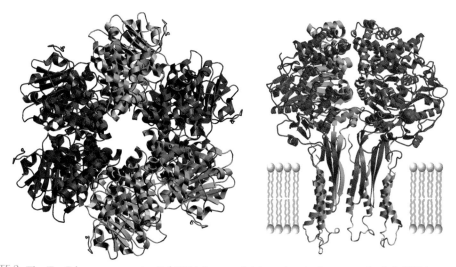

PLATE 2. The TrwB hexamer, a potential DNA "pump." A top view is shown on the left. DNA may be extruded through the pore in the center. On the right is shown a possible orientation of TwrB in the bacterial membrane (only 4 of the 6 monomers are shown for clarity). (Modified, with permission, from Gomis-Ruth et al. [2001] *Nature 409:* 637–641 [© Macmillan Magazines Ltd.]; rendering by Miquel Coll, Institut de Biologia Molecular de Barcelona, Spain.)

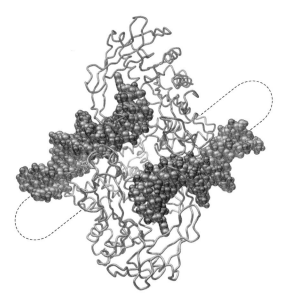

PLATE 3. A dimer of Tn*5* transposases (*gray* and *gold* backbone models) bound to the Tn*5* DNA ends (*black* and *red* space-filling models). The dashed line indicates the connectivity of the Tn*5* DNA in the transposition complex. The Tn*5* DNA ends contain terminal hairpins. (Adapted from Davies et al. [2000] *Science 289:* 77–85; rendering by Witek Kwiatkowski, Structural Biology Laboratory, Salk Institute and F.D.B.)

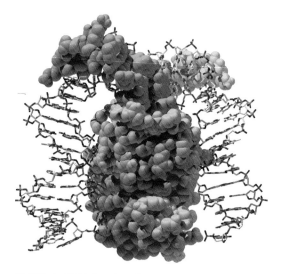

PLATE 4. DNA bending by IHF protein. The IHF heterodimer is shown by blue and yellow space-filling models. DNA is shown by the "stick" model, with CPK colors. (Adapted from Rice et al. [1996] *Cell 87:* 1295–1306; rendering by Witek Kwiatkowski, Structural Biology Laboratory, Salk Institute and F.D.B.)

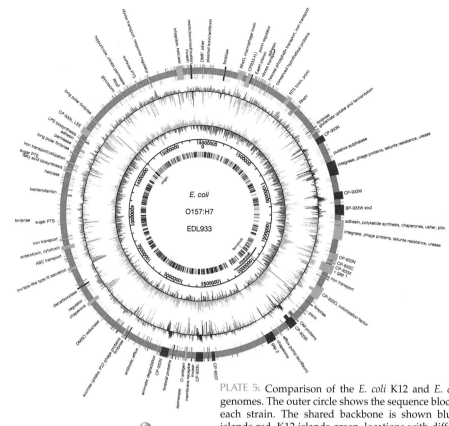

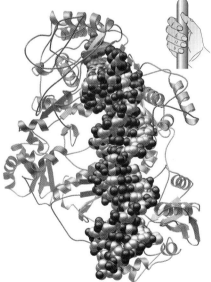

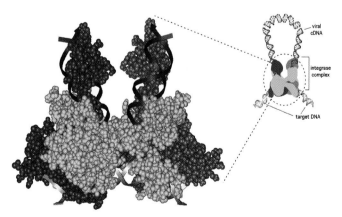

PLATE 5. Comparison of the *E. coli* K12 and *E. coli* O157:H7 genomes. The outer circle shows the sequence blocks unique to each strain. The shared backbone is shown blue, O157-H7 islands red, K12 islands green, locations with different islands in the two strains tan, hypervariable regions purple. The second circle shows the G/C content for genes, colored as for the outer circle. The third circle shows the G/C skew in third codon positions. The fourth circle is the scale (in base pairs). The fifth circle shows the distribution of Chi recombination sites; blue and purple indicate the orientation of the sites. (Reprinted, with permission, from Perna et al. [2001] *Nature 409:* 529–533 [© Macmillan Magazines Ltd.].)

PLATE 6. HIV reverse transcriptase bound to DNA and a nucleotide substrate. The p66 subunit is shown in bold, the p51 subunit silver. The DNA substrate is shown as the space-filling model (CPK colors) and the incoming nucleotide is shown as the green space-filling model. The drawing of a right hand (upper right) emphasizes that RT bound to DNA resembles a right hand grasping a mop handle. (Adapted from Huang et al. [1998] *Science 282:* 1669–1675; rendering by Witek Kwiatkowski, Structural Biology Laboratory, Salk Institute and F.D.B.)

PLATE 7. A speculative model for the organization of HIV-1 integrase bound to DNA. A dimer of integrase dimers (one composed of dark and light blue monomers, the other orange and yellow) for a tetramer binding the viral DNA ends (*black ribbons*). The diagram at the right illustrates the placement of the tetramer in the larger preintegration complex. (Modified from Gao et al. [2001] *EMBO J. 20:* 3565–3576; rendering by Kui Gao.)

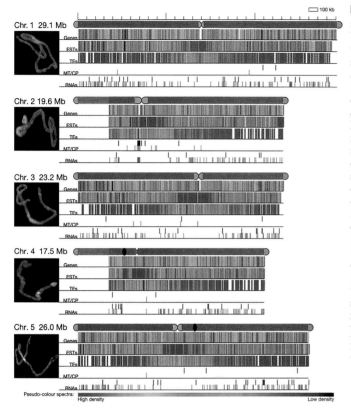

PLATE 8. The *Arabidopsis* chromosomes, showing locations of mobile DNA elements. The scale is shown at the top in units of 100 kb. Chromosome numbers and sizes are shown above images of stained chromosomes (*left*). The sequenced regions in each chromosome are shown red (*top right diagram*), telomeric and centromeric heterochromatin in blue, heterochromatic knobs in black, and rDNA repeats in magenta. Genes and expressed sequence tags (ESTs) are shown in the next two lines, color-coded for density (Genes ranged in density from 1 to 38 per 100 kb, ESTs from 1 to 200 per 100 kb. Transposable elements are shown in the next line (TEs; density ranging from 1 to 33 per 100 kb), and imports from the mitochondrial and chloroplast genomes (MT/CP) below that (*black* and *green* ticks, respectively). The bottom line shown tRNAs (*black* ticks) and small nucleolar RNAs (*red* ticks). (Reprinted, with permission, from The Arabidopsis Initiative [2000] *Nature 408:* 796–815 [© Macmillan Magazines Ltd.].)

PLATE 9. Organization of type IV secretion systems including that of *Agrobacteria*. Hypothetical structure for the type IV section apparatus is shown at the top. The thick red arrow indicates the direction of extrusion of macromolecules. "NTP" indicates ATPase activity of the indicated protein. Genes are shown schematically at the bottom connected to the subunits they are proposed to encode. Genes of similar sequence and probable function are shown in the same color. Genes marked in gray are thought to encode the "core" transfer apparatus. (Reprinted, with permission, from Covacci et al. [1999] *Science 284:* 1328–1333 [© American Association for the Advancement of Science].)

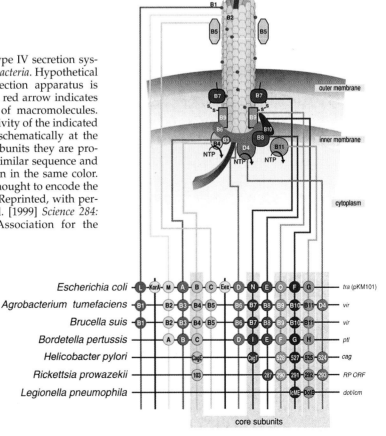

5

Microbial Genomes and DNA Exchange

THE COMPLETE DNA SEQUENCES of more than 55 prokaryotic organisms have been determined (10 Archaea and 45 bacteria), and more are on the way (some examples are summarized in Table 5.1). This gigantic collection of sequence information provides a rich new resource for studies of the mechanism and consequences of lateral DNA transfer. It has been clear for several decades that DNA can move between cells, but just in the last five years, with the determination of complete microbial sequences, have we begun to appreciate the full impact of lateral transfer on the genetic makeup of organisms.

This chapter has a different focus from the previous two chapters, which considered lateral transfer in antibiotic resistance (Chapter 3) and bacterial pathogenesis (Chapter 4). This chapter considers the general structure of prokaryotic genomes, highlighting the role of lateral transfer in their evolution. The structures of genomes and the evidence for active lateral transfer are first considered, then evidence for extensive transfer in natural settings, which bolsters the idea that lateral

TABLE 5.1. Genome Features from 24 Microbial Genome Sequencing Projects

Organism	Genome size (Mbp)	Number of ORFs (% coding)		Unknown function		Unique ORFs	
Aeropyrum pernix K1	1.67	1,885	(89%)	663	(44%)	407	(27%)
A. aeolicus VF5	1.50	1,749	(93%)	1,315	(54%)	641	(26%)
A. fulgidus	2.18	2,437	(92%)	1,722	(42%)	1,053	(26%)
B. subtilis	4.20	4,779	(87%)	1,722	(42%)	1,053	(26%)
B. burgdorferi	1.44	1,738	(88%)	1,132	(65%)	682	(39%)
Chlamydia pneumoniae AR39	1.23	1,134	(90%)	543	(48%)	262	(23%)
C. trachomatis MoPn	1.07	936	(91%)	353	(38%)	77	(8%)
C. trachomatis serovar D	1.04	928	(92%)	290	(32%)	255	(29%)
Deinococcus radiodurans	3.28	3,187	(91%)	1,715	(54%)	1,001	(31%)
E. coli K-12-MG 1655	4.60	5,295	(88%)	1,632	(38%)	1,114	(26%)
H. influenzae	1.83	1,738	(88%)	592	(35%)	237	(14%)
H. pylori 26695	1.66	1,589	(91%)	744	(45%)	539	(33%)
Methanobacterium thermoautotrophicum	1.75	2,008	(90%)	1,010	(54%)	496	(27%)
Methanococcus jannaschii	1.66	1,783	(87%)	1,076	(62%)	525	(30%)
M. tuberculosis CSU#93	4.41	4,275	(92%)	1,521	(39%)	606	(15%)
M. genitalium	0.58	483	(91%)	173	(37%)	7	(2%)
M. pneumoniae	0.81	680	(89%)	248	(37%)	67	(10%)
N. meningitidis MC58	2.24	2,155	(83%)	856	(40%)	517	(24%)
Pyrococcus horikoshii OT3	1.74	1,994	(91%)	859	(42%)	453	(22%)
Rickettsia prowazekii Madrid E	1.11	878	(75%)	311	(37%)	209	(25%)
Synechocystis sp.	3.57	4,003	(87%)	2,384	(75%)	1,426	(45%)
T. maritime MSB8	1.86	1,879	(95%)	863	(46%)	373	(26%)
Treponema pallidum	1.14	1,039	(93%)	461	(44%)	280	(27%)
Vibrio cholerae El Tor N1696	4.03	3,890	(88%)	1,806	(46%)	934	(24%)
	50.60	52,462	(89%)	22,358	(43%)	12,161	(23%)

(Reprinted, with permission, from Fraser et al. 2000 [© Macmillan Magazines Ltd.]. [25])

The Beginning of the Genomics Era (24, 90)

The era of complete genomic sequences began with the determination of the 1.8-million-base sequence of *Hemophilus influenzae* by Craig Venter and colleagues at The Institute for Genomic Research (TIGR). The sequence was determined by first fragmenting *H. influenzae* DNA into myriad small pieces of about 2 kb, then reading the sequences of 450 bp from each edge (the longest achievable sequence reads). About 24,000 such short sequences were determined and then linked up by a computer that identified and joined overlapping sequences determined from different DNA fragments. After a cleanup phase, involving targeted sequencing to fill in the remaining gaps, the circular sequence was closed, yielding the final chromosome sequence.

Computer analysis identified about 1,800 genes, although the number remains an estimate due to the difficulty of identifying genes with certainty from sequence data alone. The early analysis of the genome revealed several examples of mobile DNA, including a cryptic Mu phage (Mu is discussed in Chapter 4) and several copies of the insertion sequence IS*1016*. Another indication of lateral transfer was the absence of a pathogenicity island containing a cluster of fimbrial genes, which is present in pathogenic *H. influenzae* strains but not in the strain sequenced.

The complete sequence of *H. influenzae* also revealed that its chromosome has evolved to promote uptake of its own DNA by transformation. Evident in the sequence were 1465 copies of the 9-base uptake signal sequence (USS) that promotes DNA capture by transformation. This frequency of occurrence of the USS is much greater than would be expected by chance for a 9-bp sequence. Thus, *H. influenzae* appears to have evolved to promote uptake of DNA extruded from cells of related bacteria, perhaps helping to generate genetic diversity under adverse conditions. The detection of high numbers of USS sequences is just one example of the new discoveries that are possible by analyzing complete genome sequences.

---

transfer has been a major force in microbial genomic evolution. The chapter ends by reviewing the turbulent state of prokaryotic systematics that results from the discovery of wide-spread DNA exchange.

MOBILE DNA AND THE *E. COLI* K-12 SEQUENCE (8, 62)

The best-understood free-living organism is probably the intestinal bacterium *Escherichia coli*. For decades *E. coli* has served as the primary model for studies of genetics and molecular biology. Experiments by members of the Phage School focused primarily on *E. coli* for studies of growth of bacteriophage. The extremely efficient genetic methods available for studies of *E. coli* itself have made it the favored model for prokaryotes generally. Many of the mobile DNA systems described in earlier chapters were first studied in *E. coli*. Thus, this chapter on microbial genomes begins with a survey of the role of lateral transfer in constructing the *E. coli* genome.

The chromosome of *E. coli* K-12 strain MG1655 was sequenced by Fred Blattner and coworkers after a multiyear effort. The molecule is a circular, double-stranded DNA 4,639,221 base pairs in length. The number of genes was initially estimated to be 4288 from the large open reading frames (ORFs) present in the sequence. The estimate of the number of genes, difficult to specify precisely, has risen with further study (discussed below). Many, but by no means all, of these sequences have been verified to encode proteins that accumulate in cells, and a still smaller subset have known functions. Probable protein-coding genes account for 87.8% of the sequence, and stable RNA genes account for 0.8%. Much of the remaining 11% of the genome is probably made up of sequences important for regulation.

Table 5.2 presents a summary of the gene content of *E. coli* K-12. As might be expected, large groups of genes are devoted to central metabolic reactions required for growth. The largest single group of genes (427 or 10% of the total) are probably binding and transport proteins involved in acquisition of nutrients from the environment and export of wastes and toxic materials. Other large groups of genes are devoted to energy metabolism (243 genes or 5.7%), metabolism of biosynthetic precursors (806 genes or 18.8%, summing several categories), transcription (55 genes or 1.3%), and translation (182 genes or 4.2%).

Important for the present story is the fact that 87 genes, or 2% of the genome, are phage, transposon, and plasmid sequences. Some are active elements, others inactive fossil sequences. Determining exactly how many sequences are derived from mobile elements can be challenging, because inactive sequences are expected to accumulate muta-

TABLE 5.2. Distribution of *E. coli* Proteins among 22 Functional Groups (Simplified Schema)

Functional class	Number	Percent of total
Regulatory function	45	1.05
Putative regulatory proteins	133	3.10
Cell structure	182	4.24
Putative membrane proteins	13	0.30
Putative structural proteins	42	0.98
Phage, transposons, plasmids	87	2.03
Transport and binding proteins	281	6.55
Putative transport proteins	146	3.40
Energy metabolism	243	5.67
DNA replication, recombination, modification, and repair	115	2.68
Transcription, RNA synthesis, metabolism, and modification	55	1.28
Translation, posttranslational protein modification	182	4.24
Cell processes (including adaptation, protection)	188	4.38
Biosynthesis of cofactors, prosthetic groups, and carriers	103	2.40
Putative chaperones	9	0.21
Nucleotide biosynthesis and metabolism	58	1.35
Amino acid biosynthesis and metabolism	131	3.06
Fatty acid and phospholipid metabolism	48	1.12
Carbon compound catabolism	130	3.03
Central intermediary metabolism	188	4.38
Putative enzymes	251	5.85
Other known genes (gene product or phenotype known)	26	0.61
Hypothetical, unclassified, unknown	1632	38.06
Total	4288	100.0

(Reprinted, with permission, from Blattner et al. 1997 [© American Association for the Advancement of Science]. [8])

tions over time and so eventually become unrecognizable. Thus, the exact figures for the abundance of mobile DNA elements in *E. coli* or any other genome are strongly influenced by the stringency of the search criteria applied.

As discussed in previous chapters, *E. coli* transposons were identified by early workers in genetic studies. Many spontaneous mutations involved insertion of these transposons, and their involvement in more complicated rearrangements such as deletions, duplications, and inversions are well documented. Two multi-component clusters of IS elements were initially identified in the genomic sequence. At one position, an IS*911* element was present with a copy of IS*30* inserted into it. At a second position, there is another copy of IS*911* with an IS*30* integrated into it and, in addition, an integrated fragment of an IS*600* element. An IS*5* element is inserted in a third position into the *rfb* gene, thereby inactivating a pathway of lipopolysaccharide synthesis. Other IS elements have been identified, often in different locations, in other *E. coli* strains.

Several of the larger blocks of mobile DNA sequences correspond to so-called cryptic prophages, remnants of once-active phages that have become inactivated by mutation. The cryptic prophages turned up in early studies by the Phage School, because mutated phages newly entering a cell could sometimes revert to wild type by recombining with a cryptic prophage. Two of the observed cryptic prophages, DLP12 and Rac, are fossil lambdoid phages. The initial isolate of *E. coli* K-12 also contained an active copy of phage λ. Other cryptic phages include Qin, e14, CP4-57, CP4-6, and CP4-44. The last three are relatives of phage P4.

Also present in the genome are numerous "phage remnants," single genes that appear to have been captured from phage. In these cases, it can be difficult to specify whether the genes represent cellular precursors of genes later captured by phage or fragments of phage genomes that have been mostly eliminated by deletion.

P4 is an example of a satellite phage, a replication-defective phage that requires factors provided by a second phage (P2) to grow. Examples of satellite viruses are also known for viruses of plants and animals. The satellite viruses illustrate that genomic parasites can themselves acquire parasites, as with defective transposons.

RATES OF LATERAL DNA TRANSFER INFERRED FROM THE *E. COLI* K-12 GENOME (8, 20, 46, 70, 84)

In 1998, Jeffrey Lawrence and Howard Ochman carried out a thorough analysis of *E. coli* sequences that allowed them to propose specific sequences derived from recent lateral transfer and to estimate their age in the *E. coli* genome. Their study suggests that much of the genome is indeed derived from lateral transfer, often as large blocks of DNA.

As the first step, the sequence of *E. coli* was analyzed to identify regions of unusual sequence content. In the genetic code, most amino acids are encoded by multiple DNA triplets, allowing different DNA sequences to be used to specify the same protein sequence. In many cases, codons for the same amino acid can differ in the number of G/C versus A/T residues. Intergenic regions are even more free to vary. Thus, it is possible to identify potentially newly introduced sequences by analyzing the G/C versus A/T content for regions that depart from the *E. coli* average of 50.8% G/C, since different bacteria are found empirically to have quite different genome-wide A/T to G/C ratios. Another way of identifying newly acquired DNA sequences is to identify coding regions with unusual codon usage, since different bacteria sometimes favor different codons for the same amino acids. In the analysis of *E. coli*, some potentially transferred segments displayed favored patterns of codon usage that could be attributed to specific bacteria, providing likely donors for the newly acquired DNA. Lawrence and Ochman then analyzed sequences to identify multigene blocks of transferred DNA and scanned databases of bacterial sequences in an effort to identify the specific candidates for the donors.

By this analysis, Lawrence and Ochman estimated that 755 of the 4288 open reading frames (ORFs) of *E. coli* K-12 or 17.6% of the *E. coli* genome could be attributed to capture by DNA transfer in at least 234 lateral transfer events since the time that *E. coli* diverged from *Salmonella*. This is a much larger fraction than the 2% of genes attributable directly to phage, transposon, and plasmid sequences. This value represents a minimal estimate for several reasons. Genes transferred from bacteria with codon usage similar to that of *E. coli* will not be identified by the above methods. Similarly, because genes are expected to "ameliorate" slowly—that is, to adopt the codon usage and G/C content of the host—very old transferred sequences will no longer be recognizable. In addition, many of the acquired genes have probably since been deleted (see below). Some caution is warranted in this analysis, since it is not ruled out that some of the sequences have unusual G/C content for reasons related to their function, but the above assumptions provide at least a first-order way of identifying recently transferred DNA. Examples of the DNA segments inferred to have been acquired by horizontal transfer are shown in Figure 5.1.

Many of the inferred transferred segments are associated with tRNA genes, implicating phage in DNA transfer because many *att* sites are located near the tRNA genes. Of the 89 transfer events shown in Figure 5.1, 15 lie near one or more tRNA genes. In addition, several of the pathogenicity islands from uropathogenic *E. coli* also lie near the genes for tRNAs (Chapter 4).

Many of the transferred gene segments near tRNA genes no longer contain recognizable phage sequences. One model would be that the genes were initially transported on a bacteriophage and integrated, then phage sequences were lost by random deletion events. After enough time, only a small subset of genes would remain, perhaps those that confer evolutionary benefit to the bacteria. Alternatively, some of the sequences may have spread as satellite phages, as with the toxic shock syndrome pathogenicity island. Downstream of the genes for tRNA*leuW* and tRNA*leuX* are sequences resembling integrases of phages Sf6 and P4, respectively, supporting involvement of phages in integration at these sites.

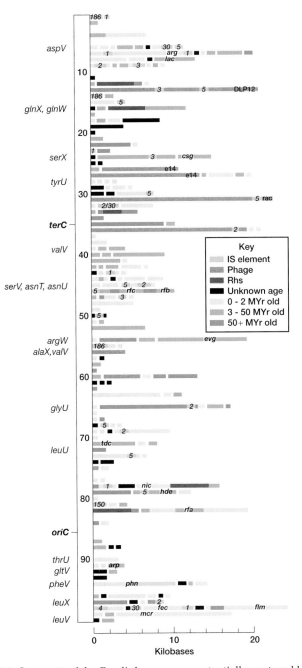

FIGURE 5.1. Segments of the *E. coli* chromosome potentially captured by horizontal transfer. The vertical bar at the left indicates the position in the *E. coli K-12* (MG1655) sequence. Horizontal bars represent individual regions inferred to have been captured by lateral transfer. The color of the horizontal bars indicates the composition and inferred duration in the *E. coli* genome. Rhs indicate rearrangement hot spots. The gene names to the left of the bar indicate tRNAs, except *terC* (terminus) and *oriC* (origin of replication). (Redrawn, with permission, from Lawrence and Ochman 1998 [© National Academy of Sciences]. [46])

Of 37 IS elements or fragments detected by Lawrence and Ochman, 25 were associated with horizontally transferred sequences. In several cases, the IS elements lie at the boundary between recently transferred DNA and bulk *E. coli* DNA. Such a structure could arise from formation of a cointegrate between transposon-containing transferred DNA and the *E. coli* chromosome (Chapter 3). This reaction would yield copies of the transposon flanking the newly transferred region, with the entire unit embedded in the *E. coli* chromosome. To create the structures with a single flanking IS element, part of the flanking DNA would have to then be removed by deletion. In other cases, however, the IS elements are placed more internally in the transferred DNA, making the steps that formed the observed structure harder to reconstruct. The degree of association with newly transferred DNA differed with different IS element classes. IS*186* was only weakly associated, whereas six of seven IS2 elements were associated, suggesting that IS2 but not IS*186* may be mechanistically involved in transfer.

By closely analyzing sequence composition and codon usage, Lawrence and Ochman were able to provide tentative dates for lateral transfer of each new DNA segment. After introduction into *E. coli*, each new DNA is expected to accumulate mutations that gradually adapt the sequence composition and codon utilization of the transferred DNA to match that of the host, thereby optimizing expression of the newly introduced genes (and possibly other functions such as translation, replication, and repair). This amelioration is hypothesized to take place at a (at least approximately) constant rate, which can be determined by comparing the sequence differences between bacteria with inferred times of evolutionary divergence. Actual implementation of this type of analysis requires a number of further refinements, such as the need to account for different amelioration rates at each of the three codon positions. Applying the full analysis to synonymous substitutions (DNA changes that did not alter the encoded proteins) in *E. coli* and *Salmonella enterica*, which diverged about 100 million years ago, revealed that the genomes differed by 47%. Thus, the synonymous substitution rate can be estimated at 0.47% per million years.

If amelioration takes place at a known rate, newly introduced sequences can be "back-ameliorated" until they match the sequence composition of a known bacterial genome. The number of changes required to recreate the hypothetical original sequence can then be used to deduce the residence time of *E. coli*, providing a date of transfer. On the basis of this kind of analysis, the oldest sequence detected was introduced into the *E. coli* genome 100 million years ago. The vast majority of transferred DNA is of relatively recent origin, giving an average age of horizontally transferred DNA of 6.7 million years. The deduced rate of accumulation is thus 64.2 kilobases per million years, or about 1.4% of the genome. For comparison, point mutations are estimated to accumulate in *E. coli* DNA at a rate of 22 kilobases of new

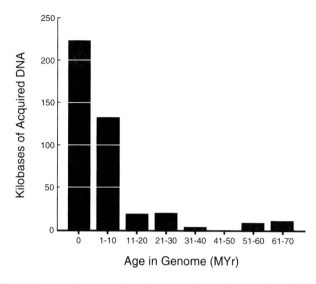

FIGURE 5.2. Graph of the age distribution of laterally transferred DNA in the *E. coli K-12* (MG1655) chromosome. Bars represent the amount of protein-coding DNA analyzed by back amelioration in 571 transferred genes. (Redrawn, with permission, from Lawrence and Ochman 1998 [© National Academy of Sciences]. [46])

sequence per million years, emphasizing the predominant contribution of lateral DNA transfer.

How much of this DNA becomes fixed in the microbial genome long-term? Much of the newly transferred DNA is composed of sequences such as IS elements, which are expected to be relatively transient over geological time. The uneven age distribution also suggests that most newly acquired DNA does not persist long-term (Fig. 5.2). Because a large fraction of sequences are relatively new, having been present for less than 10 million years, most newly acquired sequences must be lost over longer periods of time. Presumably those sequences that last longer will often confer increased fitness on the host. To isolate those sequences that have persisted long-term, Lawrence and Ochman eliminated sequences acquired within the last 1 million years. Re-analysis yielded a rate of 16 kilobases transferred per million years, suggesting that *E. coli* has gained about 1,600 kilobases of sequence since the divergence from *S. enterica* 100 million years ago.

What are the genes that have been acquired by lateral transfer and then fixed in bacterial genomes? In the previous two chapters, we discussed genes involved in antibiotic resistance and pathogenesis. Myriad further gene sets have conferred new phenotypes on the recipient bacteria. Some examples include genes for phosphonate utilization (*phn*) and lactose utilization (*lac*) in *E. coli* and genes for coenzyme B12 biosynthesis (*cbi/cob*) and citrate utilization (*tct*) in *S. enterica*. For the case of the familiar *E. coli lac* operon, analysis of base composition and codon usage suggests that it was actually acquired in two pieces. The *lacI* and *lacZ* genes are 56% GC, whereas *lacYA* is 43% GC. The ability of *E. coli* to utilize lactose (milk sugar) may have allowed the bacteria to colonize a new niche, the human gut. The general conclusion is that the vast majority of genes that are found in either *E. coli* or *S. enterica,* but not in both, were likely acquired by lateral DNA transfer since their divergence.

Despite the influx of new sequences, the *E. coli* genome has not grown in size since the divergence from *S. enterica*, implying that sequences must be lost at a comparable frequency. Recall too that the back-amelioration method of Lawrence and Ochman provides a minimal estimate of transfer frequency. Sequences that have resided in the *E. coli* genome for long periods will have ameliorated to the point of matching the host genome, and so will no longer be detectably different. Sequences transferred from bacteria with similar base composition and codon preferences will also be undetectable. Although the methods used involve a number of assumptions, not all of which are likely to hold 100% of the time, the broad conclusions are nevertheless likely to be correct. The picture that emerges is one of an extremely dynamic process, with new sequences gained and lost at a high rate.

The "islands" nomenclature, originally introduced to describe pathogenicity islands, has been broadened (loosely) to include diverse types of inserted sequences. Blattner and colleagues refer to insertions in the E. coli sequences as "K-islands" and "O-islands," reflecting the strain of origin. Potentially mobile blocks of genes important for other functions, such as plant-bacterial symbiosis (mutualism), have also been called "symbiosis islands."

THE ENTEROHEMORRHAGIC *E. COLI* O157:H7 SEQUENCE (74)

In 2001 Blattner and coworkers completed the second sequence of an *E. coli* isolate, this time the enterohemorrhagic strain O157:H7. This strain has been associated with outbreaks of severe hemorrhagic colitis and fatalities caused by the associated hemolytic uremic syndrome. The first outbreak occurred in 1982, in association with contaminated hamburger meat. At present some 75,000 cases occur annually in the United States.

The divergences between K-12 and O157:H7 are remarkable (Plate 5). The two share a recognizable "backbone" structure that is colinear except for a single 422-kb inversion. However, the homology is interrupted by hundreds of "islands" unique to each strain.

O-islands comprise 1.34 Mb of DNA in 177 clusters, and K-islands comprise 0.53 Mb in 234 clusters. Nine large O-islands, each greater than 15 kb, contain many of the known pathogenic determinants, including genes for toxins, adhesins, protein secretion, and integrases. Four of the islands are located adjacent to tRNA genes, as in the pathogenicity island paradigm. Many smaller gene blocks encode other potential virulence determinants, such as genes for iron uptake and various adhesins. Smaller blocks encode genes that may be important for evolutionary persistence in different niches, such as genes for antibiotic resistance, aromatic compound degradation, and glutamate fermentation. Fully 18 multigene regions in O157:H7 are related to phages, with only one, the Stx2 Shiga toxin phage BP-933W, known to be competent for replication. Many of the other genes resemble collections of phage gene "cassettes" without appearing to encode complete prophages. These islands appear to be captured primarily by lateral transfer, although it is not formally ruled out that some may have been present in the common ancestor of the two strains and lost during the subsequent divergence. In all, the O157:H7 chromosome contains 1,387 genes not found in K-12 out of 5,416, whereas K-12 contains 528 unique genes out of 4,405 total, an amazing degree of divergence for organisms of the same ostensible species.

Below we return to other examples of the remarkable sequence divergence in different isolates from the same bacterial species. We next turn to a study that compares the genetic divergence in many strains of *E. coli* using DNA chips.

E. COLI DIVERSITY STUDIED WITH DNA CHIPS (69)

In a related study, Howard Ochman and Isaac Jones investigated the rates of DNA transfer in *E. coli* by comparing the genetic makeup of different *E. coli* strains. They measured the length of the chromosomal DNA in each strain and used DNA array technology to determine which genes were in common with the sequenced K-12 MG1655 strain.

DNA Arrays

DNA arrays provide a method for rapidly characterizing the sequence content of a nucleic acid sample. Test DNAs, each corresponding to a single gene, are spotted in ordered arrays at high density on a flat surface. In the above example, the arrays contained representative sequences from every gene of *E. coli* MG1655. To compare the gene content of other strains, bulk DNA was isolated from each test strain, labeled, and applied to the chip. Most spots on the array became labeled, indicating that the tested gene was common to the two strains, but occasionally genes were found to be absent.

These high-density arrays provide a valuable new tool for many applications in biology and medicine. For example, with the recent completion of the human genome, it will soon be possible to create chips that "query" the activity of all of the human genes, allowing a full description of anatomy and pathology in terms of gene activity.

Characterization of five widely diverged strains of *E. coli* revealed that 90% of open reading frames (ORFs; probable genes) were common to all the strains, but the rest varied. Up to a megabase (a million base pairs) of DNA was unique to each strain. Each divergent lineage was inferred to contain several hundred unique genes. Many of the deletion/insertion events were found to be associated with phage or IS elements, which are known to be mobile. In many cases, the deleted/inserted genes accounted for particular phenotypes in the different strains, such as the *aga* operon, which confers the ability to grow on glactosamine, or the *rfa* and *rfb* operons that specify the medically important O-antigen.

The distribution of genes in different strains provided information on the ages of the genes in the genome, because genes acquired earlier in evolution should be more widespread among strains. This provides a way of checking some of the conclusions drawn from the back-amelioration method in the Ochman and Lawrence study. In general, genes predicted to be recently acquired (<50 million years ago) are sporadically distributed among strains, confirming the general conclusions. Genes acquired by lateral transfer at earlier times (>50 million years ago) were more commonly present in all strains, consistent with their presence in the progenitor of the strains studied. Thus, the comparison of genes in divergent *E. coli* strains supported the idea that the genome is experiencing high rates of gene acquisition and loss.

LATERAL DNA TRANSFER AND THE MICROBIAL GENOME SEQUENCES (21, 70)

The sequenced microbial genomes are quite heterogeneous, varying greatly in size, number of genes, and complement of mobile DNA elements. The nature of each genome can often be understood, at least in part, as a consequence of life-style of the sequenced organism. For example, the mycoplasmas, the organisms with the smallest genomes, live as intracellular parasites and so do not encode synthetic enzymes for precursors derived from the host. The genomes of the thermophiles, prokaryotes that grow at high temperatures, also have distinctive sets of genes that appear to be important for life at high temperature.

Several ways of analyzing the complete genome sequences reveal extensive lateral DNA transfer. An analysis by Lawrence, Ochman, and coworkers, as in their work on *E. coli* but now applied to 19 bacterial genomes, is presented in Figure 5.3. Potential recently transferred DNA is recognized by base composition and codon usage different from the genome average. This analysis reveals quite different amounts of recently transferred DNA in different genomes. *E. coli*, *Bacillus subtilis*, and *Synechocystis* all contain relatively large proportions of newly acquired DNA, the latter containing fully 16.6%. Others like *Borrelia burgdorferi* and *Rickettsia prowazekii* have little or no newly transferred DNA. Often the anomalous sequences are associated with mobile element remnants such as plasmids, phage, or transposons and are located near tRNA genes, consistent with phage integration.

These conclusions also hold for the Archaea, the third domain of life. The fraction of newly transferred DNA varies among archaeal genomes, as with bacteria, ranging from 9.4% to 1.3% for the genomes

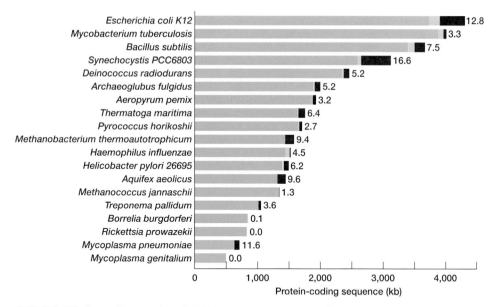

FIGURE 5.3. Laterally transferred DNA sequences in 19 prokaryotic genomes inferred by the method of Lawrence and Ochman. The lengths of the bars indicate the amount of protein-coding DNA. The numbers to the right of each bar indicate the percentage of foreign DNA. "A" indicates archaeal genomes. Green indicates "native" DNA, gray indicates mobile DNA elements, and black indicates "foreign" DNA. (Redrawn, with permission, from Ochman et al. 2000 [© Macmillan Magazines Ltd.]. [70])

studied. Mechanisms of lateral transfer have not been as closely studied for Archaea as for bacteria and eukaryotes, but active phages and transposons have been identified. Gene transfer mechanisms in the Archaea probably resemble those of the bacteria, although further studies could still yield surprises.

A related approach reinforces the findings of lateral DNA transfer and documents exchange between the bacterial and archaeal domains. Once a genomic region of anomalous structure is recognized, the genes present can be compared to those of other species. If multiple genes are present, their order and transcriptional orientation can also be compared. Lateral transfer can often be inferred if a similarly organized gene cluster is seen in two distantly related organisms, particularly if the cluster is absent in intermediate forms. Striking similarities have been found for gene clusters in thermophilic Bacteria and Archaea. For *Thermatoga*, a thermophilic bacteria, fully 24% of ORFs (inferred genes) are most similar to an archaeal gene. Many of these are organized in gene clusters also found in Archaea. Similarly for *Aquifex*, 16% of ORFs are most similar to an archaeal gene. Similar evidence of transfer can be found in Archaeal genomes. The Archaeon *Archaeoglobus fulgidis*, for example, contains genes for fatty acid metabolism that are absent in other archaeal genomes but present in bacteria. It seems most likely that these sequences were acquired by lateral transfer, although it is not ruled out that some of these genes were inherited from the distant common ancestor of Bacteria and Archaea, which may itself have lived at high temperature. According to the latter idea, these genes have been lost in progeny that live under less extreme conditions (mesophiles). Further studies have the potential to clarify this origin of genes for life at high temperature—for example, defining the mechanisms of DNA exchange among thermophiles could bolster the case for lateral transfer.

LATERAL TRANSFER IN ARCHAEA (79, 85)

Less information is available for Archaea than for Bacteria and Eukaryotes, but nevertheless, the limited data suggest that lateral transfer is frequent. Figure 5.3 shows that Archaeal genomes assayed by Lawrence and Ochman contain amounts of mobile DNA and laterally transferred sequences comparable to the bacterial genomes. Numerous examples of probable transfer events have been documented as mentioned above. A temperate bacteriophage of Archaea, SSV1, has also been isolated and studied.

One newly discovered feature of gene transfer in Archaea appears to be unique. She and coworkers reported that genes for λ-integrase family proteins were often found in two pieces, named *int* (N) and *int* (C) (Fig. 5.4). The *int* (N) sequence overlaps a tRNA gene, and a portion of the tRNA gene is repeated within *int* (C). Excision is inferred to restore the continuity of the *int* gene and thereby allow function, so that the gene is functional only in the extrachromosomal form. Sequence analysis revealed several examples of integrated plasmids or phages containing split-*int* genes in Archaea genomes. Different split-*int* cassettes flanked several unrelated blocks of genes. Apparently these split-*int* cassettes have evolved to allow stable incorporation of new sequences into the Archaeal genomes, as excision is blocked due to the lack of integrase. The number of such loci is unknown, in part because the cassettes are somewhat different from each other and so can be difficult to identify in the genomic sequences. Capture of a new source of integrase protein could potentially lead to rearrangement of loci flanked by the split-*int* cassettes. If an Archaeal genome harbors multiple split-*int* cassettes, the genome could be subject to a "burst" of genomic rearrangement in response to acquisition of a source of integrase.

In the next section, we turn to another possible effect of lateral transfer on genome structure, the clustering of genes that work together.

OPERON STRUCTURE AND LATERAL DNA TRANSFER (16, 47, 70)

The extreme genetic fluidity of the microbial genomes is reminiscent of the bacteriophage chromosomes discussed in Chapter 4, where fre-

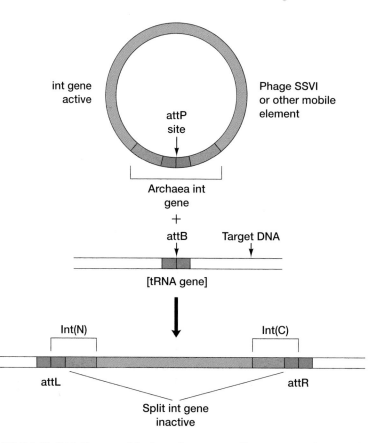

FIGURE 5.4. "Split-*int*" genes of Archaea. Integrase coding genes on phage such as SSV1 are intact and active in the circular phage genome (*top*). Integration involving an *attP* site within the *int* gene splits the genes into two segments (*bottom*), neither of which alone can supply *int* function. (Adapted from She et al. 2001. [85])

quent exchange apparently promoted the organization of genes into functional cassettes. Many bacterial genomes also seem to be loosely arranged in cassettes, in this case the operons. Diverse models have been proposed to explain the origin of operons. A relatively new model, "The Selfish Operon Hypothesis," holds that genes are grouped into functional units as a consequence of lateral DNA transfer, as with the phage.

Theories of Operons (6, 10, 16, 71)

Operons are now defined as multi-gene blocks with associated regulatory regions that allow transcription from a common promoter to make a single mRNA. Early ideas about operon structure arose in the late 1950s from studies of Jacob and Monod in the phage λ (Chapter 4) and *E. coli lac* systems. The *lac* operon comprises the *lacZYA* genes for lactose fermentation, its promoter, *Plac*, and *lacI*, the gene for lac repressor. In the absence of lactose in the growth medium, lac repressor binds to sites in and around *Plac*, called the lac operators, thereby blocking expression of *lacZYA*. Allolactose (a metabolite of lactose) binds to lac repressor, causing a conformational change that blocks binding to the lac operators. This allows *lacZYA* to be transcribed (but only if glucose is unavailable, otherwise *Plac* is subject to glucose repression). The *lacI* gene is linked to *lacZYA*, located just upstream of *Plac*.

Today the concept of the operon has been somewhat broadened. Operons can be turned on by regulatory genes as well as turned off, and in some cases there are secondary internal promoters that fine-tune the relative expression levels of different genes. A few examples are shown in Figure 5.5. Genes important for histidine synthesis are clustered together in the *his* operon under the control of a regulatory region that is activated by low levels of histidine. Genes for utilization of arabinose are clustered together in the *araBAD* operon, which is also linked to its regulatory gene *araC*. Expression of *araBAD* is induced when arabinose is present and glucose is absent.

Not all genes are joined into operons, although in *E. coli* and *Salmonella*, operons are probably more the rule than the exception. In many cases, there are insufficient data available to assess whether a locus observed in the primary DNA sequence is formally an operon or not. Further complicating the picture, some mixed operons contain genes for functions not obviously related to the majority of the genes present. However, operon structure can often be inferred from primary sequence data when a chromosomal region contains predominantly genes that are likely to act together and are oriented so that they are transcribed in the same direction.

The Selfish Operon Hypothesis of Jeffrey Lawrence and John Roth holds that clustering exists as a consequence of lateral DNA transfer. A sequence that can function immediately upon introduction into a new bacterium is more likely to be fixed by evolution than an inactive sequence. If multiple genes are required for some function; for example, the enzymes of a biosynthetic pathway, the sequence can be positively selected only if all the genes are present. The probability of lateral transfer of all the needed genes is greater if they are linked in the chromosome. As was argued for phage, if genes are frequently reshuffled in the genome by recombination, and transported to new hosts periodically, those arrangements that preserve function thanks to clustering are most likely to confer benefit to the new host and so persist. Many rounds of gene rearrangement and lateral transfer will eventually lead to clustering of genes for many functions in the genome. Such clusters are "selfish" because clustering allows the grouped genes to spread at the expense of unclustered relatives, a process that need not benefit the host cell.

Other models do not appear to explain the origin of operons as readily. An early model proposed that operons exist because genes originate in clusters, but this is now known not to be the case. Another model posits that genes are clustered for convenience of coregulation. This may well be an advantage, but there are many examples of trans-acting factors that are not constrained to act on nearby genes and are encoded at distant positions in the genome. AraC protein acts on the nearby *araBAD* operon, but also acts on other nonlinked genes for arabinose utilization. Such a model also makes it hard to understand how an operon could come together, because little benefit results until a full set of genes is joined together, presumably by random, and thus infre-

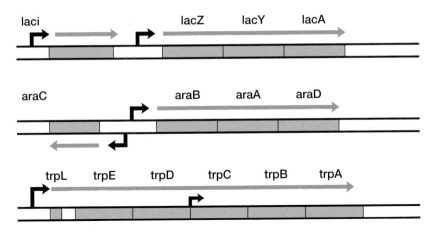

FIGURE 5.5. Some operons from *E. coli K-12* (MG1655). (*Top*) Genes for catabolism of lactose. (*Middle*) Genes for catabolism of arabinose. (*Bottom*) Genes for synthesis of tryptophan.

quent, processes. Another alternative, the Fisher model, proposes that genes are clustered because the proteins they encode interact particularly well. If the genes were located at distant sites, they would more frequently be separated by recombination, and thus the benefit would be lost. The Fisher model may well operate in some cases, but it does not comfortably explain clustering of genes whose proteins are not known to interact. The Fisher model may explain clustering of genes for ribosomal proteins, which interact extensively.

The newly available information from bacterial genome sequences allows potential tests of the Selfish Operon Hypothesis. There are several small bacterial genomes that have no known plasmid, phage, or transposon sequences of any kind. If these organisms do not engage in frequent lateral DNA transfer, there should be no pressure in favor of gene clustering. Thus, according to the Selfish Operon Hypothesis, the genes of these organisms should not be grouped into operons. This correlation appears to hold true at least in the *Campylobacter* genome. On the other hand, data against the hypothesis are provided by the *Aquifex* genome, which has engaged in extensive lateral transfer but does not appear to have a prominent operon structure, at odds with the prediction of hypothesis. Detailed analysis of operon structure in the complete genome sequences may clarify whether lateral transfer promotes gene clustering.

The pathogenicity islands of Chapter 4 and the integrons of Chapter 3 may represent extreme embodiments of the Selfish Operon Hypothesis. The same forces that favor gene clustering in chromosomal operons likely operate on pathogenicity islands and antibiotic-resistance genes. In many of these, we also see integrase genes and flanking *att* sites, as with *Staphylococcus aureus* SaPI1 or the integrons. If the Selfish Operon idea is correct, operons incorporating machinery for gene acquisition and dispersal should be particularly successful. Many pathogenicity islands may be understood as selfish operons that have gone one step further, adding machinery for mobilization as satellite phages to existing gene clusters. Possibly organization of genes in integrons promotes their dispersal. "Selfish Islands" may well represent a new higher-order unit of genetic self-interest.

GENE TRANSFER BETWEEN BACTERIA IN THE ENVIRONMENT (15, 26, 36, 39, 50, 75, 95, 97)

Lateral DNA transfer between bacteria has been found to take place at high rates in natural settings, adding important support to the idea that lateral transfer shapes prokaryotic genomes. Field studies of gene transfer are also important for evaluating the possible spread of engineered DNA from genetically modified organisms to other creatures. In this section, we review studies of bacterial gene transfer under natural conditions.

DNA transformation, the process whereby naked DNA is transferred between bacteria, provides a robust means for DNA exchange between bacteria in natural settings. As discussed in Chapter 3, some bacteria (*Haemophilus influenzae, Streptococcus pneumoniae, Neisseria gonorrhoeae*) encode elaborate machinery that takes up DNA. It might seem that transformation is unlikely in a natural setting, because free DNA would be expected to be present in only low concentrations and, furthermore, rapidly degraded. However, natural conditions are actually considerably more favorable. Direct measurements suggest that DNA is in fact relatively abundant in the environment. DNA may adhere to sand or clay particles, and so be stabilized 100- to 1000-fold. Furthermore, both bacteria and DNA may cluster in biofilms or small particles, thereby greatly increasing the local concentrations of both and the frequency of transformation.

DNA transfer by transformation has been demonstrated in several natural settings (Table 5.3). For example, Williams and coworkers assayed biofilms of *Acinetobacter caloaceticus* growing on river stones for incorporation of a DNA marked with a *his* gene and documented transfer. Other studies documented transfer by transformation in seawater, fresh water, or dirt. Environmental conditions affected the rates of gene transfer. In soil, for example, the humidity was, not surprisingly, an important variable. More curiously, churning of soil by earthworms also promoted DNA transfer.

Phage transduction is probably a more important pathway of gene transfer in natural environments (Table 5.4). A substantial fraction of bacteria isolated from the environment harbor lysogens (temperate phages), allowing some estimates of frequency to be made. Treatments that damage DNA induce many temperate phages, such as the lamb-

TABLE 5.3. Bacterial Transformation in the Environment

Bacterial host	Environmental situation	Genetic marker	Reference
P. stutzeri	marine water microcosm	chromosomal *rif*^{R}	Stewart and Sinigalliano (1991) (94)
Pseudomonas sp.	marine water and sediment microcosm	plasmid multimers	Paul et al. (1991, 1992) (72, 73)
A. calcoaceticus	ground water and soil extract	chromosomal *trp*	Lorenz et al. (1991, 1992) (51, 52)
A. calcoaceticus	ground and aquifer water	plasmid	Chamier et al. (1993) (11)
A. calcoaceticus	river epilithon	chromosomal *his*	Williams et al. (1996) (107)
A. calcoaceticus	soil microcosm	chromosomal DNA +*Km*^{R}, *Gm*^{R} cassette	Nielsen et al. (1997) (64)
E. coli	river and spring water	plasmid	Baur et al. (1996) (5)
P. stutzeri	soil microcosm	chromosomal *his* and plasmid	Sikorski et al. (1988) (87)

(Modified, with permission, from Davison 1999 [© Academic Press]. [15])

doid family, so phages can be detected by lysis of a culture after treatment with UV or mitomycin C. However, some phages are not induced by this means (such as Mu), so only a subset of phages will be recovered by screening with inducing treatments. Nevertheless, the fraction of marine bacteria containing temperate phages assayed by induction is very high. Natural bacterial isolates from Tampa Bay were screened for inducible prophages by treatment with mitomycin C. Fully 43% of cells lysed, yielding phage particles or particle-like structures. Because bacterial hosts are not known for most of these phages, it was impossible to assess what fraction of the liberated particles were biologically active. In some cases, the particles in the lysate looked incomplete and were probably inactive, but a substantial fraction of the total yielded morphologically mature phage particles.

TABLE 5.4. Transfer of Genes by Phage Transduction under Environmental Conditions

Bacterial donor	Bacterial recipient	Phage	Environmental situation	Reference
Marine bacteria	marine bacteria HSIC	T-phiHSIC	seawater-nutrient mix	Jiang and Paul (1998) (39)
P. aeruginosa	P. aeruginosa	F116	fresh water	Morrison et al. (1978) (60)
P. aeruginosa	P. aeruginosa	F116	leaf surface	Saye et al. (1987) (82)
Vibrio	Vibrio		seawater	Saye et al. (1987) (82)
Vibrio cholerae	Vibrio cholerae			Kidambi et al. (1994) (42)
E. coli	E. coli	P1	soil	Zeph and Stotzsky (1989) (109)

Studies of marine phages by Sunny Jaing and John Paul yielded the dramatic transduction rates in ocean water described in Chapter 1. They introduced a plasmid encoding resistance to kanamycin (Kmr) and streptomycin (Smr) into the marine bacteria HSIC, then infected with the marine phage φHSIC. The progeny phages were then used to infect an HSIC strain lacking the plasmid. Transfer of the plasmid was found to take place at a rate of 1×10^{-7} to 5×10^{-9} transductants per plaque-forming unit. Similar studies were carried out with mixed natural populations of marine bacteria, revealing transfer of the plasmid in some tests. Studies by other investigators of generalized transduction by a marine vibriophage yielded even higher rates. On the basis of these data, Jaing and Paul estimated the rate of transfer in the Tampa Bay Estuary to be 10^{14} times per year. The estimate in Chapter 1 extrapolates this number across all the oceans, supported by the observation of extremely high densities of phages in diverse marine environments, yielding a rate of 20 million billion transduction events *per second* in the world's oceans.

The above study was carried out in seawater-based nutrient broth, but other studies have documented transfer in more natural settings. Studies of transfer between *Pseudomonas aeruginosa* strains revealed that transduction by phage F116 or UTI could take place in test chambers in a fresh-water reservoir. Further studies revealed that a moist leaf surface was also sufficient to support transduction by F116. Under simulated field conditions, gene transfer was reported to take place even if the bacteria and phage were initially inoculated on different plants.

Perhaps the most prominent means of transferring genes among prokaryotes is conjugation, the main mechanism mediating transfer of antibiotic-resistance genes (Chapter 3). As discussed in earlier sections, conjugation involves the transfer of DNA from one cell to another mediated by a plasmid-encoded mating machinery. Many examples are known of transfer of antibiotic-resistance determinants under natural conditions (Table 5.5). For example, it is possible to carry out controlled experiments in which different bacterial strains are introduced into the intestines of laboratory rats or mice to serve as artificial transduction donors or recipients. Transfer of antibiotic-resistance genes between the introduced bacteria and native bacterial populations could be readily demonstrated. Experiments have also been carried out on human volunteers, allowing direct demonstration of transfer, although in these cases the site of transfer was not determined. In many other cases, gene transfer is inferred by the close similarity of plasmids in presumed donor and recipient bacteria. Transfer was also demonstrated between various pathogenic bacteria on food-chopping boards, a setting designed to simulate a food-processing environment. Insect guts have also been found to be efficient settings for gene transfer between bacteria, because they provide nutrient-rich environments in which the bacteria are highly concentrated. Many studies also document conjugative gene transfer between bacteria in soil and fresh water.

TABLE 5.5. Bacterial Conjugation in the Environment

Donor	Recipient	Environmental situation	Genetic marker	Reference
Animal ecosystems				
E. coli	Shigella flexneri	urinary tract?	A^R-P	Taure et al. (1989) (99)
E. coli	Salmonella enteritidis	human intestine?	A^R-P	Balis et al. (1996) (4)
Lactobacillus lactis	Enterococcus faecalis, Bacteroides sp., Bifidobacterium sp., Peptostreptococcus sp.	gnotobiotic mouse intestine	A^R-P	Gruzza et al. (1994) (32)
Human, rat, and pig intestinal bacteria	E. coli	germ-free rat intestine	A^R-P	Nijsten et al. (1995) (65)
Klebsiella sp.	Klebsiella sp.	human intestine?	A^R-P	Prodinger et al. (1996) (76)
E. coli	Shigella dysenteriae, S. flexneri	human intestine?	A^R-P	Bratoeva and John (1994) (9)
Prevotella sp. (several gram-positive and -negative bacteria)	Bacteroides fragilis	human intestine?	A^R-P, con Tn	Nikolich et al. (1994) (66)
		human intestine?	A^R-P, con Tn	Salyers and Shoemaker (1996) (80)
E. coli (Human, farm animal, and fish bacterial pathogens)	E. coli	simulated sheep rumen microcosm; meat and fish chopping board	A^R-P	Scott and Flint (1995) (83)
Bacillus thuringiensis	B. thuringiensis	lepidopterous larvae	A^R-P	Kruse and Sorum (1994) (44)
Enterobacter cloacae	E. cloacae	cutworm insect gut	B_t-P	Jarrett and Stephenson (1990) (38)
E. coli	indigenous microflora	soil microarthropod gut	A^R-P	Armstrong et al. (1990) (2)
E. coli	E. cloacae	silkworm larvae	AR , luc-P	Hoffmann et al. (1998) (35)
Erwinia herbicola				
Mesorhizobium loti	non-symbiotic soil Mesorhizobium sp.	rhizosphere or soil	AR -P	Watanabe and Sato (1998) (106)
			sym-1	Sullivan and Ronson (1998); Sullivan et al. (1995, 1996) (95–97)
Rhizobium leguminosarum	R. leguminosarum	non-rhizosphere soil?	sym-P	Louvrier et al. (1996) (53)
Sinorhizobium fredii	R. leguminosarum	non-sterile soil	sym-P	Kinkle et al. (1991) (43)
Pseudomonas fluorescens	P. fluorescens	rhizosphere soil	mob-P	Smit et al. (1993) (88)
Ralstonia eutrophus	indigenous rhizosphere bacteria	wheat rhizosphere	mob-P	van Elsas et al. (1998) (105)
Indigenous rhizosphere bacteria	Pseudomonas fluorescens, Pseudomonas putida, E. cloacae	wheat rhizosphere	mob-P	Smit et al. (1998) (89)

(Continued on following pages.)

TABLE 5.5. (Continued)

Donor	Recipient	Environmental situation	Genetic marker	Reference
Pseudomonas sp.	indigenous rhizosphere bacteria	beet rhizosphere	A^R-P	Lilley et al. (1994) (49)
Rhizosphere bacteria	P. fluorescens	beet rhizosphere	M^R-P	Lilley and Bailey (1997) (48)
P. fluorescens	P. fluorescens	wheat rhizosphere	chromosomal housekeeping genes	Troxler et al. (1997) (104)
Pseudomonas aeruginosa	P. aeruginosa			
Plant leaves				
Pseudomonas syringae and Erwinia herbicola	Erwinia amylovora	pear leaves	A^R-P	Lacy et al. (1984) (45)
P. syringae	plant epiphytic bacteria	bush bean leaves	A^R-P	Bjorklof et al. (1995) (7)
P. syringae	P. syringae	ornamental pear leaves	A^R, M^R-P; A^R Tn	Sundin et al. (1994) (98)
Leaf surface bacteria	P. fluorescens	beet leaves and roots	M^R-P	Lilley and Bailey (1997) (48)
P. putida	P. putida and leaf surface bacteria	bush bean leaves	cat, gfp-P	Normander et al. (1998) (67)
Nonpolluted water and soil				
P. putida	P. fluorescens	oligotrophic river epilithon	A^R cat-P	Bale et al. (1988) (3)
P. aeruginosa	Pseudomonas chlororaphis Pseudomonas aureofaciens P. aeruginosa	oligotrophic river epilithon	mob-P	Hill et al. (1992) (34)
B. japonicum	Bradyrhizobium sp.	rhizosphere	A^R, M^R-P:cat, M^R-P	Kinkle et al. (1991) (43)
E. coli	R. eutrophus	amended soil	M^R-P	Top et al. (1990) (101)
P. aeruginosa	indigenous soil bacteria	soil	A^R-P	Glew et al. (1993) (29)
P. fluorescens	soil bacteria	soil and earthworm cast	M^R,cat-P	Daane et al. (1996) (13)
Indigenous soil bacteria	P. putida	field soil containing pig manure	A^R-P, mob-P	Gotz and Smalla (1997) (31)
E. faecalis	E. faecalis	sewage wastewater microcosm	A^R-P: A^RTn	Marcinek et al. (1998) (55)
E. coli	indigenous seawater bacteria	L-broth-amended seawater	A^R-P	Sorensen (1993) (92)

Starved *Vibrio* and *E. coli*	*Vibrio* and *E. coli*	oligotrophic marine microcosm	A^R-P	Goodman et al. (1993) (30)
Aeromonas salmonicida	indigenous marine bacteria	marine sediment microcosm	A^R-P	Sandaa and Enger (1994) (81)
P. putida	indigenous marine bacteria	seawater and marine sediment	AR, M^R,gfp-P	Dahlberg et al. (1998) (14)
Polluted soils, sludges, and water				
Indigenous soil bacteria	*Alcaligenes eutrophus*	polluted soil	cat-P	Top et al. (1994) (102)
Indigenous bacteria	indigenous bacteria	coal tar-contaminated site	cat-P	Herrick et al. (1997) (33)
P. putida	indigenous *Pseudomonas* sp.	phenol-contaminated site	cat-P	Peters et al. (1997) (75)
Enterobacter agglomerans	*Pseudomonas* and *Comamonas* sp.	polluted soil	A^R-P: cat-Tan	De Rore et al. (1994) (18)
R. eutrophus	*P. glathei* / *B. cepacia* / *B. caryophylli* / *A. eutrophus*	polluted soil	M^R, cat-P	DiGiovanni et al. (1996) (19)
R. eutrophus	*V. paradoxus*	soil microcosm	M^R, cat-P	Neilson et al. (1994) (63)
R. eutrophus	indigenous soil bacteria	soil microcosm	cat-P	Top et al. (1998) (103)
Indigenous soil and sludge bacteria	*E. coli*	soil and sludge	mob-P	Top et al. (1994) (102)
P. putida	sludge *Pseudomonas* sp.	activated sludge unit	cat, mob-P	McClure et al. (1989, 1990) (58,59)
P. putida	*P. putida*	activated sludge microcosm	cat-P	Nußlein et al. (1992) (68)
Alcaligenes sp.	indigenous bacteria	xenobiotic-polluted fresh water	cat-Tn	Fulthorpe and Wyndham (1991, 1992) (27, 28); Nakatsu et al. (1995) (61)
(*Alcaligenes paradoxux* and *Pseudomonas pickettii*)		agricultural soil?	cat-P	Ka and Tiedje (1994) (40)
Burkholderia sp.	other *Burkholderia* sp	soil?	chromosomal cat	Matheson et al. (1997) (56)
Indigenous sludge bacteria	*E. coli*	wastewater treatment plant	mob-P	Mancini et al. (1987) (54)
P. putida	*P. putida*	polluted flow chamber biofilm	cat, gfp-P	Christensen et al. (1998) (12)
Pseudomonas sp. B13	*R. eutrophus*	aquifer	cat-Tn?	Zhou and Tiedje (1995) (110)
Pseudomonas sp. B13	*P. putida* F1	activated sludge microcosm	cat-Tn?	Ravatn et al. (1998) (77)

(M^R-P) Heavy metal-resistant plasmid; (A^R-P) antibiotic-resistant plasmid; (cat-P) catabolic plasmid; (mob-P) mobilizable plasmid; (A^RTn) antibiotic-resistant transposon; (con Tn) conjugative transposon; (cat-Tn) catabolic transposon; (sym-P) nitrogen-fixing symbiotic plasmid; (sym-1) chromosomal, nitrogen-fixing, symbiosis island; (B_t-P) *B. thuringiensis* insect toxin plasmid; (gfp) green fluorescent protein gene; (*luc*) luciferase gene; (?) possible or unknown. (Modified, with permission, from Davison 1999)[© Academic Press]. [151]

One intriguing example involves the bacterial genes for symbiotic nitrogen fixation (*nif*). Eukaryotes cannot acquire nitrogen directly from the atmosphere, instead relying on nitrogen-fixing microorganisms to synthesize nitrogen-containing compounds, and so supply the food chain with fixed nitrogen. Leguminous plants acquire nitrogen from symbiotic nitrogen-fixing bacteria, which inhabit specialized root nodules elaborated by plants for this purpose in response to infection. The bacterial genes encoding the machinery required for symbiotic nitrogen fixation are clustered, either on plasmids or on the bacterial chromosomes, much like the pathogenicity islands. These "symbiotic islands" have been found to be transferable between strains, including a compelling example of transfer under field conditions.

John Sullivan, Clive Ronson, and coworkers documented lateral transfer of a symbiotic island in the evolution of a new partnership between an exotic plant species and a native bacterium. The legume *Lotus corniculatus* was introduced into New Zealand and inoculated with a strain of the symbiotic nitrogen-fixing bacterium *Mesorhizobium loti*, which is not found in New Zealand. Seven years later, the *Lotus* plant was found to engage in symbiotic nitrogen fixation with native *Rhizobium* species that are not normally capable of fixing nitrogen. Study of the newly symbiotic bacterium revealed that it harbored the "symbiotic island" from the originally inoculated *M. loti* strain. Evidently, the native *Rhizobium* species captured the symbiotic island and formed a novel symbiosis with the *Lotus* plant. In this case, the mechanism of transfer may have involved phage-mediated transduction, as with many of the pathogenicity islands. The symbiotic island was incorporated into the chromosome of both strains and encoded a phage-like integrase as well as the *nif* genes. Moreover, in the recipient strain, the symbiotic island was found to be integrated near a tRNA *Phe* gene. Many other examples of symbiotic islands are known, and a few have been shown to be mobile under laboratory conditions. The example from *Lotus* suggests that the symbiosis islands may be transferred frequently in natural settings.

Bacterial genes for degradation of pollutants have been studied for their potential use in bioremediation of contaminated sites. For example, an operon encoding genes for degradation of chlorobenzoate (*cbaAB*) was found in a conjugative plasmid of *Alicaligenes* isolated at a contaminated site. The *cbaAB* genes were subsequently found to be associated with transposon Tn*5271*, a compound transposon integrated into the originally identified plasmid. In an environment simulating a chlorobenzoate-contaminated river, the initial host bacteria did not survive well, but the plasmid was found to transfer to a wide range of indigenous species. Tn*5271*-like sequences have also been found in diverse bacterial species from a bioremediation system treating contaminated water from a chemical landfill.

A variety of bacterial genes have been identified that are capable of degrading pollutants, leading to renewed interest in genetic engineering to facilitate bioremediation of contaminated sites.

Perhaps the most dramatic example of gene transfer in polluted environments resulted after introduction of a phenol-degrading

Pseudomonas into an Estonian river heavily contaminated from a sub-terranean oil-shale fire. Six years later, the original bacterial strain could not be detected. However, the *pheBA* degradative operon could be found in association with different bacterial species up to 22 km from the original site of introduction. In most cases, the *pheBA* genes were carried on a plasmid, but in one isolate, the genes were chromosomal. Although the study was incomplete in an important respect—samples were not taken prior to inoculation of the *pheBA* strain—it seems likely that the spread of the *pheBA* operon is an example of large-scale lateral transfer in a polluted ecosystem.

In summary, diverse studies have documented lateral gene transfer between bacteria in natural or simulated environmental settings. This work provides important perspective for considering the deliberate release of genetically modified organisms. The results summarized in this section indicate that genetic engineering is occurring naturally in the environment at a high rate. For example, the newly evolved nitro-gen-fixing symbiont provides an example of natural genetic engineer-ing occurring in just seven years.

These findings bear on evaluating the potential spread of genes deliberately engineered by humans. Engineered genes in genetically modified prokaryotes introduced into the environment probably will be transferable to new species on occasion, leading to potential spread. On the other hand, transfer and rearrangement of DNA is taking place naturally at very high rates without human intervention. Although caution is warranted, human genetic engineering would appear to be a small drop in the ocean of global DNA exchange. Many considera-tions go into evaluating possible hazards, such as the nature of the genes involved and their potential impact, but an informed analysis will require placing human activity in the context of extremely high natural rates of lateral transfer.

CAN MICROBIAL ORGANISMS BE DIVIDED INTO SPECIES? (20–22)

Can microbes be divided into discrete species? The observed high rates of lateral transfer call all of bacterial taxonomy into question. Maybe bacterial genes are just mixed too rapidly for conventional taxonomy to describe the reality. These issues have been the topic of intense debates, fueled by analysis of the complete genome sequences.

Taxonomy and Lateral Transfer

Early Linnean taxonomy grouped species by shared characteristics—creatures with more characters in common were judged to be more closely related, and those with fewer, more distantly related. More modern phylogenetic taxonomy seeks to put the field on an evolutionary basis. Phylogenetic taxonomy holds that all creatures from microbes to people are derived from a common ancestor. Single-celled organisms first arose 3.5–4 billion years ago and then radiated into the modern domains of life—Bacteria, Archaea, and Eucarya.

Taxonomic Controversies (57, 108)

Many aspects of taxonomy are subject to debate, including the grouping of organisms into separate domains. Carl Woese and others favor three domains—Bacteria, Archaea, and Eucarya (eukaryotes)—based largely on molecular data. Ernst Mayr and his supporters favor only Prokaryotes and Eukaryotes, pointing to morphological similarities among organisms. This book uses the popular tripartite classification of Woese to maintain a consistent terminology.

The relationships among organisms are expressed by their position in the taxonomic hierarchy. The highest divisions within each domain are the kingdoms (Fig. 5.6), each of which is subdivided into phyla, classes, orders, families, genera, and species. The implication is that creatures in different families, but the same order, for example, are more closely related than creatures in different orders. The different orders are expected to have diverged from one another earlier in evolution than the families within an order. A neatly branching evolutionary tree is shown in Figure 5.6A.

Molecular taxonomy has used the explosion in DNA sequence information to group organisms into taxa. The timing of divergence of different groups should be reflected in the DNA sequence (at least according to a simple analysis ignoring lateral transfer). DNA is expected to accumulate changes over time. For example, neutral changes can accumulate in protein-coding regions that alter the sequence of a codon without changing the encoded amino acid. Noncoding regions will usually be even more tolerant of substitutions. Two populations of organisms that are no longer interbreeding will slowly come to differ in sequence by accumulation of mutations. The time of the divergence between lineages can be estimated by quantitating sequence divergence, assuming that substitution rates are not greatly different between lineages (in some cases a risky assumption). Data from DNA divergence times can then be used to construct molecular phylogenetic trees. Efforts to generate molecular phylogenies, however, have met with mixed success, requiring extensive re-evaluation of the data in light of lateral DNA transfer.

A pioneering effort by R.F. Doolittle and coworkers to use molecular data to map the divergence points of higher taxa initially yielded curious results. Vertebrate DNA sequences were used to calibrate a molecular clock, because the relatively rich fossil record could provide an independent means of calibrating the DNA substitution rate. Extrapolating backward, this yielded a date for the last common ancestor of all existing life of 2 billion years ago. This result was very surprising, because probable microfossils are known to be much older.

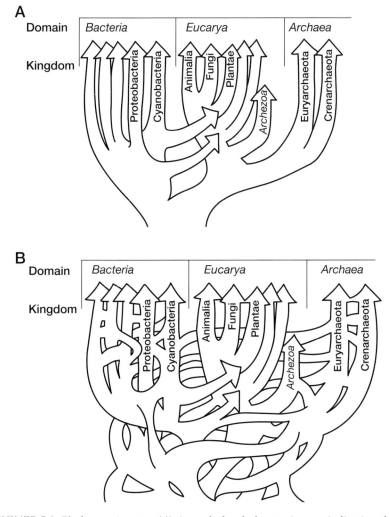

FIGURE 5.6. Phylogenetic trees. (*A*) An orderly phylogenetic tree, indicating the origin of the Bacteria, Eucarya, and Archaea. Transfer of genes between kingdoms is minimized in this view, with only the acquisition of chloroplasts and mitochondria by eukaryotes shown as transfers between domains. (*B*) A highly reticulated tree, reflecting the consequences of extensive lateral DNA transfer. (Redrawn, with permission, from Doolittle 1999 [© American Association for the Advancement of Science]. [21])

Furthermore, chemical changes in the earth's atmosphere, attributed to action of primitive microorganisms, suggest that life existed on earth over 3.5 billion years ago.

A re-evaluation of these data suggested that lateral DNA transfer had complicated the earlier analysis. The Bacteria were found to be anomalously similar to one another, an observation attributable to extensive lateral transfer leading to increased homogeneity of sequences. Some of the sequences in eukaryotes were judged to be recent acquisitions from bacteria via symbionts that are preserved today as the organelles, chloroplasts, and mitochondria (discussed in Chapter 12). "Resetting" of the molecular clock taking into account these revisions yielded a date for the last common ancestor of 3.2–3.8 billion years ago, a proposal more in keeping with other evidence. This study and many others highlight the contribution of lateral transfer to sculpting modern genomes, yielding a "tree-of-life" picture like that in Figure 5.6B.

Complete Sequences for Different Strains of the Same Species Display Remarkable Diversity (1, 8, 41, 74, 78, 86, 100)

At this writing, there are four bacterial species for which more than one complete sequence has been determined: *E. coli, Helicobacter pylori, Chlamydia trachomatis*, and *Chlamydia pneumoniae*. A comparison yields surprisingly different conclusions from case to case (Table 5.6).

As discussed earlier in this chapter, the two sequenced *E. coli* genomes, from the laboratory K-12 strain and pathogenic O157:H7 strains, were remarkably different. The two genomes differ by more than a megabase, with the pathogenic strain being larger. Each strain contains a remarkable fraction of unique genes, 26% in O157:H7 and 12% in K-12. The "backbone" of the chromosomes is roughly colinear, although here too, the two genomes differ by a large inversion. The general conclusion is that the two genomes have been greatly altered by import of new sequences and internal rearrangements since their evolutionary divergence. Whether it is appropriate to label two such

TABLE 5.6. Comparison of Genome Sequences for Strains of *E. coli, H. pylori, C. trachomatis*, and *C. pneumoniae*

Genomes	Size (Mb)	Number of genes	Strain-specific genes	Reference
E. coli K-12	4.63	4045	528	Blattner et al. (1997) (8)
E. coli O157:H7	5.44	5416	1387	Perna et al. (2001) (74)
H. pylori 26695	1.67	1552	117	Tomb et al. (1997) (100)
H. pylori J99	1.64	1495	89	Alm et al. (1999) (1)
C. trachomatis MoPn	1.07	924	~6	Read et al. (2000) (78)
C. trachomatis serovar D	1.04	894	3	Stephens et al. (1998) (93)
C. pneumoniae J138	1.23	1052	none	Shirai et al. (2000) (86)
C. pneumoniae CWL029	1.23	1052	none	Kalman et al. (1999) (41)
C. pneumoniae AR39	1.23	1052	none	Read et al. (2000) (78)

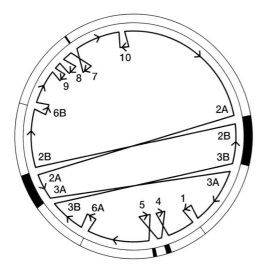

FIGURE 5.7. A comparison of the large-scale structures of the two sequenced isolates of *Helicobacter pylori*. The lines between segments indicate the insertions, inversions, and translocations relating the two sequences. The numbers near each rearrangement refer to the events tabulated in Table 5.7. Dark bars indicate regions of inverted transcriptional orientation. (Redrawn, with permission, from Alm et al. 1999 [© Macmillan Magazines Ltd.]. [1])

genetically and functionally different strains as members of the same species is open to question.

Two complete genome sequences have also been determined from the intestinal pathogen *H. pylori*. The two strains were isolated years apart from patients with different symptoms. The measured degree of divergence is remarkable (Fig. 5.7). Fully 6–7% of genes are unique to each strain. The sequences differ in length by 24,036 base pairs. The genomes differ in the number of IS elements present (17 versus 10, including defective elements). In many cases, the locations of the IS elements in the sequence differ as well, indicating recent mobility. Strikingly, the two sequences differed by ten inversions or translocations, involving sequences from 1 kb to 83 kb in length. Many of the inferred breakpoints in these rearrangements coincide with insertion elements and other repeated sequences. In three cases, genes for restriction-modification systems also thought to be frequently transferred are found at the breakpoints as well. A pathogenic island, *cagPAI*, is present in both strains sequenced but not in other nonpathogenic strains of the same species. In summary, the two sequences are strikingly divergent for genomes of the same ostensible species.

Multiple sequences are also available for genomes of the intercellular parasites *C. trachomatis* and *C. pneumoniae*, and for these the picture is quite different. Both of these species grow only in eukaryotic host cells, inhabiting a specialized vacuole in the post-Golgi exocytic vesicular compartment. Both species rely on the host for many macromo-

TABLE 5.7. Elements Associated with the Artificial End Points Required to Align the Two *H. pylori* Chromosomes (strains J99 and 26695)

Locus	Type	Size (kb)	Associated elements and genes	Strain
1	TR	1.5	lower (G + C) -content DNA; repeat element	both
2 A/B	IN	75	IS605 in 26695; genes of unknown function in J99	26695 or J99
3 A/B	IN	83	"orphan" 5S rRNA; inverted copies of repeat 7	26695
4	IN	10	insertion of DNA-restriction/modification genes	J99
5	IN	2.5	conserved C terminus of *omp* genes; Is605 left-end fragment	both
6 A/B	TR	2	conserved C terminus of *bab* genes and 5′ repeat element	both
7	IN	5.5	repeated, overlapping C terminus of histidine-rich genes	both
8	IN/TR	24	DNA-restriction/modification-gene replacement	both
10	IN/TR	21	IS605	26695
9	IN/TR	1	DNA-restriction/modification genes; duplication of response regulator	both

TR, Translocation; IN, inversion.
(Reprinted, with permission, from Alm et al. 1999 [© Macmillan Magazines Ltd.]. [1])

lecular precursors, allowing the *Chlamydia* species to survive with a genome of only about one megabase. The nature and the order of genes in the genome sequences from each species were very highly conserved, much more so than the *E. coli* or *H. pylori* isolates. For *C. pneumoniae,* strains isolated at different times in different parts of the world are very similar, displaying fully 99% identity in sequence and identical gene numbers. In *C. trachomatis*, a small region of variation was found, the "plasticity region," in which the few strain-specific genes were mostly located, although the rest of the genome was quite homogeneous between isolates.

These highly conserved genomes showed no evidence of lateral DNA transfer. The few islands of unique genes did not have G/C to A/T ratios different from the genome average, failing to support the idea that they were acquired by lateral transfer. For all of the *Chlamydia* genomes sequenced, no IS elements were detected, and the coding density is very high, suggesting that these organisms can eliminate unnecessary DNA quite efficiently. One speculation holds that the *Chlamydia* life-style as intracellular parasites provides little opportunity for contact with other bacteria, preventing acquisition of new genes by lateral transfer. Life as an intracellular parasite may also demand a particularly streamlined genome, providing the pressure to eliminate all but the most crucial sequences. The combination of these two factors, according to the hypothesis, results in a high degree of conservation of genome structure. It seems likely that most bacteria resemble *E. coli* and *H. pylori*, living in environments with large numbers of bacteria in close contact and so engaging in frequent lateral transfer. If so, the problems for phylogeny raised by the *E. coli* and *H. pylori* examples are severe.

Preserving Bacterial Phylogeny? (17, 23, 37, 91)

One effort to preserve the idea of discrete bacterial species focuses on phylogenetic analysis of genes that may be less prone to lateral transfer. The reassessment of molecular phylogeny by Doolittle and coworkers, for example, restricted the analysis to well-conserved enzymes. Genes encoding functions for the most central enzymatic processes may be particularly well conserved, since these may be more difficult to transfer if their products need to interact with many partners. This idea has been explored by James Lake and colleagues, who contrasted genes for central "informational" pathways with "operational" genes for housekeeping functions. They found that informational genes, for example, those involved in transcription, translation, and tRNA charging, were less prone to lateral transfer than operational genes such as those involved in amino acid biosynthesis, cofactor biosynthesis, intermediary metabolism, and other functions. Lake and colleagues argued that proteins encoded by informational genes interact with more total macromolecules, as in the ribosome, and so cannot readily function after lateral transfer. This leads to the complexity hypothesis, which holds that the probability of lateral transfer is greater the simpler the system involved. According to this view, a sensible prokaryotic molecular phylogeny may be possible if only the genes for the most complex informational complexes are considered.

Traditional phylogenies of microorganisms have been constructed using sequences of the small subunit of the ribosomal RNA (SSU), an informational gene whose product is expected to interact with at least 100 other proteins and RNAs, making it intuitively unlikely to be transferable without disrupting core cellular systems. This sequence also has the advantages of being readily identified and isolated, but sufficiently different to distinguish related species. However, even in this case, there is evidence that rRNA genes can be transferred, at least on rare occasions. At present, there does not seem to be any sequence class immune to lateral transfer, and therefore no fully sound basis for generating organismic phylogenetic trees from trees based on single genes.

Where does all this leave microbial taxonomy? If genes are so readily transferred between organisms, does it make sense to talk of separate species at all? The orderly tree of life, with sequentially finer branches representing ever more closely related organisms, is being replaced with a highly reticulated tree, more like a series of interconnected pipes. Such a picture is strongly supported by the sequences for two strains of *E. coli* and *H. pylori*, which document very extensive differences.

Furthermore, how does one deal with the finding that analyzing different gene sequences often gives different organismic phylogenies?

Several groups have taken the next logical step, using all of the genes in completely sequenced genomes to make phylogenetic trees, but even this has led to complications. Brend Snel and coworkers constructed a phylogeny based on gene content, scoring the presence or absence of genes in 13 microbial genomes and using this information to assess relationships. They generated a phylogenetic tree that was robust by several statistical tests, and which placed the Archaea and Bacteria in separate groups, as expected. Their tree roughly agreed with the tree constructed using similarity among the 16S ribosomal RNA sequences, although there were a number of differences in detail. Sorel Fitz-Gibbon and Christopher House took a related approach, comparing the presence or absence of protein sequence in 11 complete genomes from free-living microorganisms. Their analysis also yielded consistent trees that withstood various types of statistical perturbations. Their phylogenetic trees generally agreed with the 16S RNA studies and work of Snel and coworkers. The authors argued that microorganisms are not an "incomprehensible amalgam of genes with complicated histories" due to lateral transfer. However, a complication arose in their study in attempts to include genome sequences from parasitic microbes, resulting in inconsistent phylogenetic trees. The authors felt that the *H. pylori* genome (one of the parasites) might not be representative of the group to which it belongs, possibly a consequence of lateral transfer. Nevertheless, despite the complications, if some type of whole-genome comparison can be shown to yield consistent results across large numbers of organisms, this approach may allow the construction of a robust prokaryotic phylogeny.

Summary

New data from the microbial genome sequences highlight the importance of lateral DNA transfer in microbial evolution. *E. coli*, the workhorse microbe of modern biology, has a genome of 4.6 million base pairs, encoding about 4288 genes (initial estimate). The genome contains about 2% sequences derived from mobile elements such as transposons, phage, and plasmid remnants. At least 18% of the genome has been derived from lateral transfer from other microbial species. This analysis relies on detecting new sequences by (1) sequence differences with the host genome, including G/C content and amino acid codon usage, (2) the association with mobile element fragments, and (3) the proximity to tRNA genes, indicative of potential phage integration. It is estimated from these data that *E. coli* captures and fixes DNA at a rate of at least 16 kilobases per million years.

Even these high values can only be taken as lower limits for the amount of lateral transfer. Older lateral transfer events will often not be detected, because they have been "ameliorated" to match the *E. coli* G/C content and codon preferences. In addition, transfer from genomes with similar base composition and codon preferences will not be detected either.

Quite different amounts of laterally transferred DNA have been detected in different prokaryotic species, but most contain mobile elements and newly acquired DNA. The *Thermotoga maritima* genome even contains multigene blocks resembling segments of archaeal genomes, providing an example of probable gene transfer between the domains of life. A few small genomes, such as that of *Mycoplasma genitalium*, contain no detectable mobile DNA elements, possibly reflecting a strong pressure to eliminate unnecessary DNA.

The observed high rate of lateral transfer may explain at least some of the clustering of bacterial genes into operons. Traditionally, operon organization was thought to be for convenience of coregulating expression of related genes, but this is not very compelling because unlinked genes are often coregulated perfectly well. An intriguing alternative is that genes are clustered for convenience of lateral transfer, an idea named The Selfish Operon Hypothesis, in which groups of genes that together specify a useful function have a greater chance of fixation after transfer. If genes are randomly reassorted within the bacterial chromosomes, and occasionally transferred to new hosts, related functions may necessarily become clustered over time.

Lateral transfer has been extensively documented in natural settings. Transformation, transduction, and conjugation can all be detected in the environment. Transformation might seem unlikely, as DNA is rapidly degraded in the environment, but actually adherence of DNA to mineral surfaces can greatly increase its lifetime and potentially concentrate it with surface-bound bacteria. Phages are extremely abundant in aquatic environments and can transfer DNA between bacteria at a high rate. Conjugation has also been extensively documented in natural settings. These findings bear on planning the release of genetically modified bacteria into the environment.

The discovery of extensive lateral transfer has left bacterial systematics in a state of chaos. Different phylogenetic trees result depending on which genes are selected for analysis. At least for the case of bacteria, the tree of life appears to be highly reticulated rather than branched in an orderly way. Trees based on whole genomes may ultimately yield satisfactory phylogenies, but the jury is still out. One is left with the feeling that traditional Linnean classification fits at best uncomfortably with the realities of microbial life, but that by admitting this, we approach more closely the biological reality.

REFERENCES

1. Alm R.A., Ling L.S., Moir D.T., King B.L., Brown E.D., Doig P.C., Smith D.R., Noonan B., Guild B.C., deJonge B.L., Carmel G., Tummino P.J., Caruso A., Uria-Nickelsen M., Mills D.M., Ives C., Gibson R., Merberg D., Mills S.D., Jiang Q., Taylor D.E., Vovis G.F., and Trust T.J. 1999. Genomic-sequence comparison of two unrelated isolates of the human gastric pathogen *Helicobacter pylori*. *Nature* **397**: 176–180.

2. Armstrong J.L., Wood N.D., and Porteous L.A. 1990. Transconjugation between bacteria in the digestive tract of the cutworm *Peridroma saucia*. *Appl. Environ. Microbiol.* **56**: 1492–1493.

3. Bale M.J., Day M.J., and Fry J.C. 1988. Novel method for studying plasmid transfer in undisturbed river epilithon. *Appl. Environ. Microbiol.* **54**: 2756–2758.

4. Balis E., Vatopoulos A., Kanelopoulou M., Mainas E., Hatzoudis G., Kontogianni V., Malamou-Lada H., Kitsou-Kiriakopoulou S., and Kalapothaki V. 1996. Indications of in vivo transfer of an epidemic R plasmid from *Salmonella enteritidis* to *Escherichia coli* of the normal human gut flora. *J. Clin. Microbiol.* **34**: 977–979.

5. Baur B., Hanselmann K., Schlimme W., and Jenni B. 1996. Genetic transformation in freshwater: *Escherichia coli* is able to develop natural competence. *Appl. Environ. Microbiol.* **62**: 3673–3678.

6. Beckwith J. 1987. The operon: An historical account. In Escherichia coli

and Salmonella typhimurium: *Cellular and molecular biology* (ed. F.C. Neid-hardt et al.), vol. 2, pp. 1439–1443. American Society for Microbiology, Washington, D.C.

7. Bjorklof K., Suoniemi A., Haahtela K., and Romantschuk M. 1995. High frequency of conjugation versus plasmid segregation of RP1 in epiphytic *Pseudomonas syringae* populations. *Microbiology* **141:** 2719–2727.

8. Blattner F.R., Plunkett G.R., Bloch C.A., Perna N.T., Burland V., Riley M., Collado-Vides J., Glasner J.D., Rode C.K., Mayhew G.F., Gregor J., Davis N.W., Kirkpatrick H.A., Goeden M.A., Rose D.J., Mau B., and Shao Y. 1997. The complete genome sequence of *Escherichia coli* K-12 (comment). *Science* **277:** 1453–1474.

9. Bratoeva M.P. and John J.F.J. 1994. In vivo R-plasmid transfer in a patient with a mixed infection of *Shigella* dysentery. *Epidemiol. Infect.* **112:** 247–252.

10. Brock T.D. 1990. *The emergence of bacterial genetics.* Cold Spring Harbor Laboratory Press, Cold Spring Harbor, New York.

11. Chamier B., Lorenz M.G., and Wackernagel W. 1993. Natural transformation of *Acinetobacter calcoaceticus* by plasmid DNA adsorbed on sand and groundwater aquifer material. *Appl. Environ. Microbiol.* **59:** 1662–1667.

12. Christensen B.B., Sternberg C., Andersen J.B., Eberl L., Moller S., Givskov M., and Molin S. 1998. Establishment of new genetic traits in a microbial biofilm community. *Appl. Environ. Microbiol.* **64:** 2247–2255.

13. Daane L.L., Molina J.A., Berry E.C., and Sadowsky M.J. 1996. Influence of earthworm activity on gene transfer from *Pseudomonas fluorescens* to indigenous soil bacteria. *Appl. Environ. Microbiol.* **62:** 515–521.

14. Dahlberg C., Bergstrom M., and Hermansson M. 1998. *In situ* detection of high levels of horizontal plasmid transfer in marine bacterial communities. *Appl. Environ. Microbiol.* **64:** 2670–2675.

15. Davison J. 1999. Genetic exchange between bacteria in the environment. *Plasmid* **42:** 73–91.

16. Deckert G., Warren P.V., Gaasterland T., Young W.G., Lenox A.L., Graham D.E., Overbeek R., Snead M.A., Keller M., Aujay M., Huber R., Feldman R.A., Short J.M., Olsen G.J., and Swanson R.V. 1998. The complete genome of the hyperthermophilic bacterium *Aquifex aeolicus. Nature* **392:** 353–358.

17. de la Cruz F. and Davies J. 2000. Horizontal gene transfer and the origin of species: Lessons from bacteria. *Trends Microbiol.* **8:** 128–133.

18. De Rore H., Demolder K., De Wilde K., Top E., Houwen F., and Verstraete W. 1994. Transfer of the catabolic plasmid RP4::Tn4371 to indigenous soil bacteria and its effect on respiration and biphenyl breakdown. *FEMS Microbiol. Ecol.* **15:** 71–78.

19. DiGiovanni G.D., Neilson J.W., Pepper I.L., and Sinclair N.A. 1996. Gene transfer of *Alcaligenes eutrophus* JMP134 plasmid pJP4 to indigenous soil recipients. *Appl. Environ. Microbiol.* **62:** 2521–2526.

20. Doolittle R.F., Feng D.F., Tsang S., Cho G., and Little E. 1996. Determining divergence times of the major kingdoms of living organisms with a protein clock (comments). *Science* **271:** 470–477.

21. Doolittle W.F. 1999. Phylogenetic classification and the universal tree. *Science* **284:** 2124–2129.

22. Feng D.-F., Cho G., and Doolittle R.F. 1997. Determining divergence times with a protein clock: Update and reevaluation. *Proc. Natl. Acad. Sci.* **94:** 13028–13033.

23. Fitz-Gibbon S.T. and House C.H. 1999. Whole genome-based phylogenet-

ic analysis of free-living microorganisms. *Nucleic Acids Res.* **27:** 4218–4222.

24. Fleischmann R.D., Adams M.D., White O., Clayton R.A., Kirkness E.F., Kerlavage A.R., Bult C.J., Tomb J.F., Dougherty B.A., Merrick J.M., et al. 1995. Whole-genome random sequencing and assembly of *Haemophilus influenzae* Rd (comments). *Science* **269:** 496–512.

25. Fraser C.M., Eisen J.A., and Salzberg S.L. 2000. Microbial genome sequencing. *Nature* **406:** 799–803.

26. Fuhrman J.A. 1999. Marine viruses and their biogeochemical and ecological effects. *Nature* **399:** 541–548.

27. Fulthorpe R.R. and Wyndham R.C. 1991. Transfer and expression of the catabolic plasmid pBRC60 in wild bacterial recipients in a freshwater ecosystem. *Appl. Environ. Microbiol.* **57:** 1546–1553.

28. Fulthorpe R.R. and Wyndham R.C. 1992. Involvement of a chlorobenzoate-catabolic transposon, Tn5271, in community adaptation to chlorobiphenyl, chloroaniline, and 2,4-dichlorophenoxyacetic acid in a freshwater ecosystem. *Appl. Environ. Microbiol.* **58:** 314–325.

29. Glew J.G., Angle J.S., and Sadowsky M.J. 1993. In vivo transfer of pR68.45 from *Pseudomonas aeruginosa* into indigenous soil bacteria. *Microbiol. Releases* **1:** 237–241.

30. Goodman A.E., Hild K.C., Marshall K.C., and Hermansson M. 1993. Conjugative plasmid transfer between bacteria under simulated marine oligotrophic conditions. *Appl. Environ. Microbiol.* **59:** 1035–1040.

31. Gotz A. and Smalla K. 1997. Manure enhances plasmid mobilization and survival in *Pseudomonas putida* introduced into the field soil. *Appl. Environ. Microbiol.* **63:** 1980–1986.

32. Gruzza M., Fons M., Ouriet M., Duval-Iflah Y., and Ducluzeau R. 1994. Study of gene transfer in vitro and in the digestive tract of gnotobiotic mice from *Lactococcus lactis* strains to various strains belonging to human intestinal flora. *Microbiol. Releases* **2:** 183–189.

33. Herrick J.B., Stuart-Keil K.G., Ghiorse W.C., and Madsen E.L. 1997. Natural horizontal transfer of a naphthalene dioxygenase gene between bacteria native to a coal tar-contaminated field site. *Appl. Environ. Microbiol.* **63:** 2330–2337.

34. Hill K.E., Weightman A.J., and Fry J.C. 1992. Isolation and screening of plasmids from the epilithon which mobilize recombinant plasmid pD10. *Appl. Environ. Microbiol.* **58:** 1292–1300.

35. Hoffmann A., Thimm T., Dröge M., Moore E.R., Munch J.C., and Tebbe C.C. 1998. Intergeneric transfer of conjugative and mobilizable plasmids harbored by *Escherichia coli* in the gut of the soil microarthropod *Folsomia candida* (Collembola). *Appl. Environ. Microbiol.* **64:** 2652–2659.

36. Ichige A., Matsutani S., Oishi K., and Mizushima S. 1989. Establishment of gene transfer systems and construction of the genetic map of a marine *Vibrio* strain. *J. Bacteriol.* **171:** 1825–1834.

37. Jain R., Rivera M.C., and Lake J.A. 1999. Horizontal gene transfer among genomes: The complexity hypothesis. *Proc. Natl. Acad. Sci.* **96:** 3801–3806.

38. Jarrett P. and Stephenson M. 1990. Plasmid transfer between strains of *Bacillus thuringiensis* infecting *Galleria mellonella* and *Spodoptera littoralis*. *Appl. Environ. Microbiol.* **56:** 1608–1614.

39. Jiang S.C. and Paul J.H. 1998. Gene transfer by transduction in the marine environment. *Appl. Env. Microbiol.* **64:** 2780–2787.

40. Ka J.O. and Tiedje J.M. 1994. Integration and excision of a 2,4-dichlorophe-

noxyacetic acid-degradative plasmid in *Alcaligenes paradoxus* and evidence of its natural intergeneric transfer. *J. Bacteriol.* **176:** 5284–5289.

41. Kalman S., Mitchell W., Marathe R., Lammel C., Fan J., Hyman R.W., Olinger L., Grimwood J., Davis R.W., and Stephens R.S. 1999. Comparative genomes of *Chlamydia* and *C. trachomatis. Nat. Genet.* **21:** 385–389.

42. Kidambi S.P., Ripp S., and Miller R.V. 1994. Evidence for phage-mediated gene transfer among *Pseudomonas aeruginosa* strains on the phylloplane. *Appl. Environ. Microbiol.* **60:** 496–500.

43. Kinkle B.K., Sadowski M.J., Schmidt E.L., and Koskinen W.C. 1991. Plasmids JP4 and r68.45 can be transferred between populations of *Bradyrhizobia* in non-sterile soil. *Appl. Environ. Microbiol.* **59:** 1762–1766.

44. Kruse H. and Sorum H. 1994. Transfer of multiple drug resistance plasmids between bacteria of diverse origins in natural microenvironments. *Appl. Environ. Microbiol.* **60:** 4015–4021.

45. Lacy G.H., Stromberg V.K., and Cannon N.P. 1984. *Erwinia amylovora* mutants and *in planta* derived transconjugants resistant to oxytetracycline. *Can. J. Microbiol.* **6:** 33–39.

46. Lawrence J.G. and Ochman H. 1998. Molecular archaeology of the *Escherichia coli* genome. *Proc. Natl. Acad. Sci.* **95:** 9413–9417.

47. Lawrence J.G. and Roth J.R. 1996. Selfish operons: Horizontal transfer may drive the evolution of gene clusters. *Genetics* **143:** 1843–1860.

48. Lilley A.K. and Bailey M.J. 1997. The acquisition of indigenous plasmids by a genetically marked pseudomonad population colonizing the sugar beet phytosphere is related to local environmental conditions. *Adv. Appl. Microbiol.* **63:** 1577–1583.

49. Lilley A.K., Fry J.C., Day M.J., and Bailey M.J. 1994. *In situ* transfer of an exogenously isolated plasmid between *Pseudomonas* spp in the sugar beet rhizosphere. *Microbiology* **140:** 27–33.

50. Lorenz M.G. and Wackernagel W. 1994. Bacterial gene transfer by natural genetic transformation in the environment. *Microbiol. Rev.* **58:** 563–602.

51. Lorenz M.G., Gerjets D., and Wackernagel W. 1991. Release of transforming plasmid and chromosomal DNA from two cultured soil bacteria. *Arch. Microbiol.* **156:** 319–326.

52. Lorenz M.G., Reipschläger K., and Wackernagel W. 1992. Plasmid transformation of naturally competent *Acinetobacter calcoaceticus* in non-sterile soil extract and groundwater. *Arch. Microbiol.* **157:** 355–360.

53. Louvrier P., Laguerre G., and Amarger N. 1996. Distribution of symbiotic genotypes in *Rhizobium leguminosarum* biovar *viciae* populations isolated directly from soils. *Appl. Environ. Microbiol.* **62:** 4202–4205.

54. Mancini P., Fertels S., Nave D., and Gealt M.A. 1987. Mobilization of plasmid pHSV106 from *Escherichia coli* HB101 in a laboratory-scale waste treatment facility. *Appl. Environ. Microbiol.* **53:** 665–671.

55. Marcinek H., Wirth R., Muscholl-Silberhorn A., and Gauer M. 1998. *Enterococcus faecalis* gene transfer under natural conditions in municipal sewage water treatment plants. *Appl. Environ. Microbiol.* **64:** 626–632.

56. Matheson V.G., Forney L.J., Suwa Y., Nakatsu C.H., Sewtone A.J., and Holben W.E. 1997. Evidence for aquisition in nature of a chromosomal 2,4-dichlorphenooxyacetic acid/ketoglutarate dioxygenase gene by different *Burkholderia* spp. *Appl. Environ. Microbiol.* **63:** 2266–2272.

57. Mayr E. 1998. Two empires or three? *Proc. Natl. Acad. Sci..* **95:** 9720–9723.

58. McClure N.C., Fry J.C., and Weightman A.J. 1990. Gene transfer in activated sludge. In *Bacterial genetics in natural environments* (ed. J.C. Fry and M.J. Daly), pp. 111–129. Chapman & Hall, London.

59. McClure N.C., Weightman A.J., and Fry J.C. 1989. Survival of *Pseudomonas putida* UWC1 containing cloned catabolic genes in a model activated-sludge unit. *Appl. Environ. Microbiol.* **55:** 2627–2634.

60. Morrison W.D., Miller R.V., and Sayler G.S. 1978. Frequency of F116-mediated transduction of *Pseudomonas aeruginosa* in a freshwater environment. *Appl. Environ. Microbiol.* **36:** 724–730.

61. Nakatsu C.H., Fulthorpe R.R., Holland B.A., Peel M.C., and Wyndham R.C. 1995. The phylogenetic distribution of a transposable dioxygenase from the Niagara River watershed. *Mol. Ecol.* **4:** 593–603.

62. Neidhardt F.C., Ingraham J.L., Low K.B., Magasanik B., Schaechter M., and Umbarger H.E., Eds. 1987. Escherichia coli *and* Salmonella typhimurium: *Cellular and molecular biology.* American Society for Microbiology, Washington, D.C.

63. Neilson J.W., Josephson K.L., Pepper I.L., Arnold R.B., Di Giovanni G.D., and Sinclair N.A. 1994. Frequency of horizontal gene transfer of a large catabolic plasmid (pJP4) in soil. *Appl. Environ. Microbiol.* **60:** 4053–4058.

64. Nielsen K.M., van Weerelt M.D., Berg T.N., Bones A.M., Hagler A.N., and van Elsas J.D. 1997. Natural transformation and availability of transforming DNA to *Acinetobacter calcoaceticus* in soil microcosms. *Appl. Environ. Microbiol.* **63:** 1945–1952.

65. Nijsten R., London N., van den Bogaard A., and Stobberingh E. 1995. In-vivo transfer of resistance plasmids in rat, human or pig-derived intestinal flora using a rat model. *J. Antimicrob. Chemother.* **36:** 975–985.

66. Nikolich M.P., Hong G., Shoemaker N.B., and Salyers A.A. 1994. Evidence for natural horizontal transfer of tetQ between bacteria that normally colonize humans and bacteria that normally colonize livestock. *Appl. Environ. Microbiol.* **60:** 3255–3260.

67. Normander B., Christensen B.B., Molin S., and Kroer N. 1998. Effect of bacterial distribution and activity on conjugal gene transfer on the phylloplane of the bush bean (*Phaseolus vulgaris*). *Appl. Environ. Microbiol.* **64:** 1902–1909.

68. NuBlein K., Maris D., Timmis K., and Dwyer D.F. 1992. Expression and transfer of engineered catabolic pathways harbored by *Pseudomonas* spp. introduced into activated sludge microcosms. *Appl. Environ. Microbiol.* **58:** 3380–3386.

69. Ochman H. and Jones I.B. 2000. Evolutionary dynamics of full genome content in *Escherichia coli. EMBO J.* **19:** 6637–6643.

70. Ochman H., Lawrence J.G., and Groisman E.A. 2000. Lateral gene transfer and the nature of bacterial innovation. *Nature* **405:** 299–304.

71. Parkhill J., Wren B.W., Mungall K., Ketley J.M., Churcher C., Basham D., Chillingworth T., Davies R.M., Feltwell T., Holroyd S., Jagels K., Karlyshev A.V., Moule S., Pallen M.J., Penn C.W., Quail M.A., Rajandream M.-A., Rutherford K.M., Van Vliet A.H.M., Whitehead S., and Barrell B.G. 2000. The genome sequence of the food-borne pathogen *Campylobacter jejuni* reveals hypervariable sequences. *Nature* **403:** 665–668.

72. Paul J.H., Thurmond J.M., and Frischer M.E. 1991. Gene transfer in marine water column and sediment microcosms by natural plasmid transformation. *Appl. Environ. Microbiol.* **57:** 1509–1515.

73. Paul J.H., Thurmond J.M., Frischer M.E., and Cannon J.P. 1992. Intergeneric natural plasmid transformation between *E. coli* and a marine *Pseudomonas* species. *Mol. Ecol.* **1:** 37–46.

74. Perna N.T., Plunkett III, G., , Burland V., Mau B., Glasner J.D., Rose D.J., Mayhew G.F., Evans P.S., Gregor J., Kirkpatrick H.A., Posfai G., Hackett J., Klink S., Boutin A., Shao Y., Miller L., Grotbeck E.J., Davis N.W., Lim A., Dimalanta E.T., Potamousis K.D., Apodaca J., Anantharaman T.S., Lin J., Yen G., Schwartz D.C., Welch R.A., and Blattner F.R. 2001. Genome sequence of enterohaemorrhagic *Escherichia coli* 0157:H7. *Nature* **409:** 529–533.

75. Peters M., Heinaru E., Talpsep E., Wand H., Stottmeister U., Heinaru A., and Nurk A. 1997. Acquisition of a deliberately introduced phenol degradation operon, pheBA, by different indigenous *Pseudomonas* species. *Appl. Environ. Microbiol.* **63:** 4899–4906.

76. Prodinger W.M., Fille M., Bauernfeind A., Stemplinger I., Amann S., Pfausler B., Lass-Florl C., and Dierich M.P. 1996. Molecular epidemiology of *Klebsiella pneumoniae* producing SHV-5 β-lactamase: Parallel outbreaks due to multiple plasmid transfer. *J. Clin. Microbiol.* **34:** 564–568.

77. Ravatn R., Zehnder A.J.B., and van der Meer J.R. 1998. Low-frequency horizontal transfer of an element containing the chlorocatechol degradation genes from *Pseudomonas* sp. strain B13 to *Pseudomonas putida* F1 and to indigenous bacteria in laboratory-scale activated-sludge microcosms. *Appl. Environ. Microbiol.* **64:** 2126–2132.

78. Read T.D., Brunham R.C., Shen C., Gill S.R., Heidelberg J.F., White O., Hickey E.K., Peterson J., Utterback T., Berry K., Bass S., Linher K., Weidman J., Khouri H., Craven B., Bowman C., Dodson R., Gwinn M., Nelson W., DeBoy R., Kolonay J., McClarty G., Salzberg S.L., Eisen J., and Fraser C.M. 2000. Genome sequences of *Chlamydia trachomatis* MoPn and *Chlamydia pneumoniae* AR39. *Nucleic Acids Res.* **28:** 1397–1406.

79. Reiter W.-D. and Palm P. 1990. Identification and characterization of a defective SSV1 genome integrated into a tRNA gene in the archaebacterium *Sulfolobus* sp. B12. *Mol. Gen. Genet.* **221:** 65–71.

80. Salyers A.A. and Shoemaker N.B. 1996. Resistance gene transfer in anaerobes: New insights, new problems. *Clin. Infect. Dis.* (suppl. 1) **23:** S36–43.

81. Sandaa R.-A. and Enger E. 1994. Transfer in marine sediments of the naturally occurring plasmid pRAS1 encoding multiple antibiotic resistance. *Appl. Environ. Microbiol.* **60:** 4234–4238.

82. Saye D.J., Ogunseitan O., Sayler G.S., and Miller R.V. 1987. Potential for transduction of plasmids in a natural freshwater environment: Effect of plasmid donor concentration and a natural microbial community on transduction in *Pseudomonas aeruginosa. Appl. Environ. Microbiol.* **53:** 987–995.

83. Scott K. and Flint H. 1995. Transfer of plasmids between strains of *Escherichia coli* under rumen conditions. *J. Appl. Bacteriol.* **78:** 189–193.

84. Sharp P.M. 1991. Determinants of DNA sequence divergence between *Escherichia coli* and *Salmonella typhimurium:* Codon usage, map position, and concerted evolution. *J. Mol. Evol.* **33:** 23–33.

85. She Q., Peng X., Zillig W., and Garrett R.A. 2001. Gene capture in archaeal chromosomes. *Nature* **409:** 478.

86. Shirai M., Hirakawa H., Kimoto M., Tabuchi M., Kishi F., Ouchi K., Shiba T., Ishii K., Hattori M., Kuhara S., and Nakazawa T. 2000. Comparison of whole genome sequences of *Chlamydia pneumoniae* J138 from Japan and

CWL029 from USA. *Nucleic Acids Res.* **28:** 2311–2314.

87. Sikorski J., Graupner S., Lorenz M.G., and Wackernagel W. 1988. Natural genetic transformation of *Pseudomonas stutzeri* in a non-sterile soil. *Microbiology* **144:** 569–576.

88. Smit E., Venne D., and van Elsas J.D. 1993. Mobilization of a IncQ plasmid between bacteria on agar surfaces and in soil via contransfer or retro-transfer. *Appl. Environ. Microbiol.* **59:** 2257–2263.

89. Smit E., Wolters A., and van Elsas J.D. 1998. Self-transmissible mercury resistance plasmids with gene-mobilizing capacity in soil bacterial populations: Influence of wheat roots and mercury addition. *Appl. Environ. Microbiol.* **64:** 1210–1219.

90. Smith H.O., Tomb J.F., Dougherty B.A., Fleischmann R.D., and Venter J.C. 1995. Frequency and distribution of DNA uptake signal sequences in the *Haemophilus influenzae* Rd genome (comments). *Science* **269:** 538–540.

91. Snel B., Bork P., and Huynen M.A. 1999. Genome phylogeny based on gene content. *Nat. Genet.* **21:** 108–110.

92. Sorensen S.J. 1993. Transfer of plasmid RP4 from *Escherichia coli* K-12 to indigenous bacteria of seawater. *Microbiol. Releases* **2:** 135–141.

93. Stephens R., Kalman S., Lammel C., Fan J., Marathe R., Aravind L., Mitchell W., Olinger L., Tatusov R., Zhao Q., Koonin E., and Davis R. 1998. Genome sequence of an obligate intracellular pathogen of humans: *Chlamydia trachomatis. Science* **282:** 754–759.

94. Stewart G.J. and Sinigalliano C.D. 1991. Exchange of chromosomal markers by natural transformation between the soil isolate, *Pseudomonas stutzeri* JM300, and the marine isolate, *Pseudomonas stutzeri* strain ZoBell. *Antonie Leeuwenhoek* **59:** 19–25.

95. Sullivan J.T. and Ronson C.W. 1998. Evolution of rhizobia by acquisition of a 500-kb symbiosis island that integrates into a phe-tRNA gene. *Proc. Natl. Acad. Sci.* **95:** 5145–5149.

96. Sullivan J.T., Eardly B.D., van Berkum P., and Ronson C.W. 1996. Four unnamed species of nonsymbiotic rhizobia isolated from the rhizosphere of *Lotus corniculatus. Appl. Environ. Microbiol.* **62:** 2818–2825.

97. Sullivan J.T., Patrick H.N., Lowther W.L., Scott D.B., and Ronson C.W. 1995. Nodulating strains of *Rhizobium loti* arise through chromosomal symbiotic gene transfer in the environment. *Proc. Natl. Acad. Sci.* **92:** 8985–8989.

98. Sundin G.W., Demezas D.H., and Bender C.L. 1994. Genetic and plasmid diversity within natural populations of P*seudomonas syringae* with various exposures to copper and streptomycin bactericides. *Appl. Environ. Microbiol.* **60:** 4421–4431.

99. Tauxe R., Cavanagh T., and Cohen M. 1989. Interspecies gene transfer in vivo producing an outbreak of multiply resistant shigellosis. *J. Infect. Dis.* **160:** 1067–1070.

100. Tomb J.-F., White O., Kerlavage A.R., Clayton R.A., Sutton R.D., Fleischmann R.D., Ketchum K.A., Klenk H.P., Gill S., Dougherty B.A., et al. 1997. The complete genome sequence of the gastric pathogen *Helicobacter pylori. Nature* **388:** 539–547.

101. Top E., Mergeay M., Springael D., and Verstraete W. 1990. Gene escape model: Transfer of heavy metal resistance genes from *Escherichia coli* to *Alcaligenes eutrophus* on agar plates and in soil samples. *Appl. Environ. Microbiol.* **56:** 2471–2479.

102. Top E., Mergeay M., Springael D., and Verstraete W. 1994. Exogenous isolation of mobilizing plasmids from polluted soils and sludges. *Appl. Environ. Microbiol.* **60:** 831–839.

103. Top E.M., Van Daele P., De Saeyer N., and Forney L.J. 1998. Enhancement of 2,4-dichlorophenoxyacetic acid (2,4-D) degradation in soil by dissemination of catabolic plasmids. *Antonie Leeuwenhoek* **73:** 87–94.

104. Troxler J., Azelvandre P., Zala M., Defago G., and Haas D. 1997. Conjugal transfer of chromosomal genes between fluorescent *Pseudomonads* in the wheat rhizosphere. *Appl. Environ. Microbiol.* **63:** 213–219.

105. van Elsas, J.D., Gardener B.B., Wolters A.C., and Smit E. 1998. Isolation, characterization, and transfer of cryptic gene-mobilizing plasmids in the wheat rhizosphere. *Appl. Environ. Microbiol.* **64:** 880–889.

106. Watanabe K. and Sato M. 1998. Plasmid-mediated gene transfer between insect-resident bacteria. *Enterobacter cloacae*, and plant-epiphytic bacteria, *Erwinia herbicola* in guts of silkworm larvae. *Curr. Microbiol.* **37:** 352–355.

107. Williams H.G., Day M., Fry J.C., and Stewart G.J. 1996. Natural transformation in river epilithon. *Appl. Environ. Microbiol.* **62:** 2994–2998.

108. Woese C.R. 1998. Default taxonomy: Ernst Mayr's view of the microbial world. *Proc. Natl. Acad. Sci.* **95:** 11043–11046.

109. Zeph L.R. and Stotzsky G. 1989. Use of a biotinylated DNA probe to detect bacteria transduced by bacteriophage P1 in soil. *Appl. Environ. Microbiol.* **55:** 661–665.

110. Zhou J.Z. and Tiedje J.M. 1995. Gene transfer from a bacterium injected into an aquifer to an indigenous bacterium. *Mol. Ecol.* **4:** 613–618.

CHAPTER 6

Gene Transfer by Retroviruses

IN THIS CHAPTER, WE BEGIN OUR LOOK AT LATERAL DNA TRANSFER in eukaryotes. Many of the principles describing DNA transfer in prokaryotes hold in eukaryotes as well, although often in a modified form.

MOBILE DNA IN PROKARYOTES AND EUKARYOTES

Lateral DNA transfer has been a central force in constructing the chromosomes of both prokaryotes and eukaryotes. Mobile elements have entered the genomes in both groups, proliferating and creating heritable changes in DNA. Besides the genomic changes due to integration itself, multiplication of transposons or integrating viruses provides new regions of homology that serve as substrates for cellular homologous recombination, potentiating subsequent chromosomal rearrangements. In both prokaryotes and eukaryotes, fragments of mobile elements and integrating viruses have occasionally been coopted to form new genes and gene regulatory regions. In both groups, sets of genes have been acquired from other organisms and incorporated into the cellular genome.

Other aspects of lateral transfer differ between prokaryotes and eukaryotes. DNA entering a eukaryotic cell must traverse not only the cytoplasmic membrane, but also the nuclear membrane, whereas in prokaryotes no nuclear membrane is present. Consequently, in eukaryotes there is a greater requirement for signals directing subcellular sorting. In eukaryotes, genes often come in pieces, with coding regions (exons) separated by noncoding regions (introns). A few examples of introns are known in prokaryotes, but in eukaryotes genes in pieces are the rule rather than the exception. Thus, eukaryotic RNAs frequently need to be "spliced" to yield mature mRNAs, influencing the life-style of eukaryotic mobile DNA systems in diverse ways. There is no conjugative transfer of genes between eukaryotic cells, although as described in Chapter 12, a conjugation-like mechanism can transfer DNA from prokaryotes to eukaryotes.

Some of the virus types that transfer DNA between cells are very different in the different domains of life. Phage come in many forms, but none have replication cycles like the retroviruses, the topic of this chapter. These differences in viral life-styles result in important differences in DNA exchange.

The human body is composed of some 50 trillion cells, many of which are arranged in complex tissues, so physical access to suitable cells can be an obstruction to gene transfer in higher eukaryotes. Another major obstacle in higher eukaryotes is presented by the immune system, which identifies and destroys foreign materials entering the body. The immune system is a major force suppressing replication of retroviruses. Retroviruses, in turn, have devised strategies for immune evasion during infection.

Another difference between prokaryotes and eukaryotes is the relative amount of "junk" DNA, primarily sequences from integrated transposons and viruses. In prokaryotes the genomes are typically 85–95% gene coding regions. The genome of the eukaryotic yeast *Saccharomyces cerevisiae* is 70% gene coding regions. In contrast, gene density is much lower in human chromosomes, which are only 1.5% coding region. Much of the noncoding DNA in humans appears to be fossil retroviral and mobile element sequences. In fact, it is now clear that at least 8% of human DNA is composed of retroviral fossils. Much of our genetic heritage was contributed by genomic "parasites," a point that Darwin could never have anticipated.

This chapter begins with a description of retroviruses and the retro-viral life cycle, then treats the role of retroviruses in gene transfer. In the next chapter, we consider a new manifestation of retroviral gene transfer, the acquired immune deficiency syndrome (AIDS) caused by infection with human immunodeficiency virus (HIV). Chapters 8 and 9 cover other kinds of eukaryotic mobile elements, and Chapter 10 treats eukaryotic genomes, emphasizing the role of lateral DNA transfer in their construction.

The Provirus Hypothesis (3, 17, 19, 40, 47)

Agents that were later understood to be retroviruses first came to the attention of scientists at the beginning of the 20th century as transferable agents that could cause cancer. Vilhelm Ellerman and Oluf Bang (1908) and, separately, Peyton Rous (1911) identified filterable agents isolated from cancerous birds that when inoculated into new animals caused disease as in the previous host. The filters were fine enough that cellular material could not pass through, implicating a virus in transmission, and thus demonstrating cell-free transmission of tumors in birds. Progress in understanding the mechanism of tumorigenesis in these systems was slow with the limited techniques available. Today tumorigenesis by these viruses, known to contemporary research as avian leukosis virus (ALV) and Rous sarcoma virus (RSV), is understood in detail. In his late old age, Peyton Rous was awarded the Nobel Prize (1966) for "his discovery of tumor-inducing viruses."

The elucidation of retroviral replication provides an example of creative science at its best. By 1961 it was clear that particles of RSV contained RNA, but this appeared to conflict with data that emerged later in the decade. Surprisingly, DNA synthesis inhibitors could block multiplication and transformation by RSV, an observation at odds with the expectation that RNA viruses would replicate through RNA intermediates only. In addition, actinomycin D, which inhibits the cellular DNA-dependent RNA polymerase, when added to cells producing virions, quickly interrupted virus production. To explain these results, Howard Temin proposed as early as 1964 that the infecting RNA genomes were converted to DNA, and the DNA copy was then integrated into the genome. The integrated form was named the "provirus." The integrated copy was then hypothesized to serve as a transcription template for cellular RNA polymerase, which produced new viral RNAs.

Temin's provirus hypothesis was initially unpopular. After all, nothing like the hypothetical reverse transcriptase enzyme had ever been encountered. Stunning support for the model came in 1970 when Temin and Mizutani and, independently, David Baltimore, isolated reverse transcriptase. They reasoned that reverse transcriptase had to be brought into cells as a part of the viral particle, as was known to be the case for some RNA polymerase enzymes of RNA viruses. Both groups therefore examined viral particles for reverse transcriptase and succeeded in identifying the activity. Temin, Baltimore, and Renato Dulbecco (who separately developed methods for animal cell culture) were awarded the Nobel Prize for their discoveries.

These results showed that the integrated copy of viral DNA maintained the genetic stability of a retroviral infection in cells. As such, they provided crucial background for understanding the origin of cancer, the evolution of vertebrate genomes, and the emergence of AIDS. We consider the role of lateral transfer by retroviruses in each of these areas below, but first we review the nature of retroviruses in more detail. Diverse further types of genomic parasites have also been found to replicate via a reverse transcription step. These elements are the topic of Chapter 8.

In a paper describing his results, Rous said "...the first tendency will be to regard the self-perpetuating agent in this sarcoma of the fowl as a minute parasitic organism."

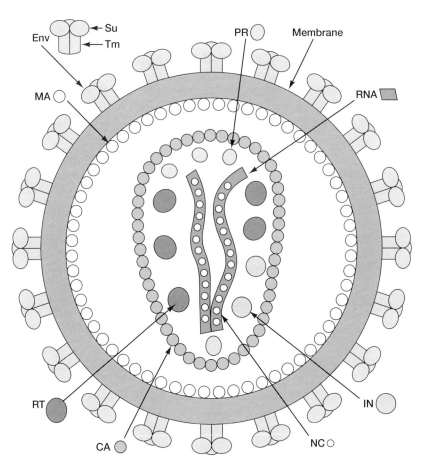

Retroviruses in mammals in the form of virus-induced mammary tumors in mice were discovered in 1936 by John Bittner.

FIGURE 6.1. A retroviral particle. The lipid bilayer is shown as a gray ring. Outside the particle are trimers of envelope protein, indicating the Su and Tm components. Just underneath the lipid bilayer lies the matrix protein (*white spheres*). The capsid protein (*green*) forms the viral capsid structure. In the case of HIV, the capsid is bullet-shaped (capsid morphology differs among retroviruses). Inside the capsid are two copies of the viral RNA (*dark green ribbons*) bound to the nucleocapsid protein (*small white spheres*). Also contained in the capsid are the viral enzymes (reverse transcriptase, *dark gray*; protease, *white*; and integrase, *light green*).

Morphological differences between cells transformed by particular RSV variants indicated to Temin that the viral genome continued to affect the transformed cell.

RETROVIRAL STRUCTURE (15, 23, 33)

Retroviruses are made up of spherical particles roughly 80–120 nm in diameter (Fig. 6.1). The viral particle is enclosed in a lipid bilayer membrane derived by budding from the infected host cell. Studding the viral membrane are viral envelope proteins (Env), which dock with receptors present on suitable target cells and mediate entry. Underlying the viral envelope is a layer of protein consisting of the viral matrix proteins (MA). Inside the envelope is a proteinaceous capsid. The major component of the capsid shell is the viral capsid (CA) protein. Inside the capsid shell is the viral genome, which consists of two complete copies of the viral genes present as single-stranded RNA. The RNA sequence of the retroviral genome matches those of the mRNA sequences. In virology such strands are called "+" strands. RNA or DNA strands complementary to the mRNA are called "−" strands. The two RNAs are bound to each other and to the virus-encoded nucleocapsid protein (NC).

Retroviral RNAs are about 7–12 kb in length.

Also contained within the core are three virus-encoded enzymes important for viral replication. The reverse transcriptase enzyme (RT) is responsible for synthesizing a cDNA copy of the viral RNA genome. The name "retroviruses" was coined in honor of this unusual reaction, since RT reverses the usual DNA-to-RNA flow described by the Central Dogma (although Crick actually left open the nature of trans-actions among nucleic acids in the original articulation, emphasizing the flow of information from nucleic acid to proteins). Note that the "c" in cDNA stands for "copy" or "complementary," reflecting the action of reverse transcriptase. The core also contains the integrase (IN) protein, which is important for integration of the viral cDNA into host cell DNA and the protease enzyme (PR), which cleaves the initially synthesized polymer of viral proteins into the individual proteins functional in the viral life cycle.

The retroviral capsids differ in structure among the different viral groups. Some of the simple retroviruses contain spherical capsids, whereas others contain more oblate capsid structures. Some capsids reside in the center of the viral particle, whereas others are usually acentric. These morphological differences provided the basis for an early scheme for classifying retroviruses into type A, B, C, or D. Some particle types are shown in Figure 6.2. Later the lentiviruses, which include HIV, were recognized as a distinct class characterized by bullet-shaped cores.

For HIV, recent work has provided a detailed picture of the viral capsid structure (Fig. 6.3). Wes Sundquist and coworkers have analyzed artificially assembled derivatives of the viral capsid, allowing them to deduce that capsids are built up of a hexameric repeating unit. It turns out that there are only a limited number of ways to assemble three-dimensional structures from hexameric repeat units, which involve introducing an occasional pentamer into the hexameric lattice to induce curvature. Analysis of laboratory-assembled capsids, incorporating the mathematical constraints worked out for networks composed of hexameric units, has provided a potentially complete model for the structure of the HIV capsid. It is presently unclear whether the capsid structures of other retroviruses will be constructed similarly, but the compelling HIV model, with slight modifications in the placement of the pentameric defects, can explain all of the capsid structures observed in retroviruses.

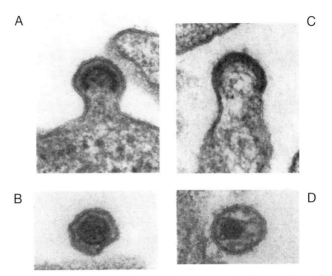

FIGURE 6.2. Electron micrographs of some retroviral particles. (*A*) Murine leukemia virus budding. (*B*) Mature murine leukemia virus. (C) Lentivirus budding. (*D*) Mature lentivirus. (Reprinted, with permission, from Coffin et al. 1997.[15])

RETROVIRAL GENOMES (15)

Simple Retroviruses

Genetic maps of some well-studied retroviruses are shown in Figure 6.4. Each is shown in the integrated or proviral state. The so-called "simple" retroviruses are represented by Moloney murine leukemia virus (Mo-MLV). For all retroviruses, each end of the viral DNA is made up of a directly repeated sequence, the long terminal repeat or LTR. These repeated sequences are important for reverse transcription, integration, and transcription.

Progressing rightward from the left LTR, we next find a sequence, called Psi, that is important in the viral RNA for packaging RNAs into assembling virions. RNAs containing Psi are selectively packaged into viral particles, even in the presence of a great excess of cellular RNAs. Viral genomes lacking Psi are encapsidated at very low levels.

Rightward of Psi begin the coding regions for the retroviral proteins. The viral MA, CA, and NC are encoded within the *gag* gene. *gag* stands for *g*roup *a*ssociated *a*ntigen, referring to early studies that grouped retroviruses by their reactivity with antibodies against proteins encoded in this region. The Gag proteins are synthesized as a single polypeptide, which is then cleaved by the action of the viral protease to yield the independent MA, CA, and NC proteins found in mature virions. Also encoded in *gag* are other proteins that generally are not conserved among retroviruses. For example, some Mo-MLVs encode a protein called p12, for its molecular weight of 12 kD. The p12

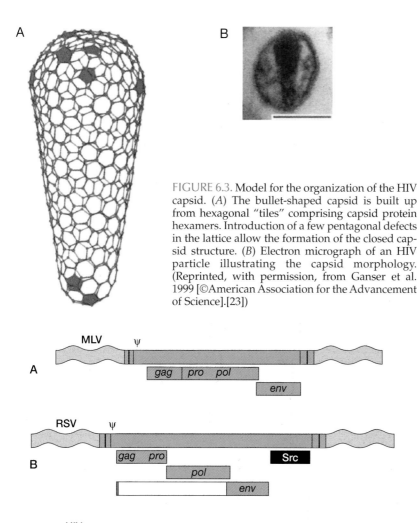

FIGURE 6.3. Model for the organization of the HIV capsid. (*A*) The bullet-shaped capsid is built up from hexagonal "tiles" comprising capsid protein hexamers. Introduction of a few pentagonal defects in the lattice allow the formation of the closed capsid structure. (*B*) Electron micrograph of an HIV particle illustrating the capsid morphology. (Reprinted, with permission, from Ganser et al. 1999 [©American Association for the Advancement of Science].[23])

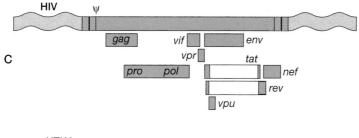

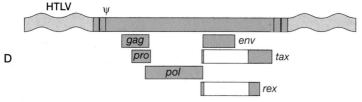

FIGURE 6.4. Some retroviral genomes. Flanking DNA is shown in gray, viral coding regions in green. The R region in the LTR is shown black. The oncogene of RSV, *src*, is also shown black. Ψ indicates the packaging site in the RNA. (Adapted from Coffin et al. 1997.[15])

protein is important for viral replication, but its detailed function is unclear.

Just downstream of the *gag* coding region is the *pol* gene, which encodes the retroviral enzymes PR, RT, and IN. Most retroviruses have evolved a mechanism to permit Gag proteins to accumulate in greater abundance than the Pol proteins. In Mo-MLV, there is a stop codon between *gag* and *pol*. About 1 time in 20, translation termination at this stop codon is suppressed, allowing translation of the *pol* genes as a Gag-Pol polyprotein precursor. Thus, translation generates a 20 to 1 ratio of Gag to Gag-Pol proteins. The PR, RT, and IN enzymes are later released from Gag-Pol by the action of protease.

Continuing rightward, we reach the *env* gene, which comprises two substituents, the transmembrane (TM) and surface (SU) proteins. *env* is synthesized from its own RNA, which differs from the genomic-length RNA by splicing out of the intervening sequences. The TM protein resides in the viral envelope. The SU protein binds to TM and is responsible for binding the cellular receptor protein.

A variation on the simple retrovirus theme is seen with Rous sarcoma virus (RSV). Like Mo-MLV, RSV contains *gag, pol*, and *env* genes. The two viruses differ in the mechanism of formation of the Gag-Pol polyprotein. In the RSV case, the *gag* and *pol* coding regions are translated in different reading frames. A special sequence, called a frameshift region, is present in a short region of overlap between the end of *gag* and the beginning of *pol*. About 1 time in 20, ribosomes translating *gag* change reading frames. Instead of terminating translation at the normal *gag* stop codon, frameshifted ribosomes are able to continue synthesis through the *pol* region. The net effect is the same as stop codon suppression in Mo-MLV, resulting in synthesis of about 1/20 the amount of Gag-Pol protein as Gag.

RSV encodes a unique protein rightward of its *env* gene, named *src*. As described in a later section, RSV is capable of causing infected cells to become transformed. Such cells multiply in an unregulated fashion, allowing them to cause tumors in suitable host animals. The *src* gene is specifically required for transformation. Src, like Env, is synthesized from a unique RNA generated from the full-length form by RNA splicing.

Complex Retroviruses

HIV-1 and a variety of other retroviruses are larger and contain more genes than simple retroviruses such as Mo-MLV, and are thus called complex retroviruses. The viruses with more complex genomes often also display more complex life-styles as well. The genomes of two complex retroviruses, HIV-1 and HTLV-1, are shown in Figure 6.4. HIV contains six genes not found in simple retroviruses. In addition to *gag, pol*, and *env*, HIV encodes *vif, vpr, vpu, tat, rev*, and *nef*. Each of these is produced from a spliced RNA form not present in simple retroviruses. The roles of these genes are discussed in detail in the next chapter.

HTLV-1, the human T-cell leukemia virus, is another complex retrovirus but with a different complement of extra genes. HTLV-1 was the first human retrovirus discovered, initially coming to attention for its association with clustered cases of T-cell leukemia, occurring in southern Japan and parts of the Caribbean. Like all retroviruses, HTLV-1 encodes *gag, pol*, and *env*, but, in addition, two other genes, *tax* and *rex*. Like *tat*, *tax* acts to promote viral transcription, but does so by a quite different mechanism. Tax binds to cellular transcription factors and thereby activates transcription of the viral genome. Tax may also act by changing transcription patterns in the cellular genome so as to create an environment more favorable for viral replication. Rex acts by binding to an internal RNA site within the viral messages and promoting their export into the cytoplasm of the infected cell, thereby optimizing and coordinating the translation of the viral messages (see Table 6.1).

Many other retroviruses are known, some of which have further unique genes suited to their hosts and modes of replication. The presence or absence of distinctive genes, and the sequences of the conserved regions, form the basis for a retroviral phylogeny.

RETROVIRAL PHYLOGENY

The underlying premise of viral phylogeny differs from that of the cellular organisms. Bacteria, Archaea, and Eukarya are believed to be descended from a common ancestor, and the finer subdivisions in phy-

TABLE 6.1. Some Retroviral Genes

Gene	Properties/function of protein
Common to all retroviruses	
gag	precursor to internal structural proteins
pro	PR enzyme
pol	precursor to RT and IN enzymes
env	precursor to envelope glycoproteins
Some accessory genes	
HTLV/BLV (e.g., HTLV-1)	
tax	transcription activator
rex	splicing/RNA transport regulator
Lentiviruses (e.g., HIV-1)	
tat	activates transcription
rev	regulates splicing/RNA transport
vif	affects infectivity of viral particles
vpr and/or *vpx*	is present in virion; has nuclear localization signal; facilitates infectivity in quiescent cells
nef	triggers CD4 endocytosis, alters signal transduction in T cells; enhances virion infectivity
vpu	integral membrane protein; triggers CD4 degradation; enhances virion release

(Reprinted, with permission, from Coffin et al. 1997.[15])

logenetic schemes reflect more recent branching on the tree of life (although this is greatly complicated by lateral DNA transfer). For many groups of viruses, however, there is no reason to think that they ever had common ancestors with other viral groups. The DNA viruses such as phage or herpesviruses, for example, may never have had a common viral ancestor with the RNA viruses or retroviruses. Different viruses could have emerged as parasites on replicating systems in altogether independent evolutionary events. Other mobile DNA elements, such as transposons or conjugative plasmids, may similarly have had multiple independent origins. Viral phylogeny carries no implication of relationships between different viral families.

The family of vertebrate retroviruses (Retroviridae) can be divided into seven genera (Fig. 6.5) based on DNA sequence relationships and presence of genes in addition to the minimal *gag, pol,* and *env.* The α retroviruses encompass a large group of bird viruses, including RSV, called the avian sarcoma-leukosis viruses. The β retroviruses include the Mason-Pfizer monkey virus. The β and γ retroviruses display a distinctive assembly pathway in which capsid structures are first formed in the cytoplasm, rather than at the cytoplasmic membrane. The γ retroviruses include mouse mammary tumor virus and human endogenous retrovirus-K (HERV-K). As described later in this chapter, the human germ line contains many copies of inactive retroviral genomes, collectively called HERV elements. The δ retroviruses contain the human and bovine leukemia viruses, and ε retroviruses include the walleye dermal sarcoma virus, an unusual retrovirus that causes benign tumors of game fish.

The lentiviruses include HIV-1, HIV-2, and the related simian immunodeficiency viruses (SIVs). This group was named "lenti" for slow, because the first member discovered, Visna-Maedi virus of sheep, caused a slowly progressing neurological disease. It is now clear, however, that members of this class can cause rapidly developing diseases as well (HIV and the lentiviruses are the topic of Chapter 7). The spumaviruses are characterized by human foamy virus (HFV), so named for the foamy appearance of infected cells in culture. This group has a number of distinctive features, such as the presence of several unique genes and a Gag protein that functions without cleavage into substituent proteins.

RETROVIRAL CELL BINDING AND ENTRY (1, 5, 15)

We now examine the retroviral life cycle, for many steps using Mo-MLV as an example (Fig. 6.6). Different viruses show important differences in their replication patterns, but these differences will be introduced only as necessary for the discussion of lateral DNA transfer.

Mo-MLV initiates infection by binding to a susceptible target cell. In this reaction, the viral envelope protein docks with a host protein

Virus

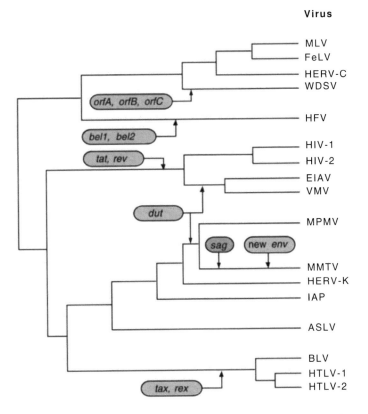

FIGURE 6.5. Retroviral phylogeny. The gene names in gray indicate where new genes were acquired in different retroviral lineages. Virus names are as follows: MLV, murine leukemia virus; FeLV, feline leukemia virus; HERV-C, human endogenous retrovirus C; WDSV, walleye dermal sarcoma virus; HFV, human foamy virus; HIV-1, human immunodeficiency virus type-1; HIV-2, human immunodeficiency virus type-2; EIAV, equine infectious anemia virus; VMV, Visna-Maedi virus; MPMV, Mason-Pfizer monkey virus; MMTV, mouse mammary tumor virus; HERV-K, human endogenous retrovirus K; IAP, intracisternal A particle; ASLV, avian sarcoma-leukosis virus; BLV, bovine leukemia virus; HTLV-1, human T-cell leukemia virus type 1; HTLV-2, human T-cell leukemia virus type 2. (Modified, with permission, from Coffin et al. 1997.[15])

exposed on the cell surface. Many different cell-surface proteins act as receptors for different retroviruses. For Mo-MLV, the receptor is a transporter protein that normally allows uptake of cationic amino acids, named CAT-1. The Mo-MLV SU protein initially docks with CAT-1, and then, in an ill-defined process, bends out of the way. Accompanying this conformational change, the TM protein extends a "fusion peptide," which inserts into the target cell membrane. Further conformational changes bring the viral and cellular membranes together and initiate fusion of the two. Probably the interaction of many individual envelope and receptor protein monomers is required to complete fusion. The fusion event can take place by either of two

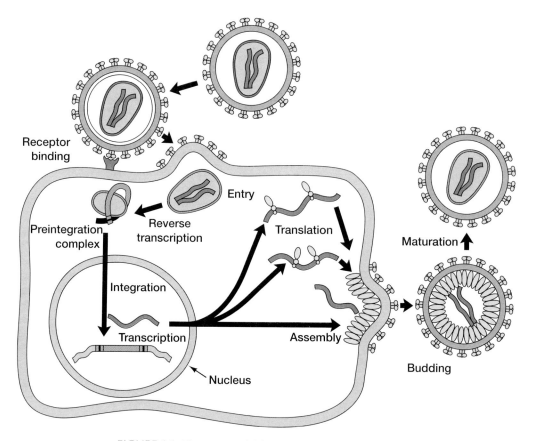

FIGURE 6.6. The retroviral life cycle. A retroviral particle (*top*) initially binds to a cell-surface receptor, leading to fusion and entry of the viral core into the cell cyto-plasm. Fusion can take place either at the membrane or after endocytosis of the bound viral particle, depending on the type of retrovirus. Following entry, the viral RNA is reverse-transcribed, yielding a double-stranded DNA copy of the viral genome. The resulting preintegration complex then migrates to the nucleus and integrates the viral cDNA into a host chromosome. The resulting provirus can then be inherited during cell division like a normal cellular gene. The provirus also serves as a transcriptional template for synthesis of viral RNA. Viral RNAs serve as mRNAs for protein synthesis, and the full-length viral RNA also can become pack-aged and serves as the genome for the next generation of virions. Assembly takes place at the membrane (in the case of type C retroviruses and lentiviruses). Budding yields the immature viral particle in which the viral proteins are present as polyprotein precursors. Proteolysis directed by the virus-encoded protease yields the mature virion.

pathways, depending on the virus involved. Fusion can take place at the cytoplasmic membrane, or the virus can first be engulfed in a process called endocytosis, and fusion can take place in an internal membrane-enclosed vesicle called an endosome. For Mo-MLV, fusion is believed to take place inside cells after endocytosis, although this conclusion is controversial.

Many studies have shown that the presence or absence of cell-sur-face receptors is a primary determinant of retroviral cell-type specificity, often called "tropism." Only cells with appropriate surface receptors can be infected. For Mo-MLV, cells of the immune system and a variety of other organs display CAT-1 on their surfaces, and Mo-MLV replicates in many of these. Another prominent example is HIV, which replicates in immune system cells bearing CD4 on the surface and in a second molecule called the "coreceptor." The nature and distribution of these receptors are key determinants of HIV transmission and disease progression, a topic discussed in the next chapter.

RETROVIRAL REVERSE TRANSCRIPTION (2, 17, 30, 42)

Fusion results in the introduction of the viral core into the cytoplasm of the infected cell. The viral core is composed of a shell of capsid protein surrounding the viral RNA and viral enzymes. Following entry, the capsid shell breaks down, before or during the process of reverse transcription, a process called "uncoating." Much of the capsid protein is released from the replication complex at this time, although the details vary with different retroviruses.

The sequences at the ends of the viral RNA differ from those at the ends of the integrated DNA (Fig. 6.7), an observation that puzzled early retrovirologists. The RNA has a repeated structure at each end, named R, and unique sequences at the 5′ end and 3′ end (U5 and U3, respectively). Sandwiched between these sequences are the coding regions for the viral proteins. In the DNA, however, the terminal structure is U3-R-U5–genes–U3-R-U5, yielding an overall direct repeat of U3-R-U5 called the long terminal repeat or LTR.

The cDNA genome including the LTRs is produced from the RNA by the mechanism of reverse transcription. The pathway involves two

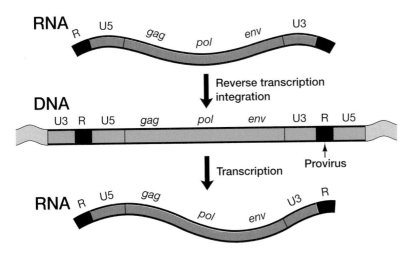

FIGURE 6.7. Information flow during retroviral replication, illustrating the differences in the LTRs between the RNA and DNA forms.

internal priming events and two strand switches (described below). Although somewhat baroque, this pathway solves an important problem for the virus. Polymerases including reverse transcriptase typically require primers for efficient initiation of polymerization. Such primers provide the point on a nucleic acid chain at which further nucleotides are added. But on a linear RNA, how is DNA synthesis to be primed, particularly in a setting where two priming events are required, one for each strand? The diverse mobile elements that use reverse transcription to replicate can be divided into families based on the means by which they solve this problem.

The pathway of reverse transcription is shown in Figure 6.8. The retroviral RNAs bind a cellular tRNA that serves as the primer for minus-strand synthesis. The edge of the tRNA-binding site, called the primer binding site (PBS), marks the edge of the U5 sequence. Reverse transcription initiates at the 3′ end of the tRNA sequence and extends to the 5′ end of the RNA, producing a short sequence complementary to the U5 and R sequences at the 5′ end of the message.

Reverse transcriptase has a second enzymatic activity required for cDNA synthesis. RT contains an RNase H activity, which cleaves the RNA strand of the RNA/DNA hybrid. As the RT activity synthesizes new DNA, the RNase H activity degrades the template RNA, chopping it into short pieces.

The activity of RNase H results in freeing the newly synthesized minus-strand DNA from the RNA template, which in turn allows the minus-strand DNA copy of R to bind to complementary R sequences at the 3′ end of an RNA molecule. This is the first template switch of reverse transcription, which allows minus-strand DNA synthesis to extend all the way across the RNA genome.

Meanwhile, the RNase H activity carries out a special cleavage to form the primer for synthesis of the second DNA strand. An RNA sequence at the edge of U3, called the polypurine tract (PPT), when cleaved yields a long RNA that can serve as a primer for plus-strand cDNA synthesis. The 3′ edge of the PPT marks the internal border of U3. Synthesis primed at the PPT extends across the minus-strand DNA into the PBS sequence. The second template jump then takes place

FIGURE 6.8. Pathway of reverse transcription. (*A*) The viral RNA is packaged with a tRNA molecule bound to the primer-binding site (PBS). After entering an appropriate target cell, reverse transcription initiates (*B*) with the extension of the tRNA 3′ end to the 5′ end of R. The viral genome is identical in sequence to the mRNA; in the jargon of virology, this is the plus strand. The complementary cDNA is the minus strand. The growing cDNA chain is then transferred (*C*) to the 3′ end of the RNA genome and extended the length of the RNA (*D*). The RNA is degraded concomitant with cDNA synthesis. An RNA fragment remains bound at the polypurine tract (PPT), which serves as the primer for plus-strand cDNA synthesis. Plus-strand synthesis extends to the tRNA primer (*E*), then jumps to the other end of the cDNA and pairs using complementary sequences in the PBS (*F*). Extension of both 3′ ends in the cDNA to the ends of the genome yields the complete linear viral cDNA (*G*). (Adapted from Varmus and Brown 1989.[49])

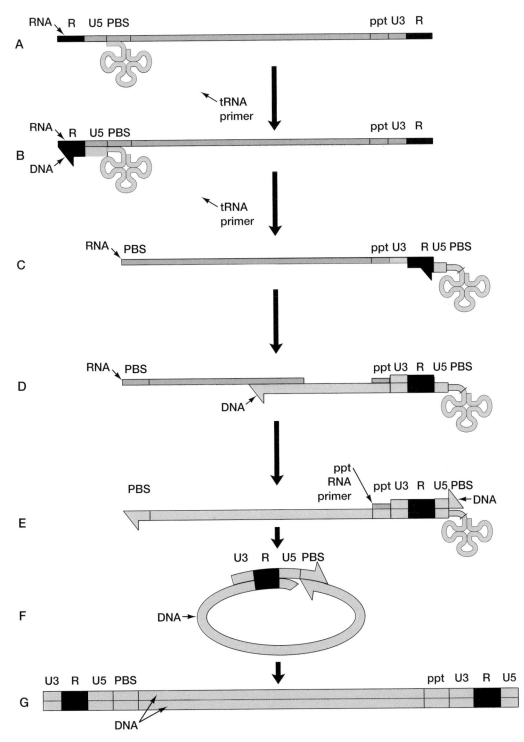

FIGURE 6.8. (*See facing page for legend.*)

using complementary PBS sequences as a bridge. This allows both the plus and minus strands to extend to the ends of their templates. The final product of reverse transcription is a double-stranded linear cDNA copy of the viral RNA genome.

It may seem strange that the RT complex jumps so readily to new templates. What prevents RT from jumping off one template without finding another? The explanation is that reverse transcription does not happen free in solution, but in a particle derived from the viral core. It seems likely that the nucleic acid strands are held in a protein scaffold that guides the strand transfers. The NC protein facilitates the required unpairing and repairing steps ("nucleic acid chaperone" activity). Thus, the reverse transcription complex, like the integration complex described below, is more a multicomponent machine than a simple enzyme acting on an extended nucleic acid substrate.

The structure of HIV-1 RT has been intensively studied because of its importance as a target for antiviral agents (Plate 6). HIV RT is active as a heterodimer. One of the two monomers, called p66 (for its molecular weight of 66 kD), contains the full polypeptides for the RT domain and the independently folded RNase H domain. The other protomer, p55, is identical in the RT domain but lacks the RNase H. RT enzymes of other retroviruses differ in the detailed composition of the mature enzyme. The two HIV RT subunits come together in an unusual asymmetric dimer. Identical copies of a single amino acid chain typically adopt the same three-dimensional folds, but the two RT domains have quite different structures. This appears to permit each of the two monomers to carry out unique functions in the RT complex.

The "Right-handed" Structure of RT

The overall shape of the HIV-1 RT complex resembles a right hand; accordingly, the different domains are named fingers, palm, connection, thumb, and RNase H domains (Plate 6). X-ray structures are available for free RT, RT bound to DNA, and RT bound to small molecule inhibitors. The structure in the presence of DNA reveals that RT "grasps" the DNA like a right hand clutching a mop handle. The chemistry of polymerization is carried out by an active site in the palm domain. Acidic residues in a conserved amino acid sequence YXDD appear to bind two metal atoms, which activate the incoming dNTP and the DNA chain at the point of joining. The right-handed structure of RT resembles that of several other well-studied polymerases, such as prokaryotic polymerases involved in DNA and RNA synthesis.

Template nucleic acid scrolls through the enzyme complex, allowing ordered polymerization to form the cDNA and concomitant degradation of the RNA by RNase H. The single-stranded RNA enters the complex from the top of the hand and translocates toward the wrist. Polymerization at the active site adds new DNA monomers directed by the RNA template sequence. The RNA/DNA complex then proceeds along the complex to the RNase H domain, where the RNA is cleaved, removing the strand complementary to the newly synthesized DNA. Polymerization along DNA to form the plus strand proceeds similarly, but the double-stranded DNA product is not a substrate for RNase H and so remains intact. The reaction cycle may well involve substantial changes in conformation of RT, since the thumb domain has been seen to adopt different conformations relative to the rest of the enzyme in some structures.

RT is quite error-prone, a property of great significance for retroviral replication. For HIV, the rate is thought to be about one misincorporation per 10,000 bases polymerized. Because the HIV genome is, in fact, about 10,000 base pairs, this means that on average every genome will have one mutation. This property allows retroviruses to evolve extremely rapidly in the face of a new selective pressure, such as a host immune response or treatment with an antiretroviral drug. We return to this point in the next chapter in our discussion of HIV therapy.

Jumping RT

RT is also able to jump between templates quite efficiently. This may be important as a damage repair reaction. If one of the two RNA molecules in a virion is damaged, RT can jump to the other and complete synthesis of the viral cDNA. This propensity to jump is also important for lateral DNA transfer. Retroviruses can pick up cellular genes and transport them between cells, a process that requires RT to jump between templates to form the transducing viral derivative.

RETROVIRAL INTEGRATION (9–12, 14, 18, 20, 24, 29, 53, 54)

Following reverse transcription, the viral cDNA is integrated into a chromosome of the host. At the termini of U3 and U5 lie short sequences, typically the terminal 10–15 bp, that are required for efficient integration. These sequences act as binding sites for the virus-encoded IN (integrase) protein, which carries out the initial DNA breaking and joining reactions mediating integration. IN is synthesized as part of the Pol polyprotein precursor and liberated from the carboxyl terminus of RT by PR.

The retroviral IN proteins differ in sequence and function from the phage λ integrase proteins, the similarity in names notwithstanding. Both carry out the integration of viral DNA into chromosomal target DNA, but the two classes of enzymes do so by different mechanisms. The retroviral integrases act chemically by a direct mechanism (single-step transesterification) rather than through a covalent protein–DNA intermediate, as with the λ integrase proteins. The viral DNA precursor is linear in the retrovirus case, but circular in the λ case. The retroviral integrase proteins are actually similar in sequence and function to many of the transposases of bacterial (Chapter 3) and eukaryotic DNA transposons (Chapter 9).

IN protein assembles with the viral cDNA concomitant with the completion of reverse transcription, yielding a large protein–DNA complex called the preintegration complex (PIC). Also present in the complex are the viral MA, RT, and potentially other proteins. A host protein, HMG IY, is present in PICs and is important for integration (at least in model reactions in vitro). Other proteins may also be associated with PICs and important for their function. The PIC can be thought of as a remodeled derivative of the viral core, differing by the conversion of the viral RNA to cDNA, release of much of the viral capsid protein, and acquisition of cellular proteins.

The DNA breaking and joining reactions mediating integration are shown in Figure 6.9. Prior to integration, the viral cDNA is cleaved by integrase at each end to remove nucleotides (usually two) from one DNA strand (Fig. 6.9A). It may seem odd that retroviruses would synthesize more DNA than necessary and then cut some of it off. However, this cleavage reaction is probably important for preparing a homogeneous substrate for subsequent reaction steps. RT is actually a rather sloppy enzyme. One consequence is that the termini of the linear viral cDNA often have heterogeneous extra nucleotides attached by an aberrant activity of RT. The terminal cleavage reaction removes the messy terminal additions and prepares a defined substrate for the next reaction step. The terminal cleavage reaction may also serve other functions as well, such as helping to form a structure that stabilizes the PIC.

Following terminal cleavage, integrase attaches the recessed ends to points in a target DNA. This reaction step is similar to the strand transfer reactions mediating transposition of elements such as Tn5 and Mu. The recessed ends in the viral DNA are joined to protruding ends in the target DNA, yielding the integration intermediate shown in Figure 6.9B. In this DNA form, the viral and target DNAs are connected, but on one DNA strand only (Fig. 6.9C). Melting of the target DNA between the points of joining yields gaps at each junction between viral and host DNA and protruding single-stranded ends on the viral side.

It is estimated that an average cell must in fact repair 10,000 sites of DNA damage (depurination) per day.

The completion of integration probably involves host DNA repair enzymes. Several repair pathways act by cutting out a damaged base and surrounding DNA, thereby generating a DNA gap. These gaps are then filled in by host polymerases, nucleases, and ligases. Cocktails of these enzymes can repair gaps in integration intermediates in vitro and probably do so in vivo as well.

Retroviral integration complexes provide one of our most detailed views of mobile DNA machinery. The organization of the RSV integrase enzyme is shown in Figure 6.10. The protein is composed of three domains. The amino-terminal domain contains a zinc-binding module. The central catalytic domain contains a sequence motif composed of three acidic amino acids (D, D-35-E). These amino acid residues come together in space to form a binding site for two metal atoms, which probably direct the chemical steps of DNA cleavage and joining. Similar metal-atom-binding sites mediate chemistry by many polymerases, RNase H, and transposase enzymes. The overall protein fold seen in the integrase catalytic domain is seen widely in enzymes that act on DNA and RNA, emphasizing its evolutionarily ancient origins. The carboxy-terminal domain comprises a barrel-like structure (5-strand β barrel, in the structural biology jargon). This domain is rich in basic amino acids, explaining the observation that this domain contributes to tight binding of the negatively charged DNA substrate.

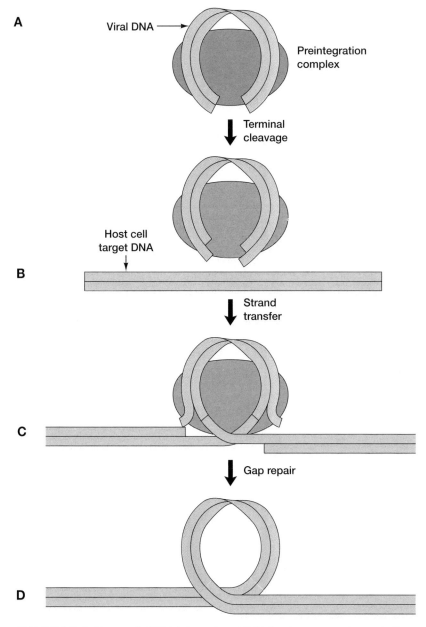

FIGURE 6.9. Pathway of cDNA integration. (*A*) Following reverse transcription, the viral cDNA (*green*) remains bound with viral and cellular proteins (*dark gray*) that direct integration. (*B*) Prior to integration, nucleotides are removed from each 3′ DNA end (typically two), probably to remove heterogeneous additional bases sometimes added by reverse transcriptase. (*C*) The recessed 3′ ends are then joined to protruding 5′ ends in the cellular target DNA. (*D*) The DNA gaps and two-base overhang at the host viral DNA junction are then repaired, probably by host cell gap repair enzymes, to yield the integrated provirus.

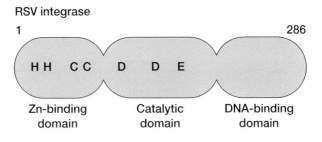

FIGURE 6.10. Schematic diagram of the RSV integrase enzyme.

Integrase is known to act as a multimer, and there are expected to be some 50–100 monomers of integrase per virion particle, so many integrase monomers may be retained in PICs. A detailed model for the organization of the larger integration complexes has been worked out for the HIV case. A fragment of HIV integrase (the catalytic and car-boxy-terminal domains) revealed by X-ray crystallography is shown in Figure 6.11A. The catalytic domain is linked to the C-domain by a long α-helix. RSV integrase differs, forming a dimer with an unusual con-formation (Fig. 6.11B). The catalytic domains form a symmetric dimer, as is seen in the HIV structure, but then the linker to C-domains and the C-domain dimer forms an asymmetric structure. Thus, a single dimer contains symmetric and asymmetric parts.

A speculative model for a full HIV integration complex is shown in Plate 7. In the model, four integrase monomers bind the LTR ends and the target DNA in an interdigitated arrangement. The two dimers fit together into a snug tetramer, with the two points of joining of the viral DNA poised for integration on either side of a DNA major groove in the target DNA. One implication of this model is that some of the inte-grase monomers act catalytically, whereas others play a purely struc-tural role.

What requirements are there for a DNA sequence to serve as an integration target? To a first approximation, any DNA sequence can serve as an integration acceptor site, although careful studies reveal slight preferences for certain primary DNA sequences over others. If a DNA-binding protein is present on a site in an integration target DNA, that site can be disfavored for integration due to simple blocking of access to the DNA. A site can be favored for integration in some cases if a protein binds to target DNA and bends the site. In the structural model in Plate 7, the target DNA fits best into the complex if a bend is introduced. A variety of studies indicate that integration is favored in bent or distorted DNA, potentially because integrase distorts its DNA substrates during the reaction cycle, so that predistorting the DNA favors reactivity.

Several studies have examined sites favored for retroviral integra-tion in vivo. As yet, there are no obvious preferences for any DNA sites

A   **B**

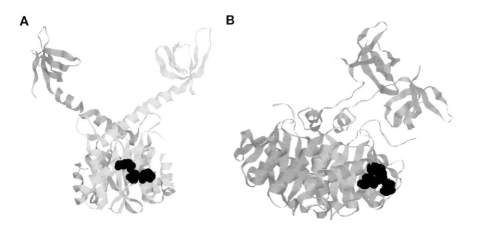

FIGURE 6.11. (*A*) X-ray structure of a fragment of HIV integrase showing the catalytic and DNA-binding domains. Rendered from coordinates generated by Chen
et al. (2000) (14). (*B*) X-ray structure of a fragment of RSV integrase showing the catalytic and DNA-binding domains. Image rendered from coordinates generated by
Yang et al. (2000) (53). Monomers in the dimer are shown in green and gray. The
three catalytic residues are shown in black.

inside cells. Instead, integration sites appear to be scattered around the
chromosome with little regard to the local features. One exception is
the centromeric regions, which are disfavored as integration target
sites. Centromeres are specialized regions of chromosomes important
for segregation of chromosomes during cell division. Centromeric
DNA is wrapped up with specialized proteins, which obstruct access
of enzymes to this region, evidently including integration complexes
as well. Many mobile elements, upon close inspection, have been
found to show target site preferences in vivo. The data set is still small
for retroviral insertion sites, leaving open the possibility that hotspots
for integration may yet be identified in larger-scale studies.

RETROVIRAL TRANSCRIPTION (15)

The process of integration results in the stable incorporation of viral
cDNA in the host chromosome. If the infected cell divides, the proviral sequences are duplicated and transmitted to daughter cells as with
normal cellular genes. Integrated proviruses encode a complete set of
the viral genes, so that transcription of the proviral DNA yields
genomes for new virions and mRNAs for use in protein synthesis.

The full-length proviral transcript serves a dual role, acting both as
the genome for the next generation of virions and as the mRNA for
gag-pol. A second RNA encoding *env* is produced by splicing of the
genomic RNA to remove the *gag* and *pol* genes. For many retroviruses,
additional proteins are produced from other spliced RNAs.

The signals directing RNA synthesis specify that the genomic RNA consists of R-U5 at the right end and U3-R on the left end. The point of initiation of transcription in the left LTR defines the boundary between U3 and R. The location of the 3′ end of the RNA transcript in the right LTR defines the boundary between R and U5. Thus, the terminal RNA structure is regenerated from the LTR DNA.

Signals controlling the rate of transcription initiation lie predominantly in the U3 region in the left LTR, just upstream of the start site of transcription. These regions are rich in binding sites for cellular proteins that promote transcription initiation (Fig. 6.12). The nature of the transcriptional signals in the LTR is another important determinant of viral tropism. In some cases, cells can support entry and provirus formation, but transcription is too inefficient to permit viral spread, hence limiting infection in such cells. The execution of retroviral transcription in most respects mimics that of cellular genes, although there are a few important retrovirus-encoded regulators. One important example is the Tat protein of HIV discussed in Chapter 7.

RETROVIRAL TRANSLATION, ASSEMBLY, AND BUDDING (46, 51)

The viral RNAs serve as templates for translation of the viral proteins. The full-length RNA serves a double role, acting both as the mRNA for Gag-Pol protein synthesis and as the genomic RNA for assembly into particles. Translation of the *gag* RNA begins at an AUG initiator codon at the left edge of the *gag* gene and proceeds rightward. Typically, translation yields the Gag polyprotein. For the Mo-MLV case, about 1 time in 20 the stop codon at the end of *gag* is read through, yielding the Gag-Pol fusion protein. The Gag and Gag-Pol proteins accumulate at the inner leaflet of the cytoplasmic membrane, facilitated in part by the attachment of a fatty myristate group to the amino terminus of Gag, which promotes binding to membranes.

Env protein synthesis proceeds differently. The Env proteins are inserted into a subcellular compartment, the endoplasmic reticulum, concomitant with translation. There the Env protein undergoes extensive modification, including the attachment of sugar chains to the protein surface. The Env protein then proceeds through an intracellular assembly line, passing through the endoplasmic reticulum and the Golgi apparatus. Ultimately, Golgi vesicles fuse with the cytoplasmic membrane of the cell, delivering Env protein to the cell surface. The mature Env protein, proteolytically cleaved to form the mature SU and TM, is found on the external side of the membrane, and so is disposed to assemble on the outside of viral particles during budding. A portion of the protein spans the membrane, however, and a short protein "tail" extends into the cell cytoplasm.

Assembly of most retroviruses takes place by coalescence of Gag and Gag-Pol proteins at the membrane, which bind together along

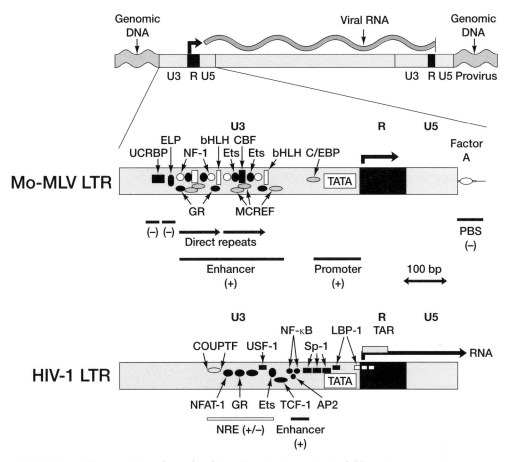

FIGURE 6.12. Transcription factor-binding sites in two retroviral U3 regions. Retroviral transcription initiates at the border between U3 and R in the upstream LTR and terminates at the border between R and U5 in the downstream LTR. The U3 region is rich in binding sites for cellular transcription factors that promote efficient transcription of the proviral genome. Some of the known binding sites in the U3 regions of Mo-MLV and HIV are shown. (Portion redrawn, with permission, from Coffin et al. 1997.[15])

their lengths (Fig. 6.13). The Gag proteins are somewhat wedge-shaped, so the accumulation of multiple Gag monomers deforms the membrane, beginning the process of viral budding. Env proteins on the other side of the membrane also become incorporated into the budding particle, potentially facilitated by contacts between the inner tail of the Env protein and the membrane-proximal part of Gag.

Meanwhile, genomic-length RNAs also accumulate at the site of viral assembly on the inside of the cytoplasmic membrane. There the NC component of Gag binds the Psi site in the RNA, facilitating RNA packaging. Psi is not present in the *env* mRNA, since splicing of the *env* message removes Psi, thereby ensuring that only the full-length genomic RNAs become packaged.

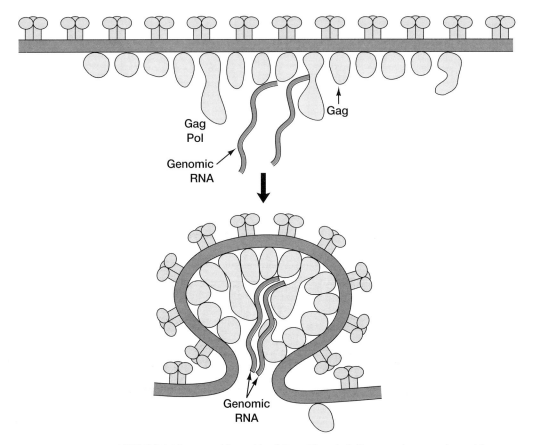

FIGURE 6.13. Assembly and budding. The viral Gag proteins associate with membranes by interactions with the amino-terminal MA domain. Env proteins are separately routed through the endoplasmic reticulum and Golgi apparatus to the cell surface. RNA is bound by the nascent particle, primarily by the NC domain of Gag. Assembly of Gag monomers deforms the membrane, allowing the virus to bud out from the membrane.

Two copies of the viral genomic RNA are packaged per virion. This is achieved by dimerization of the genomic RNA, followed by packaging of the dimeric form. The viral NC protein, acting in the context of the Gag polyprotein precursor, facilitates formation of the RNA dimer and subsequent binding of Psi. It seems likely that the RNA in the region of the dimer linkage and Psi adopts an intricate folded structure that facilitates packaging and later steps in replication.

The continued coalescence of Gag and Gag-Pol at the membrane ultimately yields a fully closed particle. Final release of the particle requires another function, called the late (L) domain, that facilitates the final pinching off of the budded particle. Intriguingly, late domains have recently been found to bind ubiquitin ligases, enzymes involved in programmed degradation of proteins. How this function facilitates budding is unknown.

The above pathway describes the lentiviruses and the type C retroviruses, but other retrovirus families follow a somewhat different assembly program. The type B viruses, such as mouse mammary tumor virus, and type D viruses, such as Mason-Pfizer monkey virus, assemble core structures first in the cytoplasm. These preassembled cores then move to the cytoplasm and bud out of the cell.

The final step of particle formation, proteolysis, takes place after budding (Fig. 6.14). The viral protease dimerizes in the environment of the budded particle and becomes activated. The protease then cleaves the Gag and Pol polyprotein precursors into the individual proteins active later in the viral life cycle. Proteolysis is also accompanied by a morphological change in viral particles. The immature particle contains a halo of proteins just under the inner leaflet of the particle membrane. Following proteolysis, these proteins condense to form the particle core. Only the mature viral particles are competent for infection, which is emphasized by the observation that HIV protease is a target for effective antiviral drugs (Chapter 7).

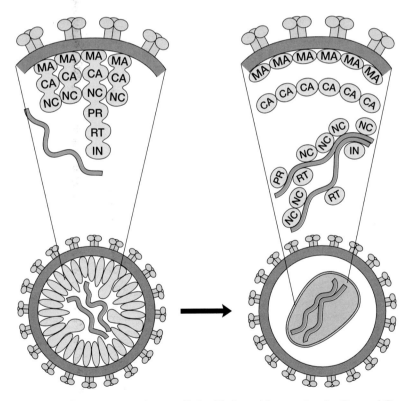

FIGURE 6.14. Maturation. The initially budded particle contains the Gag and Gag-Pol polyprotein precursors (*left*). Following budding, action of the viral protease converts the polyprotein precursors to the individual proteins active in subsequent steps of the viral life cycle (*right*). Maturation is accompanied by formation of the mature viral core.

RETROVIRAL TRANSDUCTION
AND CANCER (16, 26, 39, 40, 43)

The first indications that retroviruses could transfer host genes between cells came from studies of virus-induced cancers. In 1911, Peyton Rous found that he could transfer a factor that promoted cancer from tumors of cancerous birds to naive animals. The agent was found to pass through filters that retained cells, including bacterial cells, implicating a virus. Later genetic studies by Peter Vogt and others revealed that only a single linear segment of the RSV RNA was responsible for malignant transformation. This region was named *src*, for the sarcomas caused by viral infection. Since then, many other examples have emerged of viruses capable of causing cancer in laboratory animals, some of which were eventually found to be retroviruses.

Proto-onco-genes are genes that when activated by mutation can cause cancer—usually only a single gene copy needs to be turned on. This is in contrast to tumor suppressor genes, which normally down-modulate cell growth. If both copies of a tumor suppressor gene are mutant, cells are prone to unregulated growth.

What was *src*? This question was the focus of intense research in the decades following its discovery. *src* clearly captured much of the mechanism of cancer induction on a relatively small piece of RNA. A key breakthrough came thanks to the establishment of methods for analyzing DNA sequences, which allowed the gene for *src* to be studied in detail. Harold Varmus, Michael Bishop, and coworkers discovered that the RSV *src* gene was in fact very closely related to a normal cellular gene. The viral gene was renamed v-*src*, and the normal cellular gene c-*src*. This greatly changed the perspective on the viral origins of cancer, placing the action of viral genes for oncogenesis (shortened to *onc* genes) squarely in normal cellular pathways regulating cell growth.

A second key insight came with the finding that *src* encoded a protein that attached phosphate groups to protein tyrosine residues. This suggested early on that Src protein might be modulating the activity of other proteins by covalent phosphorylation.

A variety of other viruses had been isolated that caused cancer quickly when inoculated into naive animals. These so-called acute transforming viruses were analyzed following the discovery by Varmus and Bishop, and many of them were also found to encode homologs of normal cellular genes. In all cases, however, the v-*onc* genes were not identical to their cellular counterparts, but somewhat modified. Functional studies in which v-*onc* genes alone were used to transform cells quickly revealed that the alterations in v-*onc* genes conferred the ability to transform cells.

The changes that converted c-*src* into v-*src* provide a classic example of oncogene activation. A diagram of the *src* gene and an outline of its regulation are shown in Figure 6.15. The Src protein comprises three protein domains. The first two are named SH3 and SH2 domains, for Src *h*omology domains. The SH3 domain binds to proline-rich protein sequences, and the SH2 domain binds to proteins containing phosphorylated tyrosine residues. The SH2 domain of c-Src binds to a phosphorylated tyrosine residue near the carboxyl terminus of the same polypeptide, inhibiting the kinase activity of the enzyme. The v-Src

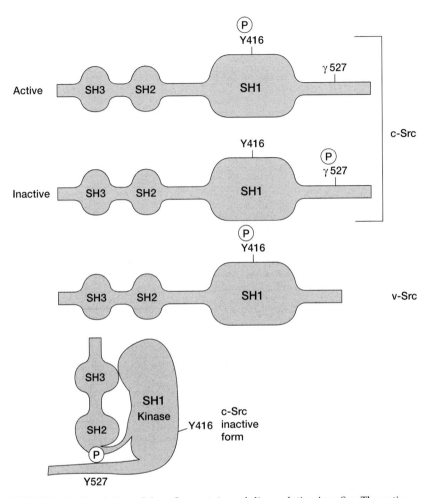

FIGURE 6.15. Regulation of the c-Src protein and disregulation in v-Src. The active form of c-Src (*top*) is phosphorylated at Y416 but not at Y527. Dephosphorylation of Y416 and phosphorylation of Y527 inactivates the enzyme (*middle*). v-Src lacks the Y527 inhibitory site and so is constitutively active. Phosphorylation of Y527 results in binding to the SH2 domain, which inhibits the kinase activity and so effects negative regulation (*bottom*). (Adapted from Coffin et al. 1997.[15])

form is mutated to remove the carboxy-terminal tyrosine residue, thereby releasing the enzyme from intramolecular negative regulation. The activated form of v-Src can thus phosphorylate tyrosine residues much more frequently than c-Src. The mechanism by which tyrosine phosphorylation of target proteins promotes cell growth is not fully clarified, although a number of substrates of potential importance have been identified. This area of "signal transduction" by protein tyrosine kinases is among the most active areas in biomedical research.

The cellular genes captured by acute transforming viruses revealed a rich variety of functions important in growth control and carcinogenesis. Many oncogenes are protein kinases. Some, like Src, are protein tyrosine kinases (e.g., Abl, v-ErbB, v-Ros, v-Sea). Others are kinases that attach phosphate groups to serine or threonine residues, such as v-Raf/Mil. Another large class of oncogenes are modified transcription factors, including v-Myc, v-Fos, and v-Jun. The v-onc protein kinases act at least in part by altering gene expression through cellular signaling pathways, whereas the v-onc transcription factors alter gene expression directly. A variety of other types of molecules implicated in growth control have also been identified as v-*onc* genes. Essentially all of the v-*onc* genes are altered so as to increase their activity, often by mutations removing functions important for negative regulation.

The structures of several representative acute transforming viruses are shown in Figure 6.16. The positioning of the v-*onc* coding sequence within the viral genome differs greatly among the acute transforming viruses. RSV is unusual, in that the *src* gene is inserted downstream of *env*, allowing normal expression of *gag, gag-pol*, and *env*. RSV is thus able to replicate while simultaneously transporting *src* to new cells. The v-*abl* gene in Abelson-MLV, in contrast, replaces most of the viral coding regions. The final protein product is actually a fusion of *abl* sequences to *gag*. The arrangement of the v-*fos* gene in FBJ-MSV is simpler, with translation yielding a relatively simple Fos protein. Other retroviruses contain quite complex gene arrangements. The Harvey MSV contains the oncogene v-*ras* embedded in another sequence, that of a VL30 retroviral remnant, which is itself embedded in the MSV genome.

If most of the acute transforming viruses contain cellular genes in place of the retroviral genes, how do they replicate? The answer is that most of the acute transforming viruses cannot replicate on their own, but multiply only in the presence of helper viruses. Initially, the acute transforming viruses were isolated by obtaining filtrates from tumors, transferring these filtrates to naive animals, then again passaging filtrates from the tumors to new hosts. In this manner, retrovirologists obtained complex mixtures of defective transforming viruses and replication-competent helper viruses. In a sense, the acute transforming viruses are a remarkable set of laboratory artifacts, virus mixtures that arose and proliferated as a result of experimental protocols.

How do pieces of cellular genes become incorporated into retroviral genomes? One reasonable pathway is shown in Figure 6.17. Initially, a retrovirus integrates upstream of a c-*onc* gene. Transcription proceeds from the left LTR, through the proviral DNA, and into the neighboring cellular gene. Retroviral proteins and normal genomes are posited to accumulate as well as the readthrough transcript. Retroviral particles form that contain one copy of the viral RNA and one copy of the readthrough fusion RNA. After infection of a new target cell, an aberrant form of reverse transcription generates a cDNA copy of the oncogene flanked by LTR sequences. The mechanism posits as one of the steps an abnormal transfer of a nascent LTR sequence to the 3′ end of the fusion c-*onc* message. The oncogene cDNA flanked by LTRs can then be integrated normally. A pathway

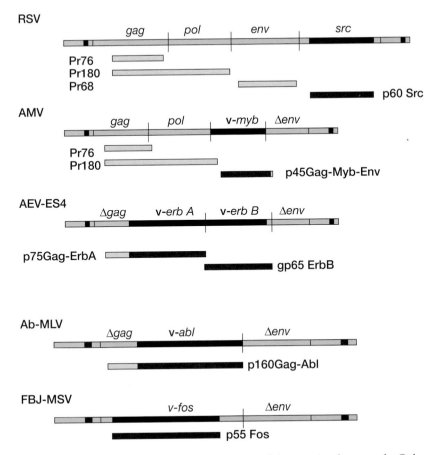

FIGURE 6.16. Some acute transforming viruses and the proteins they encode. Only Rous sarcoma virus (RSV) is capable of independent replication; the others require helper viruses to supply proteins for replication. White rectangles indicate coding regions for viral proteins, black rectangles, oncogenes. Abbreviations: AMV, avian myeloblastosis virus; AEV, avian erythroblastosis virus; Ab-MLV, Abelson murine leukemia virus; FBJ-MSV, Finkel-Biskis-Jenkins mouse sarcoma virus. (Modified and redrawn, with permission, from Coffin et al. 1997.[15])

involving a deletion of cellular DNA as an early step can also explain the observations.

INSERTIONAL ACTIVATION OF ONCOGENES (39)

Retroviral cDNA integration can also turn on oncogenes in another manner, by insertional activation (Fig. 6.19). Integration of the retroviral cDNA upstream of a gene can direct transcription initiating in the LTR outward through the oncogene, resulting in increased production of the gene product. Another means of activation involves integration of viral transcriptional control sequences that activate the oncogene, but without providing a promoter. Such "enhancer insertion" takes

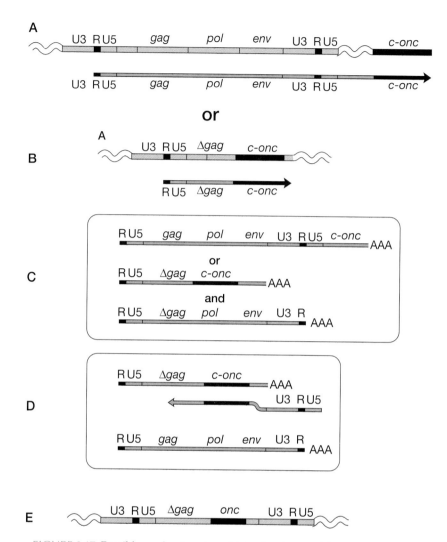

FIGURE 6.17. Possible mechanisms by which retroviruses pick up oncogenes. DNA is shown as thick light green lines, RNA as thin dark green lines. (*A*) Oncogene acquisition begins when a retrovirus integrates upstream of a proto-oncogene. A transcript is then formed containing both a retroviral LTR sequence and oncogene sequences (*B*). Such transcripts can form either by readthrough of the viral RNA or by a deletion in the DNA. These transcripts are packaged with proteins supplied by helper viruses elsewhere in the genome. (*C*) The readthrough transcript, or a deleted transcript together with an intact genome, is then packaged. Recombination during reverse transcription allows the formation of a cDNA containing the oncogene flanked by LTRs (*D* and *E*). In the presence of an intact helper virus, the resulting genome can be packaged and transferred to new cells. (Redrawn, with permission, from Coffin et al. 1997.[15])

FIGURE 6.18. Retroviral vectors. Normal retrovirus infection (*top*) involves packaging the wild-type genome, which directs synthesis of the viral proteins. For retroviral vectors (*bottom*), separate DNAs encode the viral proteins and the genome that becomes packaged. Consequently, the genome transferred to the target cell does not encode the viral proteins and so cannot spread farther. The net effect is to install only the engineered "gene x" into target cells.

Retroviral vectors (13, 34, 36, 38, 41)

Following the recognition that defective acute transforming viruses could be complemented by helper viruses *in trans* came the idea of using retroviruses as vectors to transduce any gene. Crucial to the translation of these research findings into technology was the discovery by Maxine Linial and others that retroviral RNAs contain a sequence, called Psi, that is required for packaging viral RNA into virions. This allows the RNA directing production of viral proteins to be separated from RNA to be packaged (Fig. 6.18). To make a retroviral vector, the *gag-pol* and *env* regions are expressed from an artificial DNA lacking Psi, providing a source of the viral proteins but not a packageable RNA. Another engineered DNA, the vector segment, encodes an RNA containing the correct terminal sequences, Psi, and the gene to be transduced. Expression of the two engineered DNAs in a single cell results in the production of particles containing the vector segment. Infection of such particles into new cells, followed by reverse transcription and integration, installs the vector segment in a chromosome of the new host cell. The infection cannot spread beyond this point because the vector does not encode the viral proteins required for further replication.

Such retroviral vectors are one of the most popular systems for delivering new genes into patients suffering from genetic diseases. Although the field, called gene therapy, is in its infancy, there are already a few successes. In perhaps the clearest case, A. Fischer and coworkers introduced the gene for a subunit of the γc cytokine receptor using a retroviral vector into patients lacking the gene and so suffering from severe combined immunodeficiency. The newly introduced gene complemented the defect, restoring immune function. This was probably an unusually favorable case, since cells containing the restored receptor had a growth advantage over the mutant cells, but it does illustrate the promise of gene therapy.

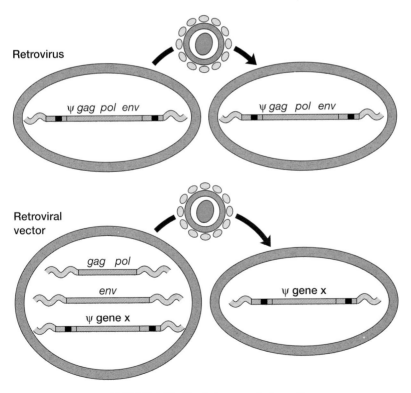

FIGURE 6.18. (*See facing page for legend.*)

place typically when the direction of transcription of the retrovirus is opposite that of the host gene. A third means of activation is the production of readthrough transcripts. In this case, a retrovirus genome inserted typically within a gene produces a readthrough transcript initiating at the left LTR, extending across the viral genome, and reading out into the oncogene. Splicing can then produce a hybrid message. Last, insertion of a retroviral cDNA can separate negative regulatory sequences from an oncogene, thereby activating its expression.

Some oncogenes can be activated by more than one mechanism. The c-*myc* gene, for example, encodes a DNA-binding protein important in growth control. An altered form of the c-*myc* gene is present in MC29, an acute transforming virus of chickens that causes myelocytomas and carcinomas. MC29 requires an ALV helper virus for repli-

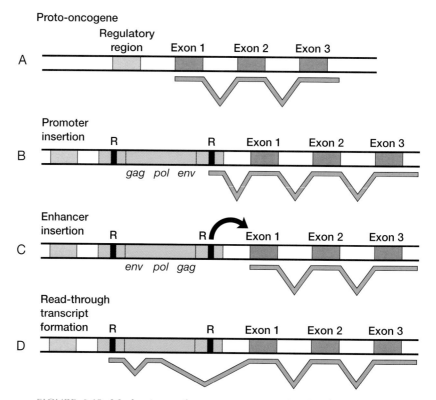

FIGURE 6.19. Mechanisms of proto-oncogene activation by provirus insertion. DNA is shown as thick light green lines, RNA as thin dark green lines. (*A*) Structure of a proto-oncogene locus, showing a transcriptional control region, three exons, and the mRNA synthesized. (*B*) Promoter insertion. Insertion of a provirus in the same orientation as the proto-oncogene can lead to production of a readthrough transcript, potentially altering the rate of transcription or the stability of the message. (*C*) Enhancer insertion. This mechanism typically involves a retrovirus in opposite orientation to the proto-oncogene, which increases the rate of transcription initiation. (*D*) Readthrough transcript formation. Transcripts initiated in the left LTR are fused by RNA splicing to the proto-oncogene exons. This can alter the rate of transcription, the stability of the message, or the nature of the encoded protein. (Adapted from Coffin et al. 1997.[15])

cation. The c-*myc* gene can also be activated by insertion of other ALVs into the c-*myc* gene, thereby causing B-cell lymphomas. Several other examples are known of oncogenes that can be activated by either of the two means.

Stepping back, the alterations in gene expression due to retroviral integration need not be restricted to activation of oncogenes. Transcriptional levels for any gene can in principle be modulated by insertion of retroviral control signals. Retroviruses represent portable gene control regions that by random integration can alter the expression of cellular genes on evolutionary time scales. Some of these changes have been fixed by natural selection and have become permanent features of the vertebrate genome.

GENE INACTIVATION BY RETROVIRAL INTEGRATION (25, 27, 31, 37, 44, 50)

Integration of retroviral cDNA can also inactivate host cell genes, although such events are often not detected because vertebrate genomes are diploid. Special circumstances are needed to detect inactivation of only one of the two alleles. Insertional activation of oncogenes, in contrast, is detected more readily, because activation of one gene copy can suffice for cellular transformation.

Several mutations in mice have nevertheless been traced to gene inactivation by retroviral integration. The genomes of mice and many other organisms contain endogenous retroviral sequences, some of which are competent for replication. These elements provide a continuous source of new cDNA genomes to act as insertional mutagens. Three well-studied examples of gene inactivation by retroviruses involved changes in the mouse coat, in part because such changes are particularly obvious to mouse handlers. One well-studied mutation is the hairless (*hr*) mutation of mice, in which mice homozygous for the *hr* mutation lack hair and also show diverse additional abnormalities. The second involves the *dilute* locus, which also affects coat color. The third is the so-called *non-agouti* locus; changes at this locus can yield coats that are black and tan, non-agouti, or light-bellied agouti. In each case, the mutation arose during inbreeding of mice in captive colonies, explaining how both copies of the mutant gene came to reside in a single animal.

Characterization of the mutant *hr* gene revealed that a Mo-MLV genome was inserted within a noncoding intron, disrupting normal RNA splicing and thereby preventing accumulation of the normal gene product. Similarly for *dilute*, integration of an Mo-MLV genome into an intron inactivated the gene.

In both cases, the insertion could be shown to be the direct cause of gene inactivation. Revertant animals were identified arising spontaneously in mouse colonies. Sequencing of the reverted alleles revealed that the retroviral coding sequences were gone, and instead a single LTR remained in its place. Evidently homologous recombination occurred between the LTRs, thereby removing the Mo-MLV genome and restoring activity to the mutant gene. Other cases of LTR–LTR recombination removing retroviral genomes and leaving solo LTRs have also been found. This appears to be quite a common fate for integrated retroviruses. The genomes of many organisms contain large numbers of solo LTR elements, almost certainly derived from integration followed by recombination between LTRs. As described in future chapters, this conclusion even holds for genomes as distant from vertebrates as plants and yeast.

Returning to coat-color mutants induced by retroviruses, the genetics of the *non-agouti* locus were more complicated than *dilute* or *hairless*. A VL30 element, a defective retroviral element, was inserted into *non-agouti*. The VL30 was unusual in containing a direct repeat of a host cell chromosomal sequence as well as the usual element sequences. The insertion caused a change from wild-type coat color to non-agouti. Recombination between the LTR elements restored only partial gene function, yielding a light-bellied agouti coat color. Recombination between the repeated fragments of host cell DNA, in contrast, yielded a black and tan coloration. Thus, the added twist of the host DNA repeats embedded within the VL30 element resulted in more possible genetic configurations and resulting coat colors.

In addition to the oncogenes, there is a second class of genes important in cancer whose activity is important in suppressing, rather than promoting, cancer. These are the tumor suppressor genes, which, when mutant, lead to an increased risk of cancer. In this case, both copies of the gene need to be inactivated for an increase in cancer rate to ensue. Although they are relatively uncommon, there are a few cases known of inactivation of the tumor suppressor gene p53 by retroviral insertions in mouse erythroleukemia cells, pre-B-cell lymphomas, and an osteosarcoma.

Several additional spontaneous mutations of mice have been linked to insertion of genomes of retrovirus or defective retroelements. The affected genes show no particular pattern (apart from their interest to mouse geneticists), suggesting that there was no special pressure for initial integration into these sites. Probably most murine genes can serve as integration sites for retroviral genomes.

Several studies have been designed to generate and detect insertional gene inactivation by retroviruses. In one clever study, Harold Varmus and coworkers reasoned that if a cell was transformed with a single copy of the *src* gene, insertion of a retroviral genome into *src* should reverse the transformed phenotype. Infection of *src*-transformed cells with Mo-MLV yielded just such events. Rudolf Jaenisch and coworkers used a similar strategy, but in mice instead of in cultured cells. They introduced Mo-MLV genomes into mice by infecting early embryos with Mo-MLV. In some cases, integration took place in cells of the germ line, giving rise to heritable proviral integrations. The mouse strains containing the insertion sites were then bred to make the altered allele homozygous, yielding a variety of new mutations. For example, the insertion in Mov13 resulted in disruption of a gene for collagen, whereas the insertion in Mov34 altered a gene important in early development.

All of the above examples in some way involved intervention by scientists, whether by transferring tumor filtrates, inbreeding mouse strains, or deliberately trying to inactivate genes by infection. In the next sections, we review the contribution of naturally occurring retroviral integration events to the formation of vertebrate genomes over evolutionary time.

ENDOGENOUS RETROVIRUSES (4, 8, 22, 28, 32, 45)

Retroviral infection of germ-line cells (egg or sperm progenitors), followed by fertilization involving the modified gamete, gives rise to an offspring that differs from the parent by insertion of a proviral genome. If the modified chromosome becomes fixed in the population in subsequent generations, the whole population will possess the integrated proviral genome. Many species harbor such endogenous proviruses, with the number known usually reflecting the effort put in by experimenters to finding them.

Endogenous proviruses can be recent insertions into the host genome or ancient molecular fossils. For recent insertions, integration sites may vary from population to population, reflecting ongoing de novo integration. Inbred strains of mice, for example, differ by the positions of a number of integrated proviruses. At the opposite extreme are the evolutionarily ancient proviral genomes, in some cases so altered by mutation as to be difficult to recognize in the DNA sequence. Some HERVs have integration sites conserved between humans and apes, indicating integration prior to the evolutionary divergence of the two lineages. Many of these are extensively deleted and mutated.

The endogenous retroviruses of mice have been particularly well studied, and at least eight types are known. Some, but not all, of these are relatives of known murine replication-competent retroviruses. In the jargon of retrovirology, viruses not known to be endogenous are called "exogenous." One relatively homogeneous group is the Mo-MLV family, present in roughly 25–100 copies per haploid mouse genome, and related to the exogenous Mo-MLVs. These proviruses clearly can contribute to pathogenesis. In the case of thymic lymphomas in AKR mice, a complex recombinant virus forms reproducibly from several endogenous proviruses during growth of the mouse. Another family of endogenous murine viruses, the mouse mammary tumor viruses (MMTV), causes mammary tumors by insertional activation of several oncogenes. Some 50 MMTV remnants are present in mouse genomes. The mammary tumors can be horizontally transferred from mother to pup by nursing, because the virus is present in the mother's milk. The remaining classes of endogenous mouse proviruses are not known to contain replication-competent members. The IAP elements are a highly abundant example, present at 1000–2000 copies per cell. IAP elements encode many of the retroviral proteins, but only rare members encoded Env. The IAP elements form distinctive particles, which bud through internal membranes into "cisternae," giving the group their name (intracisternal A-type particles). VL30 elements are another retroviral relative, first found as RNA "hitchhikers" in preparations of Mo-MLV virions. These elements do not encode any proteins, but they do contain all the *cis* signals in the RNA required for reverse transcription and integration. Thus, these elements can multiply as parasites on the Mo-MLVs. Several additional types of endogenous murine proviruses are also known.

Pigs also contain endogenous proviruses, called PERVs; this is a worrisome finding because pig organs are used as grafts in humans. Pig heart valves are widely used for surgical repair of human valves, and pig strains are being engineered to improve the usefulness of pig organs for applications in transplantation. However, recent work reveals that pigs encode replication-competent endogenous retroviruses that are capable of replicating in human cells. The viruses come in at least two classes, PERV-1 and PERV-2, both relatives of the simple type-C retroviruses. The fear is that PERV viruses produced from pig grafts in humans may adapt to their human host, thereby potentially creating new disease by zoonosis, the jumping of a pathogen between species. This is no idle fear—the AIDS epidemic clearly began with jumps of viruses from primates into people (Chapter 7). So far there is no evidence of adaptation of PERV virus to growth in human cells in transplant patients or laboratory studies. PERV viruses do replicate in human cells, but so far they do so only sluggishly. The possibility that PERVs from grafts will evolve in humans to replicate efficiently remains a major concern in the use of pig tissues. Baboons also harbor endogenous retroviruses, and baboon tissues are also used in transplantation, raising similar issues as with pigs and PERVs. Continued monitoring and open reporting are crucial to assessing the possible hazards.

Humans harbor many endogenous retroviruses, called HERVs, but none of them is known to be replication-competent. Analysis of the draft sequence of the human genome reveals that it contains fully 8% endogenous retroviral sequences, emphasizing the contribution of retroviruses to our genetic heritage (the draft human genome sequence is discussed in detail in Chapter 10). As discussed in Chapter 8, an even larger fraction of the human genome is contributed by other types of elements that replicate by reverse transcription. The HERV-1 group is related to the mammalian type-C retroviruses, HERV-2 to mammalian type-A and -B viruses. The ages of the HERV proviruses differ widely. Some are quite ancient, being present in similar locations in primate and human genomes and displaying an accumulation of mutations consistent with great age. Others appear to be newer. Some of the HERV-K elements seem to have only a few mutations that prevent replication. It is possible that further studies will reveal replication-competent HERV-K elements in the human genome, although none has been found to date; the draft of the human genome contains several that are close to intact but no strong candidates for active elements (Chapter 10).

The HERV elements in their role as mobile regions of homology represent a force for change in human chromosomes. In one study, the DNA of a pair of siblings was found to differ by a recombination between LTRs that removed a HERV-H element in one. David Page and coworkers recently discovered that a HERV element is responsible for rare cases of sterility in men. Two cases were identified in which identical deletions of the Y chromosome were associated with azoospermia factor a (AZFa), a syndrome involving spermatogenic failure and infertility. Analysis of the deletion revealed that it resulted in recombination between two 94% identical HERV15 insertions.

Just because an endogenous retrovirus is replication-incompetent does not mean that it cannot move in the genome. At least two mechanisms can mediate movement, both relatives of the normal replication cycle. As mentioned above for VL30, if an endogenous genome is transcribed and encodes all the necessary *cis* signals for packaging, reverse transcription, and integration, proteins supplied by other viruses can direct replication.

The second mechanism, retrotransposition, involves replication by means of transcription of the genome, reverse transcription, and integration, but all within a single cell. Thus, retrotransposition is a form of transposition rather than viral replication. Many transposons are now known that replicate by retrotransposition. Such elements contain homologs of the *gag* and *pol* genes, but no *env*. For the endogenous retroviruses, many defective elements are known that are largely intact but lack *env* genes. Such elements may replicate by retrotransposition rather than by normal infection. Retrotransposition has been demonstrated for IAP elements in laboratory studies, but the efficiency was low relative to normal retroviral infection. It is at present unclear whether retrotransposition of defective proviruses is a major means of viral spread.

What fraction of retroviral integration events result in altered phenotype of the host? Two lines of evidence from studies in mice allow rough estimates. For inbred strains of laboratory mice, new insertions into the mouse germ line can occasionally be identified. Comparison of the number of insertions with the number of new phenotypes indicates that about 5% of insertions result in new phenotypes. This analysis is biased, because various types of changes would likely have been systematically overlooked. For example, special tests are need to determine whether a new insertion when homozygous is lethal in an early embryo. Another type of test involves infecting mouse embryonic stem cells with an artificial virus that expresses a marker gene only upon integration into a cellular gene. Mouse ES cells can be used to generate live mice after appropriate manipulations, allowing the consequences of insertion into a gene to be assessed in the full organism. Such tests indicate that fully 40% of such insertions have phenotypic consequences in the animal. These studies document some of the potential of retroviral integration for generating evolutionary change.

RETROVIRAL SEQUENCES INCORPORATED IN HOST CELL GENES (6–8, 21, 35, 48, 52)

Incorporation of retroviral sequences has contributed to the evolution of many vertebrate genes. In this final section, we discuss several well-studied examples; analysis of data from the genome sequencing projects will likely soon yield many more.

Several cases have been uncovered of retroviral sequences contributing new transcriptional control signals. One intriguing example concerns regulation of a human amylase gene (Fig. 6.20). Amylase degrades complex starch molecules into simple sugars. Humans encode several genes for amylase. All vertebrates studied express amylase in the pancreas, where much of the metabolism of sugars is controlled. However, humans also express one of the amylase genes in the salivary gland. The salivary-expressed copy differs from other amylase genes by the integration of HERV-E proviral sequences into the 5′ gene regulatory region. A functional analysis of control sequences revealed that the retrovirus DNA contributes the signals for expression of the amylase gene in the salivary gland. This insertion may have had a direct bearing on human evolution, because the partial digestion of starch in the mouth yields a sweet taste, possibly causing humans to favor certain grains as foods.

In the amylase case, the integrated HERV element provided control signals that activated the preexisting amylase gene promoter. In the case of the phospholipase-A2-related gene, the insertion of a HERV-H element introduced a complete new regulatory region, including the promoter and start site for transcription. The mRNA is formed by splicing of 5′ HERV-H sequences to the body of the cellular gene.

A mouse gene called *slp*, for *sex-*linked *p*rotein, provides another example of a regulatory change by retroviral integration. The *slp* gene requires the presence of androgen hormones for efficient expression. The androgen-responsive regulatory elements are provided by an integrated retroviral element located 2 kb upstream of the coding region.

The henny-feathering trait of chickens provides another example. This dominant feather-color variant results from overexpression of the gene for aromatase. Analysis of the gene reveals that the 5′ end of the

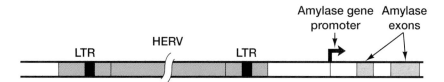

FIGURE 6.20. Insertion of a HERV element upstream of an amylase gene in humans results in parotid (salivary gland) expression. This allows partial digestion of some starches in the mouth, resulting in a sweet taste. (Adapted from Ting et al. 1992.[48])

aromatase message is formed as a fusion with a RAV-0 provirus. Evidently in this case also, integration of a retrovirus led to the formation of a fused transcription unit that overexpressed the encoded protein.

Integrated retroviruses can also supply new transcriptional stop signals to genes, as with the *dilute* and *hairless* mutations of mice. Several hundred human cDNAs with stop signals from HERV elements have been described, and studies of these are just beginning.

Recent data suggest that a captured retroviral Env protein may have been coopted for use during human placental development. Development of the human placenta involves fusion of fetal cells called trophoblasts to form giant cells with many nuclei called syncytiotrophoblasts. Many studies indicate that retroviral envelope proteins expressed on cell surfaces can similarly fuse envelope-expressing cells with receptor-expressing cells. Functional studies show that a retroviral envelope encoded by a HERV-W element is capable of causing fusion of trophoblasts in laboratory tests, and the HERV-W envelope is expressed in fetal trophoblasts. These data support the proposal that the HERV-W envelope, renamed syncytin for its ability to form cellular syncytia, has been recruited to carry out cell fusion during trophoblast development.

Endogenous retroviral sequences can also benefit the host by expressing proteins that interfere with infection by pathogenic relatives. These proviral derivatives thus qualify as host antiviral genes. The best-understood examples come from mice, which have been extensively studied as models of viral pathogenesis. The murine Fv-4 locus is an endogenous provirus that blocks superinfection by expressing a viral envelope protein. The Env protein binds to viral receptors, thereby preventing their use by incoming viruses. The presence of the Fv-4 gene in a mouse prevents the induction of leukemias by Mo-MLV.

The Fv-1 locus also blocks superinfection, but by a different and still unclear mechanism. Mo-MLVs can be of either N or B tropism, depending on whether they possess a specific amino acid at one position in the CA protein. B-tropic Mo-MLVs are able to infect BALB/c cells, but they infect NIH-3T3 cells 100-fold less efficiently. N-tropic viruses have the reverse tropism. The Fv-1 locus determines this difference, causing a defect in replication prior to integration. The Fv-1 gene is an endogenous provirus related to the HERV-L family that encodes MA and CA proteins. The Fv-1 CA-like protein may bind the incoming Mo-MLV core, possibly through interactions with the Mo-MLV CA. The mechanism is unclear but of intense interest as a unique antiviral mechanism.

These examples of retroviral sequences incorporated into host cell genes illustrate the opportunistic nature of molecular evolution. Darwinian selection operates on randomly generated changes in host cell genes. Mobile DNA elements such as retroviruses provide much of the random variation on which selection acts. The products are often surprising, looking more like historical accidents than carefully fashioned solutions to molecular problems. Many more types of mobile elements are present in eukaryotic chromosomes, essentially all of which, like the retroviruses, are implicated in genomic evolution.

Summary

Retroviruses take their name from the fact that they copy RNA sequences into DNA, reversing the typical flow of biological information. Lateral transfer is a side product of this replication mechanism, because cellular RNAs can occasionally be packaged into virions, transported between cells, and integrated into the new host. Integration can sometimes take place in germ cells, leading to heritable alteration of the progeny of the modified cell. About 8% of the human genome consists of retrovirus-derived sequences.

Retroviral particles are spherical, containing on the outside a lipid bilayer membrane derived from the virus-producer cell. Inserted into the viral membrane are the virus-encoded envelope proteins, which are responsible for binding receptors on target cells to initiate infection. Inside the membrane is a layer of matrix protein, and inside that is the viral capsid. The capsid is composed of capsid (CA) protein, which assembles to form a hollow shell. Inside the shell are two copies of the viral RNA genome, bound by the nucleocapsid protein (NC). The viral enzymes protease (PR), reverse transcriptase (RT), and integrase (IN) are also found inside the capsid shell.

The infectious cycle begins with the binding of a retroviral particle to a sensitive cell. The interaction of the Env protein and its receptor initiates a complex series of steps that result in the fusion of the viral and cellular membranes. Following entry, the core disassembles concomitant with the beginning of reverse transcription.

DNA synthesis on the viral RNA initiates from the 3′ end of a bound cellular tRNA. Reverse transcription proceeds to the end of the viral RNA template, then RT transfers between templates. Meanwhile, the RNase H activity of RT cleaves the underlying RNA into short fragments. One such fragment, annealed on the "polypurine tract," serves as the primer for synthesis of the other cDNA strand. A second template switch and full-length extension result in production of a double-stranded linear cDNA copy of the viral RNA genome.

The linear cDNA is then integrated into the host genome by the action of the virus-encoded integrase protein. The complex that carries out integration, the preintegration complex (PIC), contains integrase and other proteins that may aid the integration process. Integrase removes two nucleotides from each viral cDNA end, probably to clean up otherwise heterogeneous reverse transcription products, and then connects one strand on each cDNA end to target DNA. Host DNA repair enzymes probably connect the remaining unjoined strands, yielding the integrated provirus. During subsequent cell division, the provirus will be copied and inherited like any normal cellular gene.

The provirus also contains all the signals necessary to direct synthesis of viral RNA. The RNAs are used as templates for translation to produce viral proteins. The viral Gag and Gag-Pol proteins are synthesized as polyproteins, facilitating their assembly. A fraction of the full-length viral RNAs also assemble into budding particles to serve as viral genomes. Viral RNA and viral proteins then coalesce and bud out of cytoplasmic membrane of the infected cell. Following budding, the viral protease cleaves the Gag and Gag-Pol polyproteins to yield the mature viral proteins.

Viruses later found to be retroviruses were discovered in the first decades of the last century as filterable agents that could transmit cancer between birds. Decades of research ultimately revealed that the "acute transforming viruses" encoded altered versions of cellular genes for growth control. Introduction of these mutant genes into normal cells pushed the cells toward unregulated growth.

Many studies document further ways in which retroviral integration can alter the genome of the host organism. Retroviral integration can activate transcription by inserting transcriptional activating signals upstream of a gene. Integration into genes can result in blocking gene activity or, more rarely, production of new proteins by altering DNA splicing.

Retroviruses have been incorporated into the genomes of all animals studied, indicating that infection of the germ line is frequent, at least over evolutionary time. In animals harboring replication-competent endogenous retroviruses, ongoing infection can be a source of new mutations. In mice, for example, replication of endogenous proviruses can be associated with high-frequency malignancies. Endogenous retroviruses have also contributed to the welfare of their hosts. Sequences of endogenous retroviruses have been recruited for the formation of new genes, such as the salivary amylase gene of humans.

Fully 8% of the draft human genome sequence is contributed by endogenous retrovirus sequences. In summary, the life history of retroviruses is ideally suited to generating evolutionary change in the organisms they infect.

REFERENCES

1. Albritton L.M., Tseng L., Scadden D., and Cunningham J.M. 1989. A putative murine ecotropic retrovirus receptor gene encodes a multiple membrane-spanning protein and confers susceptibility to virus infection. *Cell* **57:** 659–666.

2. Arnold E., Jacobo-Molina A., Nanni R.G., Williams R.L., Lu X., Ding J., Clark A.D.J., Zhang A., Ferris A.L., Clark P., et al. 1992. Structure of HIV-1 reverse transcriptase/DNA comples at 7 Å resolution showing active site locations. *Nature* **357:** 85–89.

3. Baltimore D. 1970. RNA-dependent DNA polymerase in virions of RNA tumor viruses. *Nature* **226:** 1209–1211.

4. Barbulescu M., Turner G., Seaman M.I., Deinard A.S., Kidd K.K., and Lenz J. 1999. Many human endogenous retrovirus K (HERV-K) proviruses are unique to humans. *Curr. Biol.* **26:** 861–868.

5. Bates P., Young J.A., and Varmus H.E. 1993. A receptor for subgroup A Rous sarcoma virus is related to the low density lipoprotein receptor. *Cell* **74:** 1043–1051.

6. Best S., Le Tissier P., Towers G., and Stoye J.P. 1996. Positional cloning of the mouse retrovirus restriction gene *Fv*1. *Nature* **382:** 826–829.

7. Blond J.L., Lavillette D., Cheynet V., Bouton O., Oriol G., Chapel-Fernandes S., Mandrand B., Mallet F., and Cosset F.L. 2000. An envelope glycoprotein of the human endogenous retrovirus HERV-W is expressed

in the human placenta and fuses cells expressing the type D mammalian retrovirus receptor. *J. Virol.* **74:** 3321–3329.

8. Boeke J.D. and Stoye J.P. 1997. Retrotransposons, endogenous retroviruses, and the evolution of retroelements. In *Retroviruses* (ed. J.M. Coffin et al.), pp. 343–435. Cold Spring Harbor Laboratory Press, Cold Spring Harbor, New York.

9. Brown P.O. 1997. Integration. In *Retroviruses* (ed. J.M. Coffin et al.), pp. 161–203. Cold Spring Harbor Laboratory Press, Cold Spring Harbor, New York.

10. Brown P.O., Bowerman B., Varmus H.E., and Bishop J.M. 1987. Correct integration of retroviral DNA in vitro. *Cell* **49:** 347–356.

11. Bushman F.D., Fujiwara T., and Craigie R. 1990. Retroviral DNA integration directed by HIV integration protein in vitro. *Science* **249:** 1555–1558.

12. Carteau S., Hoffmann C., and Bushman F.D. 1998. Chromosome structure and HIV-1 cDNA integration: Centromeric alphoid repeats are a disfavored target. *J. Virol.* **72:** 4005–4014.

13. Cavazzana-Calvo M., Hacein-Bey S., de Saint Basile G., Gross F., Yvon E., Nusbaum P., Selz F., Hue C., Certain S., Casanova J.L., Bousso P., Deist F.L., and Fischer A. 2000. Gene therapy of human severe combined immunodeficiency (SCID)-X1 disease. *Science* **288:** 669–672.

14. Chen J.C.-H., Krucinski J., Miercke L.J.W., Finer-Moore J.S., Tang A.H., Leavitt A.D., and Stroud R.M. 2000. Crystal structure of the HIV-1 integrase catalytic core and C-terminal domains: A model for viral DNA binding. *Proc. Natl. Acad. Sci.* **97:** 8233–8238.

15. Coffin J.M., Hughes S.H., and Varmus H.E. 1997. *Retroviruses.* Cold Spring Harbor Laboratory Press, Cold Spring Harbor, New York.

16. Collet M.S., Purchio A.F., and Erikson R.L. 1980. Avian sarcoma virus-transforming protein pp60src shows protein tyrosine kinase activity specific for tyrosine. *Nature* **285:** 167–169.

17. Cooper G.M., Temin R.G., and Sugden B. 1995. *The DNA provirus: Howard Temin's scientific legacy.* American Society for Microbiology, Washington, D.C.

18. Craigie R., Fujiwara T., and Bushman F. 1990. The IN protein of Moloney murine leukemia virus processes the viral DNA ends and accomplishes their integration in vitro. *Cell* **62:** 829–837.

19. Ellerman V. and Bang O. 1908. Experimentelle Leukamie bei Huhnern. *Zentralbl. Bakteriol. Parasitenkd. Infektkrankh. Hyg. Abt. I Orig.* **46:** 595–609.

20. Farnet C. and Bushman F.D. 1997. HIV-1 cDNA Integration: Requirement of HMG I(Y) protein for function of preintegration complexes in vitro. *Cell* **88:** 1–20.

21. Feutcher-Murthy A.E., Freeman J.D., and Mager D.L. 1993. Splicing of a human endogenous retrovirus to a novel phospholipase A2 related gene. *Nucleic Acids Res.* **21:** 135–143.

22. Friedrich G. and Soriano P. 1991. Promoter traps in embryonic stem cells: A genetic screen to identify and mutate developmental genes in mice. *Genes Dev.* **5:** 1513–1523.

23. Ganser B.K., Li S., Klishko V.Y., Finch J.T., and Sundquist W.I. 1999. Assembly and analysis of conical models for the HIV-1 core. *Science* **283:** 80–83.

24. Hansen M.S.T., Carteau S., Hoffmann C., Li L., and Bushman F. 1998. Retroviral cDNA integration: Mechanism, applications and inhibition.

Genet. Eng. **20:** 41–62.

25. Hicks G.G. and Mowat M. 1988. Integration of Friend murine leukemia virus into both alleles of the p53 oncogene in an erythroleukemic cell line. *J. Virol.* **62:** 4752–4755.

26. Hunter T. and Sefton B.M. 1980. Transforming gene product of Rous sarcoma virus phosphorylates tyrosine. *Proc. Natl. Acad. Sci.* **77:** 1311–1315.

27. Jenkins N.A., Copeland N.G., Taylor B.A., and Lee B.K. 1981. Dilute (d) coat colour mutation of DBA/2J mice is associated with the site of integration of an ecotropic MuLV genome. *Nature* **293:** 370–374.

28. Johnson W.E. and Coffin J.M. 1999. Constructing primate phylogenies from ancient retrovirus sequences. *Proc. Natl. Acad. Sci.* **96:** 10254–10260.

29. Katz R.A., Merkel G., Kulkosky J., Leis J., and Skalka A.M. 1990. The avian retroviral IN protein is both necessary and sufficient for integrative recombination in vitro. *Cell* **63:** 87–95.

30. Kohlstaedt L.A., Wang J., Friedman J.M., Rice P.A., and Steitz T.A. 1992. Crystal structure at 3.5 Å resolution of HIV reverse transcriptase complexed with an inhibitor. *Science* **256:** 1783–1790.

31. Landau N.R., St. John T.P., Weissman I.L., Wolf S.C., Silverstone A.E., and Baltimore D. 1984. Cloning of terminal transferase cDNA by antibody screening. *Proc. Natl. Acad. Sci.* **81:** 5836–5840.

32. Lander E., Linton L.M., Birren B., Nusbaum C., Zody M.C., Baldwin J., Devon K., Dewar K., Doyle M., FitzHugh W., et al. 2001. Initial sequencing and analysis of the human genome. *Nature* **409:** 860–921.

33. Li S., Hill C., Sundquist W., and Finch J. 2000. Image reconstructions of helical assemblies of the HIV-1 CA protein. *Nature* **407:** 409–413.

34. Mann R., Mulligan R.C., and Baltimore D. 1983. Construction of a retrovirus packaging mutant and its use to produce helper-free defective retrovirus. *Cell* **33:** 153–159.

35. Mi S., Lee X., Li X., Veldman G.M., Finnerty H., Racie L., LaVallie E., Tang X.Y., Edouard P., Howes S., Keith J.C.J., and McCoy J.M. 2000. Syncytin is a captive retroviral envelope protein involved in human placental morphogenesis. *Nature* **403:** 785–789.

36. Miller M.D. 1997. Development and application of retroviral vectors. In *Retroviruses* (ed. J.M. Coffin et al.), pp. 437–474. Cold Spring Harbor Laboratory Press, Cold Spring Harbor, New York.

37. Mowat M., Cheng A., Kimura N., Bernstein A., and Benchimol S. 1985. Rearrangements of the cellular p53 gene in erythroleukaemic cells transformed by Friend virus. *Nature* **314:** 633–636.

38. Naldini L., Blomer U., Gallay P., Ory D., Mulligan R., Gage F.H., Verma I.M., and Trono D. 1996. In vivo gene delivery and stable transduction of nondividing cells by a lentiviral vector. *Science* **272:** 263–267.

39. Rosenberg N. and Jolicouer P. 1997. Retroviral pathogenesis. *Retroviruses* (ed. J.M. Coffin et al.), pp. 475–586. Cold Spring Harbor Laboratory Press, Cold Spring Harbor, New York.

40. Rous P. 1911. A sarcoma of the fowl transmissible by an agent seperable from the tumor cells. *J. Exp. Med.* **13:** 397–411.

41. Shank P.R. and Linial M. 1980. Avian oncovirus mutant (SE21Q1b) deficient in genomic RNA: Characterization of a deletion in the provirus. *J. Virol.* **36:** 450–456.

42. Skalka A.M. and Goff S.P. 1993. *Reverse transcriptase.* Cold Spring Harbor Laboratory Press, Cold Spring Harbor, New York.

43. Spector D.H., Varmus H.E., and Bishop J.M. 1978. Nucleotide sequences related to the transforming gene of avian sarcoma virus are present in DNA of uninfected vertebrates. *Proc. Natl. Acad. Sci.* **75:** 4102–4106.

44. Stoye J.P., Fenner S., Geeenoak G.E., Moran C., and Coffin J.M. 1988. Role of endogenous retroviruses as mutagens: The hairless mutation of mice. *Cell* **54:** 383–391.

45. Sun C., Skaletsky H., Rozen S., Gromoll J., Nieschlag E., and Page D.C. 2000. Deletion of *azoospermia factor a* (*AZFa*) region of human Y chromosome caused by recombination between HERV15 proviruses. *Hum. Mol. Genet.* **9:** 2291–2296.

46. Swanstrom R. and Wills J.W. 1997. Synthesis, assembly, and processing of viral proteins. In *Retroviruses* (ed. J.M. Coffin et al.), pp. 263–334. Cold Spring Harbor Laboratory Press, Cold Spring Harbor, New York.

47. Temin H. and Mizutani S. 1970. RNA-dependent DNA polymerase in virions of Rous sarcoma virus. *Nature* **226:** 1211–1213.

48. Ting C.N., Rosenberg M., Snow C., Samuelson L., and Meisler M. 1992. Endogenous retroviral sequences are required for tissue-specific expression of a human salivary amylase gene. *Genes Dev.* **6:** 1457–1465.

49. Varmus H.E. and Brown P.O. 1989. Retroviruses. In *Mobile DNA* (ed. E. Berg and M.M. Howe), pp. 53–108. American Society for Microbiology, Washington, D.C.

50. Varmus H.E., Quintrell N.E., and Ortiz S. 1981. Retroviruses as mutagens: Insertion and excision of a non-transforming provirus alters expression of a resident transforming provirus. *Cell* **25:** 23–26.

51. Vogt V.M. 2000. Ubiquitin in retrovirus assembly: Actor or bystander? *Proc. Natl. Acad. Sci.* **97:** 12945–12947.

52. Wilkinson D.A., Mager D.L., and Leong J.A. 1994. Endogenous human retroviruses. In *The retroviridae* (ed. J.A. Levy), vol. 3, pp. 465–534. Plenum Press, New York.

53. Yang Z.N., Mueser T.C., Bushman F.D., and Hyde C.C. 2000. Crystal structure of an active two-domain derivative of Rous sarcoma virus integrase. *J. Mol. Biol.* **296:** 535–548.

54. Yoder K. and Bushman F.D. 2000. Repair of gaps in retroviral DNA integration intermediates. *J. Virol.* **74:** 11191–11200.

Lateral DNA Transfer and the AIDS Epidemic

THE ACQUIRED IMMUNE DEFICIENCY SYNDROME (AIDS) epidemic provides an extreme example of the consequences of retroviral infection. The human immunodeficiency virus (HIV), like many virulent viruses, apparently originated with a recent jump of an animal virus into humans. The human endogenous retrovirus (HERV) elements, in contrast, infected a progenitor of modern apes and humans long ago, ultimately integrating into germ-line cells to become quiescent endogenous proviruses. HIV differs in that there are no known endogenous HIV proviruses, so HIV is not involved in lateral DNA transfer in the sense of heritably modifying the germ line, at least not yet. With HIV we see an earlier step in virus–host evolution, the expansion of a virulent infectious agent by transfer between previously uninfected individuals.

The retroviral replication strategy—involving reverse transcription and cDNA integration into the host chromosome—has profound consequences for HIV-induced disease. Integrated HIV genomes are capable of adopting a quiescent state in some infected cells, which has the unfortunate consequence of permitting a fraction of viral genomes to elude the host immune system and evade antiviral therapies. Such latent viruses can reseed infection after otherwise successful treatment with antiviral drugs. We review the rise of AIDS in this section, emphasizing the importance of viral DNA transfer mechanisms in the epidemic. The chapter begins with a description of the epidemic and the disease course, then turns to the molecular nature of HIV replication, pointing out the features that differ from simple retroviruses discussed in Chapter 6. The zoonotic origin of HIV is then discussed, a contentious issue following (improbable) allegations that the disease was spread in early polio vaccines prepared using African green monkey kidney cells. The chapter ends with some new hopes from contemporary research on treatment strategies and vaccines.

THE AIDS EPIDEMIC

As of 2000, the United Nations estimated that 36.1 million people were infected with HIV, 1.4 million of whom were children (Fig. 7.1). Three million people died of HIV infection in 2000 alone. It is estimated that 21.8 million people have died between the start of the epidemic and 2000, 4.3 million of whom were children.

A total of 47 million people have been infected with HIV since the start of the epidemic. AIDS is now the fourth leading cause of death in the world.

The hardest-hit region is sub-Saharan Africa, where 25.3 million people are believed to be living with HIV infection (Fig. 7.2). In 2000 alone, there were 3.8 million new infections, down slightly from 4 million new infections in 1999. A staggering 8.8% of the adult population is believed to be infected. The rates are even higher in the worst-affected countries—up to 25% of the population may be infected in Zimbabwe, Uganda, and Botswana. As a result, the total population of these countries may actually begin to decline due to AIDS. Transmission of the disease in sub-Saharan Africa is primarily heterosexual, and women comprise 55% of the HIV-positive adults in sub-Saharan Africa, as opposed to the developed world where the majority of HIV-positive adults are male. Worldwide the people affected are disproportionately young adults in the most productive parts of their lives. The rate of death in this age group has created a parallel epidemic of

**Adults and children estimated to be living
with HIV/AIDS as of end 2000**

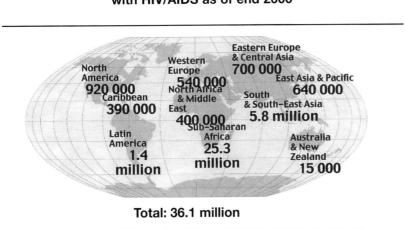

Total: 36.1 million

FIGURE 7.1. HIV infections worldwide as of the year 2000. (Redrawn from Joint United Nations Programme on HIV/AIDS [http://www.unaids.org/].)

orphans, estimated at 11 million, further tearing at the fabric of life in these countries.

In North America the epidemic has a different character. Almost a million people are infected, for an HIV-positive rate of 0.56%. However, in North America men represent 80% of infected people, reflecting the early spread of the epidemic in gay males. In North America and the

**Children (<15 years) estimated to be living
with HIV/AIDS as of end 2000**

Total: 1.4 million

FIGURE 7.2. HIV infections in children as of the year 2000. (Redrawn from Joint United Nations Programme on HIV/AIDS [http://www.unaids.org/].)

developed world generally, the rate of death has declined in the last several years due to the introduction of effective anti-retroviral therapy. As discussed below, it is unclear whether this improvement can be sustained given the high rate of evolution of drug-resistant HIV strains.

The improvement in the welfare of patients in the developed world is unfortunately matched by an expansion of the epidemic in new regions. The virus swept into Southeast Asia in the late 1980s and is presently exploding in India. Disease in the developing world cannot generally be treated with effective therapy due to the expense and lack of necessary infrastructure. In most of the world, the AIDS epidemic continues to expand unchecked.

TRANSMISSION OF HIV (17)

HIV can be transferred between individuals by blood, birth, or sex. Early in the epidemic in the developed world, many cases resulted from transmission between gay men. Treatment of hemophiliacs with clotting factors purified from contaminated human blood provided another early route of infection. Intravenous drug abusers also became widely infected due to exchange of blood while sharing contaminated needles. Worldwide, transmission is primarily heterosexual. HIV can also be transmitted from an infected mother to newborns during the birth process, when the newborn comes into contact with infected maternal fluids, although in a minority of cases infection also takes place in utero. Transmission from mother to child can also occur during nursing, since HIV is present in breast milk of infected mothers.

DISEASE COURSE (3, 17, 25)

Following initial infection, the virus spreads rapidly in the body (Fig. 7.3). CD4-positive cells in blood and tissues become infected, widely seeding lymphoid-related tissues. The disease proceeds through three phases: acute infection, clinical latency, and decline to AIDS.

In the first weeks, the concentration of the viral RNA in blood plasma can reach as high as 10 million copies per milliliter. For some patients, this initial acute phase is accompanied by flu-like symptoms. In many cases, however, initial infection has few symptoms, although a characteristic rash on the body trunk has recently been recognized.

After a few weeks, the concentration of virus falls, concomitant with the rise of the host immune response. Both humoral and cell-mediated responses against HIV develop within weeks of infection. As the intensity of the immune response increases, the concentration of virus in blood falls, probably because the cell-mediated response kills HIV-infected cells. Once the immune response sets in, a new equilibri-

The immune system comprises two branches, the humoral or antibody-based branch and the cell-based branch. The immune system and the role of mobile DNA in its function is the topic of Chapter 11.

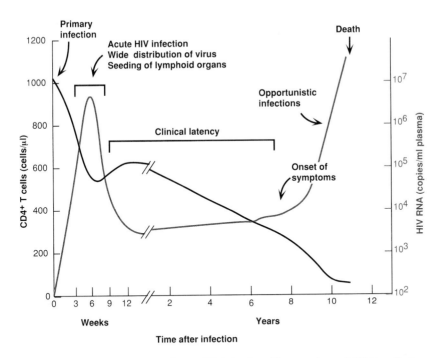

FIGURE 7.3. Typical course of an HIV infection. The number of CD4+ cells is shown on the left vertical axis; the number of copies of HIV RNA per ml of blood plasma is shown on the right. Time after infection is shown on the horizontal axis; note the break between the rapid development of the acute infection and the later slow decline. Responses of individual patients often depart considerably from the average. (Reprinted, with permission, from Fauci and Desrosiers 1997.[18]).

um is reached, in which virus is synthesized and destroyed at equivalent rates. For the infected person, this is the start of the clinical latency period. There are few or no symptoms during this phase, and patients may not realize that they are infected. The clinical latency phase can be very short in some cases, but in others it can last for a decade or more. In a few cases, the different rates of disease progression can be attributed to specific features of the virus or infected patient, but most of the variation remains unexplained.

The Dynamics of HIV Infection (5, 31, 45, 56)

In the early years of the epidemic, the clinical latent period was thought to involve very little virus replication, but this picture has changed dramatically. HIV was initially believed to enter cells and become latent, remaining inactive for years. With the development of better ways of detecting HIV in infected people, it was realized that the clinical latent period involves highly active viral replication coupled with equally high rates of clearance. Remarkably, some 10^{10} virions are produced per day in an infected person, and a similar number are destroyed. The half-life of an HIV virion is only 5–8 hours. CD4-positive cells infected with HIV are also quickly destroyed, with a half-life of only about 2.5 days, so that roughly 10^9 CD4-positive cells are destroyed and replaced each day. Over a 10-year illness, some 4×10^{13} viral particles are produced!

The latent period ends with an accelerating decline of CD4-positive cells and concomitant rise in virus concentration in blood. Tissues rich in CD4-positive cells, such as lymph nodes, become increasingly morbid. The reason for the decline is still unclear, although one possibility is that the capacity to replace CD4-positive cells becomes reduced and ultimately exhausted. Initial disease symptoms include a wasting syndrome in which patients can lose up to 10% of their body weight, swollen lymph glands, and occasional neurological disease. Depression is also common, but it is unclear whether this derives from direct damage to the nervous system or because of the patient's awareness of his situation.

Once CD4-positive cells have declined to 200 per microliter or fewer, an HIV- positive person is defined as having the acquired immune deficiency syndrome, or AIDS. CD4-positive cells are themselves components of the immune system, so with the decline in CD4 cell concentration comes a decline in immune function. As a consequence, the patient is increasingly afflicted with infections. Many of these cause little or no disease in healthy people. Normally benign microbes from cattle and cats, for example, become life-threatening pathogens for HIV-infected people. The fungus *Pneumocystis pneumonia* (PCP) can cause lethal pneumonia, and cytomegalovirus can cause blindness. Even baker's yeast (*Saccharomyces cerevisiae*) can be a pathogen in AIDS patients. Several characteristic cancers cause dangerous tumors and death. Ultimately, the burden of opportunistic infections and other symptoms causes death in nearly all untreated patients.

Curiously, the cancers associated with HIV infection do not seem to be due to transduction or insertional activation of oncogenes. HIV-induced tumors do not typically harbor any integrated HIV genomes. In the case of one cancer, Kaposi's sarcoma, the disease has been attributed to a herpesvirus, either co-transmitted with HIV or activated by immune decline. For the non-Hodgkin lymphoma associated with HIV, cancer may be promoted by the abnormal environment associated with high-level production of CD4 cells. Most HIV-infected cells have such short life spans that they are probably destroyed before they have a chance to proliferate as a cancer.

HIV INFECTION AT THE MOLECULAR LEVEL (6, 10, 15, 36, 54)

The HIV genome is larger than that of most retroviruses, encoding nine genes (Fig. 7.4). As discussed in the previous chapter, many of the functions of HIV Gag and Pol are similar to those of other retroviruses. The properties of Env are atypical, in that HIV entry requires not only the receptor CD4, but also additional cellular coreceptors, which has important consequences for the disease. Many of the six genes in addition to *gag, pol*, and *env* in HIV are shared among other lentivirus-

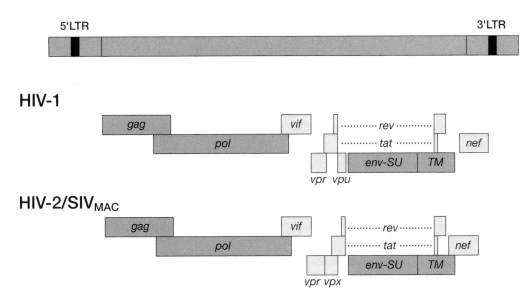

FIGURE 7.4. Genetic maps of HIV-1 and HIV-2/SIV$_{MAC}$. The genes found in all retroviruses (*gag, pol,* and *env*) are shown as green, the auxiliary genes are shown as gray. (Modified, with permission, from Fields et al. 1996.[20])

es. Studies in which these supernumerary genes were deleted from monkey viruses (simian immunodeficiency viruses, SIVs) have confirmed that most are important for efficient replication and disease progression in animals. An exception is *vpu*, which is not present in SIV and so could not be studied by this means. In this section, we review molecular aspects of infection unique to HIV—first the function of the HIV coreceptors and then function of the lentiviral-specific genes.

Coreceptors (1, 11, 12, 19, 37)

The requirements for entry into cells differ between HIV and other retroviruses, with crucial consequences for disease progression. The HIV envelope must interact not only with CD4, but also with another molecule on cell surfaces called a coreceptor. There are two major variants of HIV, each of which favors infection of different types of cells. R5 viruses use the coreceptor CCR5 for entry and favor infection of macrophages, which are rich in cell-surface CCR5. The X4 viruses, on the other hand, use CXCR4 as the coreceptor and favor infection of T cells that are correspondingly rich in this coreceptor. The detailed picture is complicated, in that there are also viruses that use both coreceptors, viruses that use still additional coreceptors, and cell types expressing multiple coreceptors.

The different strains of viruses have different consequences in the infected patient. In many cases, the X4 viruses are most prominent late in the course of the disease. These viruses are often particularly cytopathic, so the rise of X4 viruses may contribute to the depletion of CD4-positive cells during late-stage disease. The R5 viruses, on the other hand, are particularly important in initial transmission. This was illustrated dramatically by the observation that several individuals who had been multiply exposed to HIV but had not become infected in fact contained mutations in both copies of the *CCR5* gene.

vif (41, 51)

Starting from the left side of the HIV genome and proceeding rightward, the first lentivirus-specific gene encountered is *vif*, named for *viral infectivity factor*. *vif* has been shown to be required for replication in simian and caprine lentiviruses, but its role in stimulating viral growth is unclear. Studies in cell culture, however, provide some tantalizing clues. When HIV viruses lacking *vif* are produced from some cell types, designated nonpermissive, they are greatly reduced in infectivity. Other permissive cell types show no phenotype for *vif*-negative viruses. The nature of the target cell is unimportant. Further studies reveal that the *vif*-negative phenotype is genetically dominant, indicating that *vif* overcomes a negative factor synthesized by *vif*-nonpermissive cells. The mechanism of action of this negative factor is presently unknown but is the topic of intense study. As described in Chapters 11 and 13, a number of cellular systems that control transposons and viruses are known, and new ones are being discovered in diverse organisms. Possibly studies of *vif* are revealing the outlines of a new antiviral system.

vpr (9, 28, 29, 49)

Progressing rightward across the HIV genome, we come to *vpr*, named for *viral protein R* (an early name from the time when the potential gene was first recognized in the DNA sequence). The Vpr protein appears to have several effects on HIV growth, although the picture is not fully clarified. Expression of Vpr in cells can arrest cell division at a specific point in the replication cycle (G_2). It has been proposed that arrest at this point in the cell cycle allows optimal expression of the HIV genome, thereby maximizing the efficiency of production of new virions. The Vpr protein is also associated with preintegration complexes and may help to direct them to the nucleus. Simple retroviruses, in contrast to HIV, gain access to the nuclear chromosomes when the nucleus is disassembled during cell division. Nuclear localization is particularly important for HIV replication, however, because key target cell types such as macrophages do not divide; thus, preintegration complexes must cross the nuclear membrane for integration guided by Vpr.

tat (23, 55)

The next unique HIV gene is *tat*, named for transactivator. Tat protein promotes accumulation of the HIV RNAs by increasing the efficiency of transcriptional elongation. In the absence of Tat, most RNA chains initiated at the HIV promoter are terminated a few nucleotides from the point of initiation. Tat acts by binding to the nascent RNA at a site called TAR, for tat recognition. The Tat–TAR complex then binds a complex of host proteins including cyclin T and the kinase Cdk9. The kinase, once recruited to the HIV promoter, phosphorylates RNA polymerase, which results in an increased efficiency of transcriptional elongation. Thus, Tat protein promotes production of the mature HIV RNAs and all the viral proteins.

rev (32, 46)

Rightward of *tat* is *rev*, the *reg*ulator of *v*irion proteins. The action of Rev protein divides the HIV life cycle into early and late phases. Initially, the RNAs produced from a newly integrated provirus are spliced, as with normal cellular RNAs. The fully spliced HIV RNAs encode only the Tat, Rev, and Nef proteins, which accumulate with translation of the early RNAs. The newly synthesized Rev protein progressively changes the types of HIV RNAs produced by promoting the export of incompletely spliced RNAs. Rev protein acts by binding to a RNA site in the envelope gene called the RRE, for *rev r*esponse *el*ement. The Rev/RRE complex then binds the host cell factor CREM-1, which promotes export of the Rev/RRE complex out of the nucleus along pathways normally used by cellular RNAs. The Rev protein, having delivered its cargo for translation in the cytoplasm, then shuttles back into the cell nucleus. The second phase of the HIV life cycle proceeds as Rev escorts unspliced HIV RNAs into the cytoplasm, allowing translation of proteins including Gag, Pol, and Env, accumulation of full-length genomic RNAs to act as new genomes, and subsequent assembly and budding.

vpu (42)

The next gene rightward is *vpu*, *v*iral *p*rotein *u*, like *vpr* named for the lettered open reading frame first seen in the HIV DNA sequence. The Vpu protein provides one of the three functions that reduces the concentration of CD4 on the cell surface. Vpu appears to program degradation of CD4 inside cells, before CD4 has an opportunity to reach the cell surface. Details are somewhat controversial, but Vpu may act at least in part by binding to CD4 and thereby routing it to sites of protein degradation at proteosomes. Vpu may also act by intercalating into membranes and thereby forming a channel for ions, promoting viral release from cells.

In addition to Vpu, the viral envelope protein also reduces the amount of CD4 on cell surfaces, probably by binding CD4 and retaining the protein inside cells. Nef, discussed below, also reduces the concentration of cell-surface CD4. Removal of cell-surface CD4 is apparently crucial for viral replication in preventing viruses from reinfecting the cells they just emerged from, or for preventing Env proteins on a virion from fusion with CD4 on itself or on another virion.

nef (7, 13, 30, 33, 50, 53)

Proceeding rightward we reach the last of the additional HIV genes, *nef*, named (confusingly) for *ne*gative *f*actor at a time when its effects were thought to be primarily transcriptional repression of the HIV promoter. More recent work has indicated that other functions of *nef* are probably more significant. *nef* is clearly important for infectivity in vivo, since SIV infection of monkeys or HIV infection of humans is much less pathogenic when the *nef* gene is deleted.

Four different mechanisms appear to mediate the action of Nef. Nef expression reduces the concentration of CD4 on cell surfaces by linking CD4 protein to β-COP, an adapter protein that promotes uptake of CD4 by cellular endocytotic mechanisms. Once internalized, CD4 is transported to lysosomes, where it is degraded. Nef also down-regulates cell-surface MHC-I, an immune system molecule that allows detection of virus-infected cells. The net effect is to abrogate the immune response against infected cells, thereby promoting virus production. Nef also alters signaling inside T cells, activating the cells and thereby creating an environment more favorable for HIV replication.

Last, Nef promotes production of the cytokines MIP1α, MIP1β, and RANTES in infected macrophages. Cytokines are immune signaling molecules, and the three produced act by recruiting cells to sites of infection. Production of these cytokines from macrophages recruits CD4-positive T cells to the vicinity of infected cells, thereby facilitating infection of the recruited T cells. This model is particularly attractive because the HIV coreceptors are in fact the receptors for the three cytokines produced, ensuring recruitment of just the right types of cells for further spread of the infection.

The importance of Nef has been dramatically illustrated in studies of humans and monkeys infected with *nef*-deleted viruses. Several cases are known of humans infected with *nef*-deleted virus, and their disease progression has been slow or absent. In monkeys, *nef*-deleted virus has greatly reduced pathogensis. This led to initial hopes that *nef*-deleted viruses might be usable as attenuated virus vaccines, but recent studies reveal that *nef*-deleted viruses do retain pathogenic potential after long-term infection. The challenge of developing an HIV vaccine remains unmet.

CONTROL OF THE HIV LIFE CYCLE

Why does HIV have a biphasic transcriptional program? Perhaps the only reason is that the early phase must be completed, generating Tat and Rev proteins from fully spliced RNAs, before it is possible to carry out the second phase. However, there may be benefits to this two-step program. Nef is also expressed from a multiply spliced RNA together with Tat and Rev, so the two-phase life cycle allows some Nef to accumulate in advance of virion production. Thus, the benefits of Nef action—receptor clearance, MHC I removal, target cell recruitment, and host cell activation—will all precede virion budding and release. The two-stage life cycle may thereby allow the creation of a more favorable setting for virion production once it begins.

THE PUZZLING LACK OF ENDOGENOUS LENTIVIRUSES

Although the lentiviruses are widespread among vertebrates, no endogenous lentiviruses have been detected. Why were HERV proviruses incorporated into the primate germ line, while SIV/HIV proviruses were not? Generally, it is risky to explain evolutionary pathways retrospectively, but the following points are potentially relevant. The lentiviruses mainly replicate in immune system cells and not reproductive cells, reducing the likelihood of infecting the germ line. Lentivirus infection often leads to death of the infected cells, whether by cytopathic effects of the virus or by immune clearance of infected cells, obviously reducing the chances of stable incorporation in the germ line. It may require a special combination of germ-line tropism and nondestructive replication to allow a retrovirus to become endogenous.

THE ORIGIN OF THE HIV EPIDEMIC (26, 35, 52, 57, 59)

The HIV epidemic began by transmission of SIVs from African primates to humans. Many types of viruses and mobile elements have been transferred between species, with profound effects on the recipient, but few cases are as clear or as tragic as HIV.

There are two types of HIVs, HIV-1 and HIV-2, that jumped species independently. The HIV-2 genome differs in sequence from that of HIV-1 and contains a slightly different complement of auxiliary genes (Fig. 7.4). HIV-2 lacks the *vpu* gene but contains *vpx*, which apparently arose by a duplication of *vpr*. In HIV-2 the multiple roles of HIV-1 *vpr* are apparently divided between HIV-2 *vpr* and *vpx*, with the latter assisting nuclear localization while the former arrests the host cell cycle.

The two types of HIV arose by cross-species transmission of different SIVs. The SIVs are widespread, with at least 18 distinct types known that infect at least 20 species of primates. In most cases, the SIVs are benign in their natural hosts, but are often pathogenic after transmission to another primate species. For example, transmission of SIVsm (SIV of sooty mangabeys) to macaques results in an AIDS-like disease that provides a key animal model for studies of AIDS.

Transmission of SIVsm to humans gave rise to HIV-2, as indicated by several lines of evidence. The genomes of the two viruses are very closely related, more so than either is to any other HIV or SIV (Fig. 7.5). The collection of viral genes is identical, including the presence of *vpr* and *vpx* and the lack of *vpu*. Sooty mangabeys are abundant in their African ranges and infected at a high rate (22% in some troops). The monkeys are eaten by local people and sometimes kept as pets, providing ample opportunity for cross-species transmission. The geographical location of infected mangabeys correlates with the regions of highest infection with HIV-2 in humans. Dramatically, in some cases the sequences of HIV-2 found in humans closely matched the sequences of SIVsm in local mangabeys, implying frequent local transmission. Available data suggest that SIVsm has been transmitted to humans independently as many as six times.

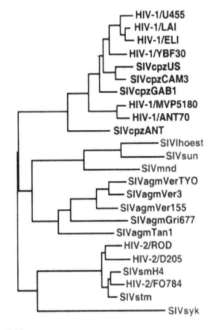

FIGURE 7.5. Phylogenetic relationships among primate retroviruses generated by comparing pol sequences. Note that HIV-1 clusters with SIVcpz, and HIV-2 clusters with SIVsm. (Reprinted, with permission, from Hahn et al. 2000 [© American Association for the Advancement of Science]. [26])

HIV-1, in contrast, resulted from the transmission of an SIV from chimpanzees (SIVcpz) into humans. In phylogenetic trees based on viral sequence, the HIV-1 and SIVcpz sequences group together, indicating their close relationship (Fig. 7.5). In addition, the complement of genes is identical, with *vpu* present and *vpx* absent. Several SIVcpz sequences have been isolated from chimpanzees, and comparison to the sequences of HIV indicates that the virus jumped into people independently at least three times.

How HIV-1 entered the human population is unknown, but that it did so is hardly surprising. Chimpanzees are eaten in Africa, and this practice offers many opportunities for chimpanzee blood to contaminate wounds in human hunters. An alternative proposal, that HIV-1 entered the human population due to SIV contamination of early polio vaccines, is very unlikely. The cells used to grow the virus for use as a vaccine do not support SIV replication, and, furthermore, the cells used for vaccine preparation did not come from mangabeys or chimpanzees.

When did HIV enter the human population? This question is impossible to answer for cases where the virus infected a local population but did not give rise to a sustained epidemic. The origin of the modern epidemic, however, can be tentatively reconstructed. The earliest clinical samples showing clear evidence of HIV-1 infection are from a sailor who died in 1959, providing direct evidence for HIV infection by this date. The rate of change of HIV sequences can be calculated, then the "molecular clock" run backward until the existing sequences coalesce into a few nodes, each of which is deduced to represent a separate jump of an SIV into humans. The rate of "ticking" of the molecular clock can be deduced from measurements of the rate of change of HIV after introduction into new populations. The recent explosion of HIV in Thailand, for example, appears to have started with the introduction of a single strain or closely related set of strains of HIV-1 in the mid-1980s. The virus has changed at a steady rate thereafter, allowing calibration of the molecular clock. Such an analysis yields a date of 1931 (1915–1941) for the jump of HIV to humans that started the modern epidemic. Probably the expansion of international travel, the increased mobility of human populations, and the introduction of routine syringe use allowed HIV infection to spread at a rate not formerly possible, seeding the modern epidemic. This date also counters the theory that contaminated cultures used for polio vaccines started the HIV epidemic. These cultures were prepared in the late 1950s, indicating that the polio vaccine is unlikely to have been the source of the HIV epidemic.

Why did some SIVs jump into the human population but not others? Many factors could have been involved, such as abundance of the particular SIV-infected primates in the vicinity of humans and the degree of interaction between them. Recent data suggest that the ability of different SIVs to replicate in human cells may also have been a major factor. Recall that the viral *vif* gene appears to act by overcoming the action of a host antiviral activity. It turns out that relatively few primate *vif* proteins are active in human cells. In most cases, SIVs replicate poorly on human cells, but their replication is boosted if artificially supplied with human *vif*. Two SIVs are clear exceptions, SIVsm and SIVcpz, exactly those two viruses that appear as HIVs in humans. A similar argument has been made for a role for Vpr in dictating which SIVs could cross into humans. These observations support the idea that the activity of the auxiliary genes determined which SIVs could spread in humans.

THERAPY OF HIV (2, 16)

An enormous effort has been devoted to developing small-molecule inhibitors of HIV replication for treatment of AIDS patients. The effort has been partially successful, in that the welfare of many patients has been improved. However, the drugs can only be tolerated by some patients, viral variants insensitive to the drugs have arisen in many cases, and, worst of all, the drug regimes are too expensive for use in the developing world, where the majority of HIV infections are found. In this section, we review the pharmacological basis for treatment of HIV. Unfortunately, the life-style of HIV—replication via reverse transcription and integration—results in profound obstacles for treatment.

Inhibiting the replication of any virus poses a difficult challenge, because most of the machinery for viral replication is stolen from the host. All viruses use host systems for translation and production of macromolecular precursors. Small viruses like HIV also use the host transcriptional apparatus to produce mRNAs, host enzymes to modify viral proteins after synthesis, and host machinery for subcellular sorting. Chemical inhibitors that interfere with these steps would be toxic to the host, since the host cell would be poisoned along with the virus. Thus, virus-encoded functions are generally the most promising inhibitor targets.

Reverse Transcriptase Inhibitors (22, 25)

The reverse transcriptase enzyme (RT) is one of the hallmarks of retroviruses and an obvious target for inhibitors. The first drug approved for HIV treatment, azido-thymidine or AZT, targeted RT. AZT acts as a chain terminator. When an AZT molecule is incorporated into a growing reverse transcript, the cDNA chain cannot be extended further because the point of incorporation of the next residue is blocked by the azide group (Fig. 7.6). Thus, incorporation of one AZT unit into an HIV genome is lethal to the virus if not removed.

But why is AZT not just as poisonous to the host? Cellular DNA polymerases must also incorporate deoxynucleotides to replicate

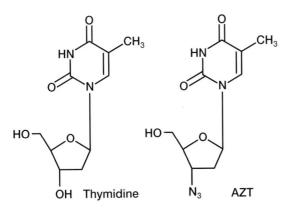

FIGURE 7.6. Comparison of thymidine and azido-thymidine (AZT). The 3′ hydroxyl in thymidine, which is the point of growth in a polynucleotide chain, in AZT is substituted with N_3 (azide), which cannot be coupled to another nucleotide. Consequently, AZT acts as a chain terminator during reverse transcription.

DNA in the course of DNA repair and cell division. Why does the incorporation of AZT by cellular polymerases not also poison the patient? The answer is that AZT is actually quite toxic, inhibiting particularly a DNA polymerase found in human mitochondria. However, empirical screening of many nucleotide analogs revealed that AZT was incorporated at least somewhat more efficiently by RT than by cellular polymerases. Several additional nucleotide analogs, which also act as chain terminators, have been approved for treatment of HIV.

A second class of RT inhibitors, the non-nucleoside RT inhibitors, binds to a different site on the enzyme. These compounds were identified by screening large libraries of small molecules for inhibitors of RT in vitro. The non-nucleoside RT inhibitors bind in a pocket beneath the active site for polymerization. The RT–inhibitor complex is impaired for polymerization, although the detailed mechanism of action is unclear. Mutations that reduce the sensitivity of RT to each of these classes of inhibitors lie in different parts of the enzyme.

Protease Inhibitors (8, 16)

The development of a second class of inhibitors against the protease enzyme (PR) has greatly improved the welfare of HIV-infected patients. The development of protease inhibitors was greatly aided by previous studies on inhibitors of related enzymes, because drugs against the HIV enzyme could be designed on the basis of the structures of known protease substrates and inhibitors. An iterative process of empirical testing, analysis of enzyme–drug complexes, and new synthesis led to the development of potent inhibitors highly active against virus in patients (Fig. 7.7). The protease inhibitors all retain some resemblance to the peptide backbone found in the natural biological substrates. The pro-

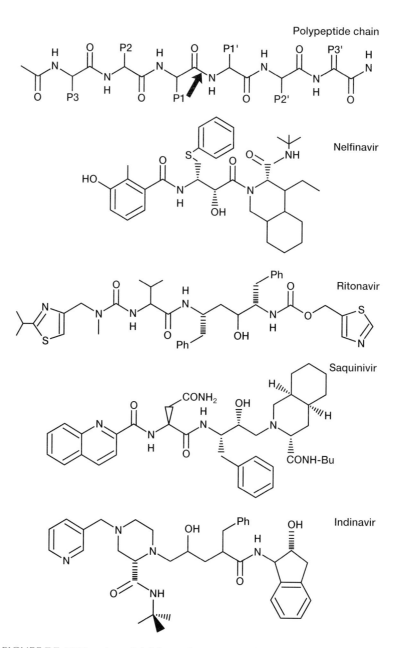

FIGURE 7.7. HIV protease inhibitors, drawn to emphasize the structural similarity with the polypeptide substrate of protease. (Modified, with permission, from Emini and Fan 1997. [16])

tease inhibitors act on immature virions, preventing the required pro-teolysis after budding that generates the mature viral proteins from the polyprotein precursors. Immature viruses are not infectious and are quickly cleared in patients.

Fusion Inhibitors (14, 34)

As discussed below, even with RT and PR inhibitors, the ability to treat patients is still quite limited, leading to intense efforts to develop additional therapies. One promising target is the gp41 component of the viral envelope. After the initial binding of envelope and receptor, gp41 promotes the fusion of the viral and cellular membranes. The gp41 protein is believed to undergo a massive conformational change during fusion. A hydrophobic peptide becomes exposed during this process, which is believed to intercalate into a lipid bilayer on one of the partners and so promote membrane fusion. A peptide matching part of gp41, named T20, is effective at blocking fusion, probably by interfering with the gp41 conformational change. T20 is in use for treating patients who lack other therapeutic options, but peptide-based drugs have the disadvantage of requiring administration by daily injections of large amounts of material. Intense efforts are under way to develop orally available small molecules active against the same target.

Integrase Inhibitors (27, 43, 44, 47)

The HIV IN protein, the third of the HIV-encoded enzymes, has not yet been exploited for inhibitor development. The integrase enzyme is an attractive target, because there are no known cellular enzymes that carry out similar reactions, so it may be possible to inhibit integrase selectively. Integrase may be a difficult target for inhibitors, however, because it only acts four times, twice to cleave each cDNA end and twice to join each cDNA end to target DNA. RT, in contrast, must turn over thousands of times to copy the viral RNA.

At present, many inhibitors of integrase have been identified, a few of which have activity against HIV in cell culture. As yet there are no clinically useful inhibitors, although efforts are under way in many companies and laboratories to achieve this goal. One exciting development is the identification of several binding sites for integrase inhibitors on the enzyme surface (Fig. 7.8). If more than one of these sites can be targeted by clinically useful inhibitors, it may be possible to develop multidrug therapy of integrase alone.

Drug Evasion by Mutation (5, 8, 31, 48)

Why are the drugs used to treat HIV infection only partially effective? As touched on above, many are toxic, with side effects ranging from anemia to nausea to kidney stones. A substantial fraction of patients cannot tolerate the treatment regimens at all. In addition, the treatments are expensive and challenging to administer. Perhaps the worst problem is the rapid development of resistance mutations, a consequence of the dynamics of HIV infection.

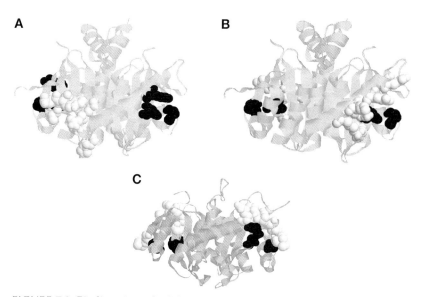

FIGURE 7.8. Binding sites of inhibitors on retroviral integrase catalytic domains, illustrating that three different sites are seen in the three cases. Active-site residues are shown in black and the drug-binding sites in gray space-filling representations. The integrase catalytic domain backbone is shown as the green ribbon. (*A*) The binding site for tetraphenyl arsonium catechol on the HIV integrase catalytic domain. Based on the structure reported in Molteni et al. (2000)(44). (*B*) Binding site for a diketoacid inhibitor on the HIV integrase catalytic domain. Based on the structure reported in Goldgur et al. (1999)(24). (*C*) Binding site for the Y3 inhibitor on an avian retrovirus catalytic domain. Based on the structure reported in Lubkowski et al. (1998).(40).

The very large numbers of virions present in an infected person, combined with the high error rate of RT, leads to the rapid generation of viruses with reduced sensitivity to the HIV drugs. In an infected person, some 10^{10} virions are produced and destroyed per day. Combined with this, the HIV RT enzyme is quite error-prone, adding an incorrect dNTP to a growing DNA chain about once every 10^4 bases polymerized. Because the HIV genome is about 10^4 base pairs, each HIV genome will have on average one mutated base due to RT errors. The average genome will have one error, but others will have none, two, or more errors. If one in 1000 viruses is competent for replication (an informed guess), then 10^7 viable virions are produced, each containing on average one altered base. Thus, all viable single-base substitutions will arise in an infected individual *each day*, as well as many of the two-base substitutions.

Against this background of constant change, it is hardly surprising that treatment with many antiviral agents is ineffective. For the case of the RT inhibitors, for example, a single amino acid substitution in RT can greatly reduce the sensitivity to many of the drugs. Because of the extremely large numbers of virions present in an infected person, the mutations conferring resistance will already be present in the patient before treatment begins. Patients treated with a single antiviral drug often display a rapid drop in the viral concentration in blood, but with-

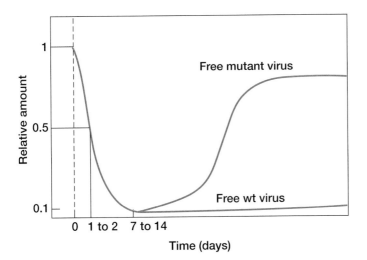

FIGURE 7.9. Evolution of drug-resistant HIV variants in response to monothera-py. The vertical axis indicates the relative amount of virus in blood plasma, the horizontal axis indicates the time in days. Upon initiation of therapy, the concentration of virus in blood falls, but quickly the level of virus rebounds and all the viruses present are resistant to the inhibitor. (Modified, with permission, from Coffin 1995 [American Association for the Advancement of Science].[5])

in a few weeks the virus population rebounds (Fig. 7.9). Analysis of the viruses present after rebound reveals that all of them are resistant to the drug used for treatment. These types of observations were in fact some of the earliest data that led to the discovery of the large HIV populations in patients. Obviously, therapy with a single drug provides transient benefits at best.

Combination Therapy (8, 16)

Treatments using multiple HIV drugs at the same time are much more effective. For example, combination of a nucleoside RT inhibitor, a non-nucleoside RT inhibitor, and a protease inhibitor demands mutations in three different parts of the viral genome to evade the drugs. Evasion of the protease inhibitors typically requires multiple amino acid substitutions in PR for high-level resistance, further increasing the "genetic barrier" that the virus must surmount to replicate in the presence of the drugs. Many patients undergoing triple combination therapy have responded well, with their virus concentrations in blood declining below the level of detection for several years. The introduction of triple combination therapy has provided great benefit to many patients, at least in the developed world.

Another treatment success has come in suppressing the transmission of HIV from mother to infant during birth. Treatment of the mother with antiviral agents before delivery has led to a greatly reduced rate of transmission during delivery. In the developed world, this has led to a heartening decline in HIV infection in infants born to infected mothers.

As was seen with antibiotic resistance in Chapter 3, however, pathogens evolve in response to drug selection. Not only do resistant HIV variants appear after monotherapy, but triply resistant viruses are also emerging. Worse, as the population becomes more experienced with the antiretroviral drugs, de novo infections increasingly involve drug-resistant virus. For a patient infected with a virus resistant to AZT, therapy with AZT and another drug is no better than monotherapy. It is clear that triple drug-resistant viruses are present in an increasing fraction of new infections. This frightening development may foreshadow the future of the epidemic.

REVERSING EVASION BY INTEGRATION (4, 21, 38, 39, 58)

Can HIV be eradicated from an infected person completely? Obviously, this would eliminate concerns about rebound of resistant virus. Much of the answer to this question turns on the nature of HIV as an integrating genetic element.

Studies of the behavior of infected cells during triple combination therapy suggest that there are at least two populations (Fig. 7.10). One declines quickly after initiating treatment, with a typical half-life of around 6 hours. A second population declines more slowly, with a half-life of 1–4 weeks. The first was initially thought to consist of blood CD4$^+$ T cells, and the second, macrophages. However, this interpretation has recently been complicated by the finding that gut immune tissue is one of the largest immune organs in the body and one of the most affected by SIV and probably HIV infection. Regardless of the sites involved, the hope has been that if triple combination therapy suppresses viral replication long enough, and infected cells decline in abundance at a steady rate, eventually all the infected cells should disappear. The patient so treated would then be cured.

Unfortunately, the discovery of reservoirs of integrated quiescent virus has dashed this hope. In 1997 several research groups isolated "resting" CD4$^+$ cells from blood of patients who had been treated successfully with triple combination therapy for up to 30 months. Resting CD4$^+$ cells were then isolated from blood and stimulated to resume cell division. Although the cells were not producing virus prior to stimulation, about one in a million cells was found to yield virus following activation. The implication is that a provirus can reside in resting cells in a quiescent or latent state. Such proviruses are not transcribed and so do not produce proteins that can be recognized by the immune system. This special class of infected cells can then persist for very long times in the body. From time to time, the resting cells are expected to be reactivated, perhaps as a part of an immune response against a new microbial invader. The latent virus would then also resume replication, thereby reseeding an otherwise successfully treated infection.

Several approaches have been suggested to eradicate the reservoir of latent virus. The immune modulator interleukin-2 (IL-2) has the

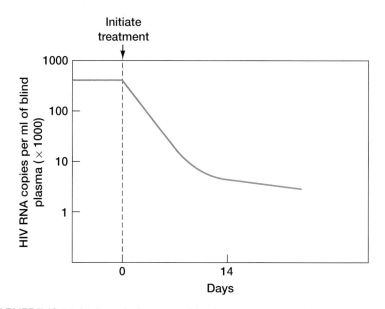

FIGURE 7.10. Multiphase decline in viral load in response to combination therapy, indicating the presence of different populations of virus. After initiating treatment, the concentration of virus in blood plasma falls rapidly, often to levels 100-fold lower than those initially present. At later times, the level of virus continues to fall, but more slowly, suggesting that virus from other compartments is slowly repopulating the blood. Curves such as these raised the hope that prolonged therapy might completely eliminate virus from patients, but later discoveries of long-lived cells containing latent virus set back these hopes. (Modified, with permission, from Ho et al. 1995 [©Macmillan].[31])

potential to activate resting T cells. One approach to eradicating virus involves the treatment of patients on triple combination therapy with IL-2 as well, with the hope of activating resting T cells and causing latent virus to reveal itself. The hope is that a normal immune response might then eliminate the uncloaked virus. Unfortunately, IL-2 is also toxic, limiting the dose that can be given to patients. Treatment of patients with IL-2 to date has yielded only disappointing results.

A related approach may be more promising. Several indirect observations suggest that the immune response against HIV may not be optimal in patients on triple combination therapy because the immune system is not exposed to enough viral proteins. According to this view, exposure of patients on effective therapy to viral antigens may boost the immune response, perhaps allowing control of virus if therapy is ended. Therapeutic vaccination is one way of applying such therapy, although results to date have been unimpressive. In contrast, striking preliminary data have been achieved with a related approach: structured treatment interruptions. In this protocol, patients are periodically taken off therapy, allowing a transient increase in virus production. With renewed triple combination therapy, viral suppression was regained. Excitingly, in a few cases, subsequent interruptions led to much smaller increases in virus production. Similar results have been achieved in some cases with triple combination therapy of macaques infected with SIV. Although much remains uncertain, it seems at least possible that this means of uncloaking hidden virus will allow HIV to be controlled in some patients.

Summary

Many host–parasite relationships reach a benign equilibrium after prolonged coevolution. Earlier on, the relationship is often much more pathogenic, as with the lethal acquired immune deficiency syndrome (AIDS) caused by HIV.

HIV is transmitted by contaminated blood, by birth, and by sex. After the initial infection, the virus spreads, seeding cells throughout the body. The virus titer in blood increases rapidly, then falls concomitant with the development of an immune response against the virus. The virus titers can remain stable at this new level for many years. This clinical latent stage is a very dynamic equilibrium, with some 10^{10} virus particles produced and destroyed per day in an infected person. Eventually the number of CD4$^+$ cells declines, perhaps because the ability to replace CD4 cells killed by infection becomes exhausted. The erosion of immune function due to loss of these cells permits myriad microbes to invade the afflicted person, which together with other maladies, eventually causes death.

HIV differs from the simple retroviruses in encoding several genes in addition to *gag*, *pol*, and *env* that are also important for replication. HIV encodes two proteins, Tat and Rev, that regulate HIV mRNA production. Tat promotes gene expression, whereas Rev promotes export of incompletely spliced mRNA from the nucleus. The Nef, Vif, Vpr, and Vpu proteins are important for infection in vivo, although their mechanisms of action are less clear. Vif may be important for overcoming a cellular system that inhibits viral replication. The Vpr protein appears to help incoming viral genomes localize to the host cell nucleus, and also to stop the host cell cycle at a point favorable for HIV gene expression. Nef and Vpu appear to carry out several functions that promote replication, such as removing CD4 from cell surfaces, but much is uncertain.

The HIV Env protein is also distinctive, requiring for cell entry not just the receptor, CD4, but also a coreceptor, which can be any of several cell-surface molecules that normally bind chemokine molecules involved in signaling. Different HIV Env proteins function together with different coreceptors. The importance of one coreceptor, CCR5, is emphasized by the finding that people mutant in both copies of the *CCR5* gene are resistant to infection by HIV.

The two types of HIV, HIV-1 and HIV-2, appear to have entered the human population by independent jumps of SIVs from nonhuman primates. HIV-1 is closely related to a SIV of chimpanzees, SIVcpz, and HIV-2 is closely related to SIVsm from sooty mangabeys. Altogether, there appear to have been at least nine independent jumps of SIVs into humans.

Intensive studies of anti-HIV drugs emphasize how difficult it can be to eradicate a mobile element integrated in large numbers of cells. Inhibitors against the HIV RT and protease are in wide use today. Intensive contemporary studies focus on developing inhibitors against new targets such as integrase and gp41. Treatments with combinations of RT and protease inhibitors in many cases provide suppression of viral load and considerable clinical benefit. However, therapy is often not successful because patients cannot tolerate the side effects of the drugs or because the virus mutates to become insensitive. In cases where patients have been taken off the drug regimen, virus usually rebounds, revealing that latent virus has remained "archived" in protected sites. A very promising recent finding is that, in a few cases, cycling patients on and off the drugs has led to greatly reduced virus rebound, as though much of the hidden virus has revealed is presence to the immune system and so been eradicated. These findings raise the exciting question of whether such "structured treatment interruptions" will allow some patients to control HIV infection.

REFERENCES

1. Berger E.A., Murphy P.M., and Farber J.M. 1999. Chemokine receptors as HIV-1 coreceptors: Roles in viral entry, trophism, and disease. *Annu. Rev. Immunol.* **17:** 657–700.

2. Bushman F.D., Landau N.R., and Emini E.A. 1998. New developments in the biology and treatment of HIV. *Proc. Natl. Acad. Sci.* **95:** 11041–11042.

3. Cao H. and Walker B.D. 2000. Immunopathogenesis of HIV-1 infection. *Clin. Dermatol.* **18:** 401–410.

4. Chun T.-W., Stuyver L., Mizell S.B., Ehler L.A., Mican J.A.M., Baseler M., Lloyd A.L., Nowak M.A., and Fauci A.S. 1997. Presence of an inducible HIV-1 latent reservoir during highly active antiretroviral therapy. *Proc. Natl. Acad. Sci.* **94:** 13193–13197.

5. Coffin J.M. 1995. HIV population dynamics in vivo: Implications for genetic variation, pathogenesis, and therapy. *Science* **267:** 483–489.

6. Coffin J.M., Hughes S.H., and Varmus H.E. 1997. *Retroviruses.* Cold Spring Harbor Laboratory Press, Cold Spring Harbor, New York.

7. Collins K.L. and Baltimore D. 1999. HIV's evasion of the cellular immune response. *Immunol. Rev.* **168:** 65–74.

8. Condra J.H. and Emini E.A. 1997. Preventing HIV-1 drug resistance. *Sci. Med.* **4:** 14–23.

9. Connor R.I., Chen B.K., Choe S., and Landau N.R. 1995. Vpr is required for efficient replication of human immunodeficiency virus type-1 in mononuclear phagocytes. *Virology* **206:** 935–944.

10. Cullen B.R. 1998. HIV-1 auxilary proteins: Making connections in a dying cell. *Cell* **93:** 685–692.

11. Deng H., Liu R., Ellmeier W., Choe S., Unutmaz D., Burkhart M., Di Marzio P., Marmon S., Sutton R.E., Hill C.M., Davis C.B., Peiper S.C., Schall T.J., Littman D.R., and Landau N.R. 1996. Identification of a major co-receptor for primary isolates of HIV-1 (see comments). *Nature* **381:** 661–666.

12. Doms R.W. and Trono D. 2000. The plasma membrane as a combat zone in the HIV battlefield. *Genes Dev.* **14:** 2677–2688.

13. Dyer W.B., Geczy A.F., Kent S.J., McIntyre L.B., Blasdall S.A., Learmont J.C., and Sullivan J.S. 1997. Lymphoproliferative immune function in the Sydney Blood Bank Cohort, infected with natural nef/long terminal repeat mutants, and in other long-term survivors of transfusion-acquired HIV-1 infections. *AIDS* **13:** 1565–1574.

14. Eckert D.M., Malashkevich V.N., Hong L.H., Carr P.A., and Kim P.S. 1999. Inhibiting HIV-1 entry: Discovery of D-peptide inhibitors that target the gp41 coiled-coil pocket. *Cell* **99:** 103–115.

15. Emerman M. and Malim M.H. 1998. HIV-1 regulatory/accessory genes: Keys to unraveling viral and host cell biology. *Science* **280:** 1880–1884.

16. Emini E.A. and Fan H.Y. 1997. Immunological and pharmacological approaches to the control of retroviral infections. In *Retroviruses* (ed. J.M. Coffin et al.), pp. 637–706. Cold Spring Harbor Laboratory Press, Cold Spring Harbor, New York.

17. Fan H.Y., Conner R.F., and Villarreal L.P. 2000. *AIDS: Science and Society.* Jones and Bartlett, Boston, Massachusetts.

18. Fauci A.S. and Desrosiers R.C. 1997. Pathogenesis of HIV and SIV. In

Retroviruses (ed. J.M. Coffin et al.), pp. 587–635. Cold Spring Harbor Laboratory Press, Cold Spring Harbor, New York.

19. Feng Y., Broder C.C., Kennedy P.E., and Berger E. 1996. HIV-1 entry cofactor: Functional cDNA cloning of a seven-transmembrane G protein-coupled receptor. *Science* **272:** 872–876.

20. Fields B.N., Knipe D.M., and Howley P.M., Eds. 1996. *Fields' Virology,* 3rd edition. Lippincott-Raven, Philadelphia, Pennsylvania.

21. Finzi D., Hermankova M., Pierson T., Carruth L.M., Buck C., Chaisson R.E., Quinn T.C., Chadwick K., Margolick J., Brookmeyer R., Gallant J., Markowitz M., Ho D.D., Richman D.D., and Siliciano R.F. 1997. Identification of a reservoir for HIV-1 in patients on highly active antiretroviral therapy. *Science* **278:** 1295–1300.

22. Fischl M.A., Richman D.D., Grieco M.H., Gottlieb M.S., Volberding P.A., Laskin O.L., Leedom J.M., Groopman J.E., Moldvan D., Schooley R.T., and et al. 1987. The efficacy of azidothymidine (AZT) in the treatment of patients with AIDS and AIDS-related complex. A double-blind, placebo-controlled trial. *N. Engl. J. Med.* **317:** 185–191.

23. Garber M. and Jones K.A. 1999. HIV-1 Tat: Coping with negative elongation factors. *Curr. Opin. Immunol.* **11:** 460–465.

24. Goldgur Y., Craigie R., Cohen G.H., Fujiwara T., Yoshinaga T., Fujishita T., Sugimoto H., Endo T., Murai H., and Davies D.R. 1999. Structure of the HIV-1 integrase catalytic domain complexed with an inhibitor: A platform for antiviral drug design. *Proc. Natl. Acad. Sci..* **96:** 13040–13043.

25. Gorbach S.L., Bartlett J.G., and Blacklow N.R. 1998. *Infectious diseases.* W.B. Saunders, Philadelphia, Pennsylvania.

26. Hahn B.H., Shaw G.M., De Cock K.M., and Sharp P.M. 2000. AIDS as a zoonosis: Science and public health implications. *Science* **287:** 607–614.

27. Hazuda D.J., Felock P., Witmer M., Wolfe A., Stillmock K., Grobler J.A., Espeseth A., Gabryelski L., Schleif W., Blau C., and Miller M.D. 2000. Inhibitors of strand transfer that prevent integration and inhibit HIV-1 replication in cells. *Science* **287:** 646–650.

28. He J., Choe S., Walker R., Di Marzio P., Morgan D.O., and Landau N.R. 1995. Human immunodeficiency virus type 1 viral protein R (Vpr) arrests cells in the G2 phase of the cell cycle by inhibiting p34cdc2 activity. *J. Virol.* **69:** 6705–6711.

29. Heinzinger N.K., Bukrinsky M.I., Haggerty S.A., Ragland A.M., Kewalramani V., Lee M.-A., Gendelman H.E., Ratner L., Stevenson M., and Emerman M. 1994. The Vpr protein of human immunodeficiency virus type 1 influences nuclear localization of viral nucleic acids in nondividing host cells. *Proc. Natl. Acad. Sci.* **91:** 7311–7315.

30. Herna Remkema G. and Saksela K. 2000. Interactions of HIV-1 NEF with cellular signal transducing proteins. *Front. Biosci.* **5:** 268–283.

31. Ho D.D., Neumann A.U., Perelson A.S., Chen W., Leonard J.M., and Markowitz M. 1995. Rapid turnover of plasm virions and CD4 lymphocytes in HIV-1 infection. *Nature* **373:** 123–126.

32. Hope T.J. 1999. The ins and outs of HIV. *Rev. Arch. Biochem. Biophys.* **365:** 186–191.

33. Kestler H.W., Ringler D.J., Mori K., Panicali D.L., Sehgal P.K., Daniel M.D., and Desrosiers R.C. 1991. Importance of the nef gene for maintenance of high virus loads and for development of AIDS. *Cell* **65:** 651–662.

34. Kilby J.M., Hopkins S., Venetta T.M., DiMassimo B., Cloud G.A., Lee J.Y.,

Alldredge L., Hunter E., Lambert D., Bolognesi D., Matthews T., Johnson M.R., Nowak M.A., Shaw G.M., and Saag M.S. 1998. Potent suppression of HIV replication in humans by T20, a peptide inhibitor of gp41-mediated virus entry. *Nat. Med.* **4:** 1302–1307.

35. Korber B., Muldoon M., Theiler J., Gao F., Gupta R., Lapedes A., Hahn B.H., Wolinsky S., and Bhattacharya T. 2000. Timing the ancestor of the HIV-1 pandemic strains. *Science* **288:** 1789–1796.

36. Landau N.R. 1999. HIV recent advances in AIDS research: Genetics, molecular biology and immunology (editorial). *Curr. Opin. Immunol.* **11:** 449–450.

37. Liu R., Paxton W.A., Choe S., Ceradini D., Martin S.R., Horuk R., MacDonald M.E., Stuhlmann H., Koup R.A., and Landau N.R. 1996. Homozygous Defect in HIV-1 coreceptor accounts for resistance of some multiply-exposed individuals to HIV-1 infection. *Cell* **86:** 367–377.

38. Lori F., Maserati R., Foli A., Seminari E., Timpone J., and Lisziweicz J. 2000. Structured treatment interruptions to control HIV-1 infection. *Lancet* **355:** 287–288.

39. Lori F., Lewis M.G., Xu J., Varga G., Zinn D.E., Crabbs C., Wagner W., Greenhouse J., Silvera P., Yalley-Ogunro J., Tinelli C., and Lisziewicz J. 2000. Control of SIV rebound through structured treatment interruptions during early infection. *Science* **290:** 1591–1593.

40. Lubkowski J., Yang F., Alexandratos J., Wlodawer A., Zhao H., Burke, Jr., T.R., Neamati N., Pommier Y., Merkel G., and Skalka A.M. 1998. Structure of the catalytic domain of avian sarcoma virus integrase with a bound HIV-1 integrase-targeted inhibitor. *Proc. Natl. Acad. Sci.* **95:** 4831–4836.

41. Madani N. and Kabat D. 1998. An endogenous inhibitor of human immunodeficiency virus in human lymphocytes is overcome by the viral Vif protein. *J. Virol.* **72:** 10251–10255.

42. Marassi F.M., Ma C., Gratkowski H., Straus S.K., Strebel K., Oblatt-Montal M., Montal M., and Opella S.J. 1999. Correlation of the structural and functional domains in the membrane protein Vpu from HIV-1. *Proc. Natl. Acad. Sci.* **96:** 14336–14341.

43. Molteni V., Rhodes D., Rubins K., Hansen M., Bushman F.D., and Siegel J.S. 2000. A new class of HIV-1 integrase inhibitors: The 3, 3, 3′,3′-tetramethyl-1,1′-spirobi(indan)-5,5′,6,6′-tetrol family. *J. Med. Chem.* **43:** 2031–2039.

44. Molteni V., Greenwald J., Rhodes D., Hwang Y., Kwiatkowski W., Bushman F.D., Siegel J.S., and Choe S. 2001. Identification of a small molecule binding site at the dimer interface of the HIV integrase catalytic domain. *Acta Crystallogr. Sect. D Biol. Crystallogr.* **57:** 536–544.

45. Perelson A.S., Neumann A.U., Markowitz M., Leonard J.M., and Ho D.D. 1996. HIV-1 dynamics in vivo: Virion clearance rate, infected cell life-span, and viral generation time. *Science* **271:** 1582–1586.

46. Pollard V.W. and Malim M.H. 1998. The HIV-1 Rev protein. *Annu. Rev. Microbiol.* **52:** 491–532.

47. Pommier Y., Pilon A.A., Bajaj K., Mazumder A., and Neamati N. 1997. HIV-1 integrase as a target for antiviral drugs. *Antivir. Chem. Chemother.* **8:** 463–483.

48. Preston B.D. 1997. Reverse transcriptase fidelity and HIV-1 variation. *Science* **275:** 228–229.

49. Re F. and Luban J. 1997. HIV-1 Vpr: G2 cell cycle arrest, macrophages and

nuclear transport. *Prog. Cell Cycle Res.* **3:** 21–27.

50. Rhodes D., Ashton L., Solomon A., Carr A., Cooper D., Kaldor J., and Deacon N. 2000. Characterization of three nef-defective human immunodeficiency virus type 1 strains associated with long-term nonprogression. *J. Virol.* **74:** 10581–10588.

51. Simon J.H.M., Gaddis N.C., Fouchier R.A.M., and Malim M.H. 1998. Evidence for a newly discovered cellular anti-HIV-1 phenotype. *Nat. Med.* **4:** 1397–1400.

52. Stivahtis G.L., Soares M.A., Vodicka M.A., Hahn B.H., and Emerman M. 1997. Conservation and host specificity of Vpr-mediated cell cycle arrest suggest a fundamental role in primate lentivirus evolution and biology. *J. Virol.* **71:** 4331–4338.

53. Swingler S., Mann A., Jacque J., Brichacek B., Sasseville V.G., Williams K., Lackner A.A., Janoff E.N., Wang R., Fisher D., and Stevenson M. 1999. HIV-1 Nef mediates lymphocyte chemotaxis and activation by infected macrophages. *Nat. Med.* **5:** 997–1003.

54. Trono D. 1998. When accessories turn out to be essential. *Nat. Med.* **4:** 1368–1369.

55. Wei P., Garber M.E., Fang S.M., Fischer W.H., and Jones K.A. 1998. A novel CDK9-associated C-type cyclin interacts directly with HIV-1 Tat and mediates its high-affinity, loop-specific binding to TAR RNA. *Cell* **92:** 451–462.

56. Wei X., Ghosh S.K., Taylor M.E., Johnson V.A., Emini E.A., Deutsch P., Lifson J.D., Bonhoeffer S., Nowak M.A., Hahn B.H., Saag M.S., and Shaw G. 1995. Viral dynamics in human immunodeficiency virus type 1 infection. *Nature* **373:** 117–122.

57. Weiss R.A. 2001. Polio vaccines exonerated. *Nature* **410:** 1035–1036.

58. Wong J.K., Hezareh M., Gunthard H.F., Havlir D.V., Ignacio C.C., Spina C., and Richman D.D. 1997. Recovery of replication-competent HIV despite prolonged supression of plasma viremia. *Science* **278:** 1291–1295.

59. Zhu T., Korber B.T., Nahmias A.J., Hooper E., Sharp P.M., and Ho D.D. 1998. An African HIV-1 sequence from 1959 and implications for the origin of the epidemic. *Nature* **391:** 594–597.

Genes Floating on a Sea of Retrotransposons

MUCH OF THE DNA IN EUKARYOTES is derived from retrotransposons, mobile elements that employ reverse transcription in their replication. In the draft sequence of the human genome, fully 29% of the DNA is identifiable as retrotransposon or retrovirus remnants. Another 13% comprises sequences probably created by reverse transcription but not themselves encoding reverse transcriptase (RT). About 50% of the genome is not identifiable as anything; much of this DNA was also probably created by reverse transcription but has become unrecognizable due to mutation.

This chapter surveys the retroelements, a heterogeneous group employing several different replication strategies. Chapter 9 describes the DNA transposons of eukaryotes, and Chapter 10 surveys the eukaryotic genome sequences and the record of DNA dynamics they provide. The main types of eukaryotic transposons are compared at the beginning of Chapter 9.

Lateral transfer of transposons in eukaryotes is less frequent than in prokaryotes. Different eukaryotic organisms do have different complements of retrotransposons, consistent with the idea that they have been acquired by lateral transfer. However, an alternative explanation is that the earliest eukaryotic cells had all existing retrotransposons, and that they were differentially lost in subsequent lineages. There are many inactive retrotransposon sequences in eukaryotic genomes derived from ancient, once-active elements. Accumulation of mutations and deletions over time can make these sequences unrecognizable, so eventually the element is judged to be lost. However, as described below, both lateral transfer and inheritance by descent are thought to be responsible for the modern distribution of retrotransposons.

INTRODUCTION TO RETROELEMENTS AND THEIR RELATIVES (1, 3)

Early studies revealed that DNA of many eukaryotes, particularly metazoans, is largely composed of repetitive sequences. Copies varied in length from a few base pairs to several thousand, and ranged in frequency from a few to millions. Later sequencing studies revealed that some of the longer elements encoded proteins related to reverse transcriptases (RT), although in most cases, the RT genes were nonfunctional and often distantly related to the previously known retroviral RTs. Proposals that repetitive DNA arose from RT-encoding elements were initially greeted with skepticism. This view has changed greatly, however, with the direct demonstration of retrotransposition for many of the element families.

Genetic elements that encode or exploit RT enzymes are introduced in this section (summarized in Table 8.1). Those elements that are implicated in altering eukaryotic genomes over time are discussed in more detail in subsequent sections. The retroviruses, the first known members of this family (Table 8.1, line 1), were discussed in Chapters 6 and 7.

The pararetro-viruses package a circular DNA genome inside viral particles.

Another group of viruses, the pararetroviruses, also replicate by reverse transcription, but differ by not integrating their cDNA (Table 8.1, line 2). Pararetrovirus replication involves transcription of a DNA template, then reverse transcription of the resulting RNA to make new genomes. Unlike the retroviruses, replication is carried out without integrating into the host cell chromosome. This group includes the human pathogen hepatitis B virus (HBV) and the plant cauliflower mosaic virus (CaMV). The pararetroviruses do not heritably alter the host cell genome as a required step in replication, and so are not considered further here.

The long terminal repeat (LTR) retrotransposons (Table 8.1, line 3) closely resemble the retroviruses, containing, as the name implies, LTR

TABLE 8-1. Retroelements and Their Relatives

Element type (other names)	Examples	Description	Recent review or reference
(1) Retroviruses	HIV, Mo-MLV, RSV, HTLV	RNA viruses that replicate by means of an integrated DNA provirus	Coffin et al. (1997) (12)
(2) Pararetroviruses	HBV, cauliflower mosaic virus	RNA viruses that replicate using reverse transcription but not integration	Hollinger (1996) (26)
(3) LTR retrotransposons	Ty1/*copia* family, Ty3/*gypsy* family	retrotransposons containing LTRs as in the retroviruses; two main families	Bennetzen and Kumar (1999) (1)
(4) Non-LTR retrotransposons	LINES, L1, I elements, ingi, Cin4	retrotransposons lacking LTRs, probably prime reverse transcription from a nick in cellular DNA	Kazazian (2000) (31)
(5) Retrotranscripts	Alu, SINES, processed pseudogenes	elements that show evidence of formation by reverse transcription, probably involving reverse transcriptase from non-LTR retrotransposons	Lander et al. (2001) (39)
(6) Retrointron	Mobile group I, II introns	mobile, intronic sequences, group II encodes RT	Lambowitz et al. (1999) (38)
(7) Retroplasmid	Mauriceville plasmid	plasmids encoding RT, potentially replicating through an RNA intermediate	Wang and Lambowitz (1993) (60)
(8) Retrons (bacterial retroelements)	msDNA	unique molecules containing RNA and DNA; found in bacteria only; function unknown	Travisano and Inouye (1995) (57)
(9) Telomerase	enzyme found in most eukaryotes	enzyme that synthesizes repeats at ends of chromosomes; uses an RNA internal guide sequence as the template for DNA synthesis	Stewart and Weinberg (2000) (55)

sequences at each end of the integrated DNA genome. LTR retrotransposons also resemble retroviruses in the proteins they encode, including Gag and Gag-Pol. They differ in lacking an envelope, and so do not have an obligatory extracellular phase. The steps of transcription, reverse transcription, and integration all take place in a single cell, resulting in the appearance of the integrated DNA in a new location in the cellular genome. Thus, the LTR retrotransposons are transposons instead of viruses, despite their close relationship with retroviruses. Members of this group are widespread in plants and animals, but absent in the Bacteria and Archaea.

The non-LTR retrotransposons (also called poly(A) retrotransposons) are a fascinating group that are also widespread in eukaryotes (Table 8.1, line 4). Much of the repeated DNA present in human cells derives either from these elements or from their action. The non-LTR retrotransposons lack LTR sequences and so cannot prime reverse transcription using the retrovirus-like strategy: Instead, they use a unique strategy described below.

The action of retrotransposons has led to the formation of several new classes of sequences in eukaryotic genomes, the SINES, processed pseudogenes, and "active pseudogenes" (Table 8.1, line 5). These sequences do not encode RT and generally do not produce any functional gene products related to retrotransposition. The machinery responsible for their formation is probably contributed by the non-LTR retrotransposons, acting occasionally on cellular RNAs instead of element RNAs. SINES are short sequences (typically 100–200 bp) that appear to be parasites on LINE elements. *Alu* elements, a type of SINE, comprise fully 10.6% of the draft sequence of the human genome. Processed pseudogenes are related to normal cellular genes but differ by loss of introns relative to the active gene copy and the presence of poly(A) sequences, both indications of their creation by reverse transcription of mRNAs.

Next on the list of retrotransposons are the mobile introns (Table 8.1, line 6), sequences that double as both introns and, for some, retrotransposons. Mobile introns reside in cellular genes but allow gene expression because the element RNAs are spliced out efficiently. The mobile introns encode much of the machinery necessary for their own removal, part of which is contributed by intron-encoded ribozymes. These elements can move by any of several means, depending on which type of the mobile intron is involved (Group I or Group II) and the nature of the host organism. To date, the mobile introns have not been found in vertebrates but are present in plants, fungi, and bacteria.

Several further types of elements are formed by reverse transcription. The retroplasmids are a group of elements found only in mitochondria of certain lower eukaryotes (Table 8.1, line 7). "Retrons" are another group, present in some bacteria and consisting of an odd RNA–DNA hybrid molecule (Table 8.1, line 8). These sequences are present as multiple copies per cell and contain single-stranded DNA,

Repeated sequences in higher eukaryotic genomes can be divided into long interspersed nuclear elements (LINES) and short interspersed nuclear elements (SINES). Although these names derived simply from the lengths of the repeat units, it is now clear that they correspond to different element families. The non-LTR retrotransposons are major components of the LINE group, and Alu *in human is the most abundant of the SINES.*

hence the name msDNA. Little is known about the biological role of these strange sequences.

Finally, the cellular telomerase enzyme is also a reverse transcriptase (Table 8.1, line 9). Telomerase enzymes are responsible for synthesizing the ends of the linear chromosomes of most eukaryotes. Telomerase adds characteristic short repeated sequences to the chromosome ends, with the detailed sequence varying from organism to organism. It turns out that telomerase is a RNA–protein complex. The sequence of the telomeric repeat is specified by an "internal guide" RNA within the complex that acts as a template for DNA synthesis, formally a reverse transcription step. The sequence of one of the telomerase proteins displays distant though clear similarity to known reverse transcriptases. Telomerase is discussed further in Chapter 10.

LATERAL TRANSFER OF RETROTRANSPOSONS (1, 19, 21, 27, 28, 46)

A variety of studies discussed below indicate that some retrotransposons are transferred between species, but how might this take place? After all, there is no known extracellular form of these elements, making their transfer more mysterious than that for viruses or conjugative plasmids. Lateral transfer in these cases is difficult to study because the transfer events occur only rarely, but plausible mechanisms can be proposed. Many of these mechanisms also apply to the DNA transposons discussed in the next chapter.

One means of transfer could be by hitchhiking on some other type of mobile element. A retrotransposon might integrate into the genome of a large DNA virus, for example, and then become transferred between cells by viral infection. If the cell survived the virus infection, the element might be transcribed from the viral genome and retrotranspose into the host chromosome. Many viruses infect multiple host species, providing a means for lateral transfer between taxa.

Several naturally occurring examples have been reported of mobile element sequences in the genomes of DNA viruses. The baculoviruses, large DNA viruses of insects, are particularly well documented as vehicles for transporting transposons. Different isolates have been found to harbor a *copia* retrotransposon, a *Tc1*-like DNA transposon, and a *piggyback* DNA transposon (specific transposon families are discussed below and in Chapter 9). Particularly interesting is another case, a *TED*-like retrotransposon in a baculovirus genome, because in this case it could be shown that the inserted element was expressed as a late gene in the virus, potentially allowing retrotransposition late during baculovirus infection.

In another example, some isolates of Marek's disease virus (a bird herpesvirus) were found to contain integrated LTRs of reticuloendotheliosis virus (REV), a retrovirus. REV could also integrate into Marek's disease virus in experimental co-infections of cultured cells.

Another plausible mechanism involves transposon transfer by parasites. As described in the next chapter, parasitic mites have been proposed to mediate the transfer of the P-element DNA transposon between *Drosophila* species. Mites feed on fly eggs but do not kill all the eggs they attack, providing an opportunity for transfer of DNA during feeding. Bacterial parasites are also candidates, particularly because they can take up DNA by transformation and introduce it into many types of cells by conjugation. Moreover, transfer of DNA into eukaryotic cells by conjugation is well documented (Chapter 12), bolstering this idea.

For those elements that can become incorporated into viral particles, DNA might also be transferred by viral infection. The *gypsy* retrotransposon turns out to be a retrovirus, as discussed below, and feeding *gypsy* particles to *Drosophila* larvae results in occasional transfer of *gypsy* genomes. Several members of the Ty3-*gypsy* family isolated from plants also encode potential envelope sequences. This provides a means for transfer not only of members of the Ty3-*gypsy* family, but for any retroelement genome RNA that can become incorporated into a *gypsy* particle. Such a "pseudotyping" mechanism probably explains the observation that some defective LTR retrotransposons have actually expanded in eukaryotic genomes, necessarily implying that they recruited replication machinery from other elements. Several of the plant LTR retrotransposons are present in many copies but do not encode functional reverse transcriptases. In these cases, the elements have presumably multiplied by hijacking the reverse transcriptase of other retrotranspsons for replication of their own genomes. Thus, even highly defective retroelement sequences may be able to multiply if their genomes encode the sequences required in *cis* for association with replication machinery encoded by other elements.

Other methods of transfer involve transient exposure of genomes to chromosomes containing transposons. Yeasts, for example, can grow as haploids (one copy of each chromosome), mate to form diploids (two copies), then sporulate to form haploids again. If mating occurs between a strain that contains a retrotransposon and one that lacks it, retrotransposition can transfer the element to a previously transposon-free chromosome. Subsequent sporulation can recreate haploid progeny, both of which contain the element. Thus, the yeast retrotransposons can be thought of as "venereal" diseases transferred by mating.

A related mechanism involves "wide crosses" between plant species. Many plants use relatively nonspecific means for transferring pollen between species, such as insects or wind. This results in occasional fertilization of one species with pollen of another. Such wide crosses rarely give rise to mature plants, but the DNA may be incorporated transiently, providing an opportunity for retrotransposition from the incoming genome to the host cell. It may even be that such crosses activate retrotransposition, increasing the transfer frequency in a manner analogous to hybrid dysgenesis (Chapter 9). Eventually, the incoming plant genome may be eliminated, but the retroelement may persist.

These possible transfer mechanisms are much more speculative than for prokaryotic elements or for retroviruses. The data supporting lateral transfer of transposons are nevertheless strong in some cases, and plausible in many others. We now turn to studies of specific types of retrotransposons and their lateral transfer.

LTR RETROTRANSPOSONS
(1, 4, 6, 8, 10, 25, 29, 33, 35, 44, 53, 54, 61, 63, 64)

How does one establish that a hypothetical retrotransposon actually uses an RNA intermediate in its transposition cycle? DNA transposons were discovered first, leaving early workers on retrotransposons faced with the job of disproving transposition through a DNA intermediate. This problem was solved beautifully by Jef Boeke, David Garfinkel, Cora Styles, and Gerry Fink, who studied the Ty1 LTR retrotransposon of yeast. They inserted DNA encoding an artificial intron into the Ty1 element and followed its fate during a cycle of retrotransposition (Fig. 8.1). For a DNA transposon, progeny elements should contain all the sequences of the starting transposon after a cycle of transposition, but the expectation for a retrotransposon is different. The element will first be transcribed to make the RNA transposition intermediate, which is then spliced like a typical cellular mRNA. The RNA is then reverse-transcribed and integrated, as with retroviruses. In a retrotransposon, the artificial intron should be lost during a cycle of transposition, and that is exactly what was observed. The Ty1 elements in new locations were found to have the intron cleanly removed, firmly establishing that Ty1 replicates through an RNA intermediate. The same strategy has since been used to confirm that a number of other candidate retrotransposons actually do transpose through RNA intermediates.

Genetic maps of representative LTR retrotransposons are shown in Figure 8.2. These elements fall into two major groups, Ty1-*copia* and Ty3-*gypsy*. These names are derived from some well-studied elements in each family: Ty1 and Ty3 elements from the yeast *Saccharomyces cerevisiae* and *copia* and *gypsy* from the fly *Drosophila melanogaster*. It is striking that the element families are distinct across such a large evolutionary distance. The Ty1 element of yeast is more similar to *copia*-like elements of plants than either is to Ty3, whereas Ty3-*gypsy* elements of plants and yeast are also more closely related to each other. The LTR retrotransposons all contain LTRs at each end of the integrated retrotransposon, with the characteristic division into U3, R, and U5 sequences (as with retroviruses, see Chapter 6). There is only one form of RNA produced, which serves both as the genome and as the mRNA for protein synthesis. Many elements contain a binding site for a tRNA

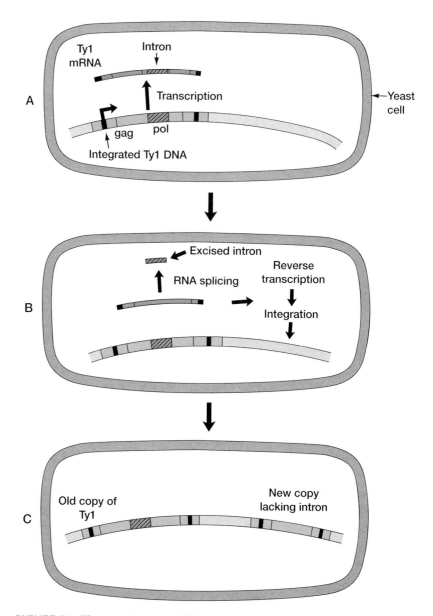

FIGURE 8.1. The experiment establishing that Ty1 elements in yeast replicate by retrotransposition. (*A*) An integrated Ty1 DNA element acts as a transcriptional template directing production of Ty1 RNA. The artificial Ty1 element studied contained an engineered artificial intron. (*B*) The Ty1 RNA is spliced to remove the engineered intron, then the RNA is reverse-transcribed and integrated. (*C*) After a cycle of retrotransposition, the newly formed Ty1 element is identical to the parental copy but lacks the engineered intron. This establishes that replication must have gone through an RNA intermediate. (Adapted from Boeke et al. 1985.[4])

primer (primer-binding site, PBS) just downstream of the left LTR for initiation of minus-strand cDNA synthesis. Upstream of the right LTR is a polypurine tract, which serves as the site of initiation of plus-strand synthesis after cleavage of the genomic RNA. Each of these functions parallels previously identified functions of retroviruses, although some

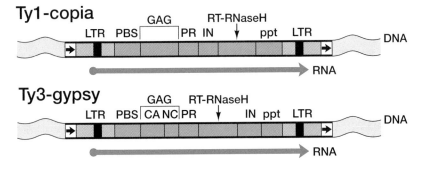

FIGURE 8.2. Genetic structure of the two main families of LTR retrotransposons. Some members of the Ty3/*gypsy* class contain in addition an *env* gene between IN and the PPT, and can replicate as retroviruses. (Adapted from Bennetzen and Kumar 1999.[1])

of the LTR retrotransposons do show variations on this theme.

LTR retrotransposons differ from the retroviruses in the detailed arrangement and processing of the element-encoded proteins. Both the Ty1-*copia* and Ty3-*gypsy* elements have *gag*-like regions. Gag of Ty1-*copia* is predominantly a single protein, whereas that of Ty3-*gypsy* is processed into at least two proteins by proteolytic cleavage. The Ty1-*copia* group is unique in having a *pol* gene order *PR-IN-RT-RNase H*. The Ty3-*gypsy* group and all retroviruses have instead the order *PR-RT-RNase H-IN*.

A distinctive feature of some of the LTR retrotransposons is their selective choices of target sites for integration. In one study, the loca-

---

Distinguishing between LTR Retrotransposons and Retroviruses (6, 53)

The distinction between LTR retrotransposons and retroviruses can be rather fuzzy, as illustrated by recent studies of the *gypsy* element. *gypsy* was initially discovered because of its ability to cause mutations in fruit flies. It was found to appear in different locations in the genomes of different fruit fly strains, so it was assumed to be a transposon. The *gypsy* element has typical LTR retrotransposon structure, but has, in addtion, a reading frame downstream of *gag* and *pol* with the potential to encode a membrane-spanning protein. Studies of fly strains containing active elements revealed that much of this *gypsy*-encoded envelope-like protein was found in extracellular particles, reminiscent of viruses. These particles were then purified by sucrose gradient fractionation and fed to larvae from fly strains lacking *gypsy* elements. The larvae devoured the sweet sucrose fractions and became infected with *gypsy* elements, establishing *gypsy* as a retrovirus instead of a retrotransposon.

Further studies revealed that fly strains in which *gypsy* is active show high levels of expression of *gypsy* proteins in follicular cells of ovaries. *gypsy* transposition in the germ line may thus be explained by production of virus in ovary cells and subsequent infection of oocytes. It is presently unclear whether *gypsy* ever actually moves by retrotransposition.

The plant LTR element Tat1 also encodes a potential envelope protein, raising the possibility that it is a retrovirus, although it has not yet been shown to form infectious extracellular particles. Env-like open reading frames have also been identified in a number of other plant LTR elements. The yeast relative Ty3, in contrast, lacks an envelope protein and clearly retrotransposes within a single cell. Thus, the Ty3-*gypsy* family apparently contains both retroviruses and retrotransposons.

tions of 32 de novo insertion sites for the yeast element Ty1 were mapped and none was found to reside in cellular genes. This shows a remarkable degree of selectivity, because the yeast chromosomal DNA is in fact 70% coding regions. Yeasts often grow as haploids, containing only a single copy of the cellular genes, so highly selective integration is likely important for survival—integration of Ty1 into a required cellular gene in a haploid strain would be fatal for both the host and the transposon. For Ty1, target sites are distributed over a few hundred base pairs upstream of Pol III-transcribed genes. For Ty3, another yeast LTR retrotransposon, integration takes place exactly at the start point of transcription of Pol III-transcribed genes. Both types of elements probably recognize the Pol III transcription complex or distinctive local chromatin in some fashion, perhaps by binding to it directly. Model studies using retroviral integrases, in which targeted integration was achieved in vitro by tethering retroviral integrase proteins to specific target DNA sites, suggest that such specificity could be mediated by integrase.

The regions upstream of Pol III-transcribed genes are relatively benign for retroelement insertion. Because the specific DNA sites recognized by Pol III transcription proteins are largely internal to the gene, insertion just upstream does not disrupt gene activity.

Ty5, another yeast LTR retroelement, has solved the same problem by a different means. Ty5 integrates into another set of benign sites, those associated with heterochromatic DNA sequences. These sites are bound by a characteristic group of proteins, the silent information regulators (Sir genes), which reduce gene expression and probably play additional roles. Disruption of some of the *sir* genes by mutation blocks Ty5 targeting to telomeres, establishing the importance of these proteins. These same proteins act at another location in the yeast genome, the silent gene cassettes used in determining yeast mating type. These sites are also preferential sites of Ty5 integration, confirming that Ty5 recognizes a distinct set of protein–DNA complexes for selecting benign integration sites.

Transposons of the slime mold *Dictyostelium discoideum* also integrate near tRNA genes, but with the added twist that different transposons have different favored positions. The DRE1 element (named for the *d*irectly *r*epeated *e*lements—the LTRs) always integrates 50 bp (+ or −3) upstream, whereas another transposon, Tdd3 (transposon of *D. discoideum* 3) always integrates 100 bp (+ or −20) downstream. Probably in this case, too, these locations mark benign integration sites.

Retroviruses, in contrast, integrate into most chromosomal locations. The integration target, the vertebrate genomes, are diploid and much less densely populated with genes—the human genome is only about 1.5% gene exons. The *gypsy* element of flies may provide an intermediate example, integrating into a genome with gene density intermediate between that of yeast and humans and displaying specificity between that of retroviruses and Ty elements.

Lateral Transfer of LTR Retrotransposons (19, 20, 22, 30, 54, 62)

One of the clearest cases of transfer of a retrotransposon in nature involves the *copia* element of insects. *copia* is abundant in all *Drosophila melanogaster* strains studied. Studies of the related *Drosophila willistoni*, however, reveal that only a subset of strains contain the element. Jordan and colleagues investigated this by isolating and sequencing *copia* elements from *D. willistoni* strains and found that they were more than 99% identical with those of *D. melanogaster*, despite the fact that the host species are believed to have diverged 50 million years ago. In contrast, *copia* elements from *D. simulans*, a closer relative of *D. melanogaster*, display only 90% sequence similarity. These findings strongly support the idea that *copia* transferred recently between *D. melanogaster* and *D. willistoni*. *D. simulans*, in contrast, appears to have inherited the element from its common ancestor with *D. melanogaster*.

Further analysis reveals the probable direction of transfer. *D. melanogaster* is cosmopolitan, whereas *D. willistoni* is found only in the New World, from Florida south through the Caribbean and Latin America (Fig. 8.3). Because the distribution of *copia* elements is universal in *D. melanogaster* and patchy in *D. willistoni*, it seems likely that the element transferred from *D. melanogaster* to *D. willistoni*.

copia, unlike *gypsy*, does not encode an Env protein, so *copia* could not transfer between species by a simple retrovirus-like mechanism. However, it may be possible for *copia* RNA genomes to be packaged into particles synthesized from another element such as *gypsy*. Studies of mobilization of *gypsy* in fact revealed simultaneous activation of *copia* elements. This may indicate that *gypsy* proteins have mediated movement of *copia* genomes (although other explanations are also possible). Whether or not *gyspy* was directly involved, such a pseudotyping mechanism provides an attractive model for the mechanism of *copia* lateral transfer.

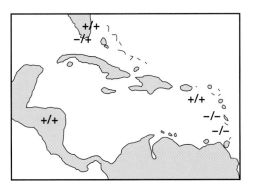

FIGURE 8.3. Distribution of the *copia* transposon in populations of *D. willistoni* from the Caribbean region. The +/+ designation indicates that *copia* was found (or not in −/−) in each of two different assays. (Redrawn, with permission, from Jordan et al. 1999 [©National Academy of Sciences].[30])

Studies of the Ty1-*copia* family in plants have also led to proposals for lateral DNA transfer. Richard Flavell and coworkers sequenced DNA from 31 Ty1-*copia* transposons from eight species of flowering plants. The transposons were found to be diverse, even within a single plant species. Comparisons of the sequences revealed that some elements had their closest relatives in distantly related species of plants, and that the molecular phylogeny of the transposons was different from that of the hosts, suggesting lateral transfer. A similar analysis of reverse transcriptase sequences from echinoderms (sea urchins, sea stars, and their relatives) reached a similar conclusion.

More broadly, the fact that the Ty1-*copia* and Ty3-*gypsy* families are distinct in fungi, plants, and animals is also consistent with the hypothesis that extensive lateral DNA transfer has maintained the distinctive character of each family.

THE NON-LTR RETROTRANSPOSONS

(5, 9, 14, 18, 31, 32, 34, 39, 41, 43, 45, 52, 59)

At least 20% of human DNA is made up of LINE 1–3 non-LTR retrotransposons.

The second large class of retroelements is the non-LTR retrotransposons, also sometimes called poly(A) retrotransposons or long interspersed nuclear elements (LINEs). As the name implies, these elements differ from the LTR-retrotransposons and retroviruses by the lack of LTRs. Well-studied examples include the human LINE elements (also called L1 elements), the *Drosophila* I factors, the *Bombyx mori* R2Bm, and *Neurospora crassa* TAD elements. The non-LTR retrotransposons can be further divided into two groups, elements that integrate into sites with specific DNA sequences versus those that integrate into many different sequences.

Structure of Non-LTR Retrotransposons

The genetic map of a representative non-LTR retrotransposon is presented in Figure 8.4 (top). Transcription proceeds from left to right as drawn. At the left edge of the element is an untranslated region (UTR) that is quite variable among different elements. This region contains binding sites for cellular transcription factors that control the rate of RNA synthesis. Proceeding rightward we reach the Orf1 (open reading frame 1) gene, which probably contributes Gag-like function to reverse transcription particles. This sequence is quite variable among poly(A) retroelements and, in general, is not close in sequence to the retroviral *gag* genes. In some insect elements, this region encodes zinc-binding sequences resembling retroviral NC protein, whereas in the human LINE elements, this region shows no such similarity.

Rightward of Orf1p is the Orf2p coding region. The first encoded function is the endonuclease (EN) coding region, which is required for insertion of the non-LTR retrotransposon sequences into the cellular DNA but is not related to the retroviral or phage integrase enzymes. Within the non-LTR retrotransposon family, the EN protein sequences

A LINE element

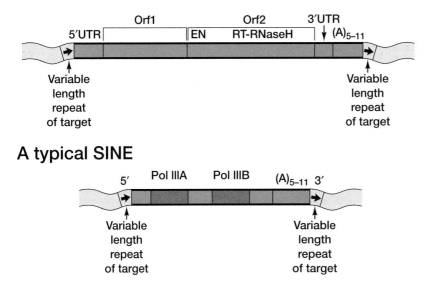

A typical SINE

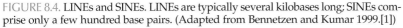

FIGURE 8.4. LINEs and SINEs. LINEs are typically several kilobases long; SINEs comprise only a few hundred base pairs. (Adapted from Bennetzen and Kumar 1999.[1])

differ between those elements that integrate sequence-specifically and those that integrate at many positions. The elements that integrate specifically, such as the silkworm R2Bm element, have EN proteins that resemble the type IIS restriction enzymes of prokaryotes (discussed in Chapter 13). The EN proteins from sequence-nonspecific elements are homologous to apurinic/pyrimidinic (AP) nucleases, which are involved in cleaving damaged DNA during DNA repair. The specificity of these EN enzymes determines the integration specificity of the non-LTR retrotransposon.

To the right of EN within Orf2p lies the RT gene. This enzyme is recognizable as an RT, but is quite distant in sequence from the retroviral enzymes. Sequences of the non-LTR retrotransposon RTs resemble one another more closely than other RTs, indicating that these enzymes form an ancient and conserved branch of the RT family. The RT enzymes are one of the most divergent sets of enzymes known, providing an argument that RT is among the evolutionarily oldest enzymes.

Flanking the RT in many, but not all, Orf2p proteins is an RNase H enzyme. This enzyme likely is important in removing the RNA template during DNA synthesis, as with retrovirus reverse transcription. Rightward of the coding regions is another untranslated region, and then typically a stretch of 5–11 A residues reminiscent of the poly(A) tails attached to mRNA molecules.

Transposition of Non-LTR Retrotransposons (14, 18, 31, 45)

The structure of integrated non-LTR retrotransposons provides clues to their replication mechanism. At least 95% of the non-LTR retrotransposons are truncated at the 5′ end (left end in Fig. 8.4). The positions of truncations vary from element to element. The 3′ (right) end, in contrast, is almost always intact and terminated with a poly(A) sequence. Flanking each integrated non-LTR retrotransposon is a short repeat of the target DNA sequence. This is similar to retroviruses and retrotransposons, but, unlike the retroviruses, the repeats differ in length from element to element, and even on occasion include short deletions. These observations on the structures of these elements, together with biochemical studies of the element-encoded enzymes, have led to a model for the mechanism of replication (Fig. 8.5).

Starting from an integrated non-LTR retrotransposon, retrotransposition begins with transcription of the element genome. The element RNA is then exported to the cytoplasm and translated to make the Orf1 and Orf2 proteins. The element RNA probably remains associated with the Orf1p and Orf2p generated by its own translation, so that each element copy avoids donating proteins to other competing elements.

The non-LTR retrotransposon particle then migrates to the host chromosomes and carries out integration. The reaction begins with formation of a single-stranded nick in the host DNA by the EN activity of Orf2p. The 3′ DNA target end generated by nicking is then used as a primer for reverse transcription. T-rich target DNA sites are favored by the EN domain, permitting annealing of the poly(A) tail to the element RNA. This reaction is probably facilitated by the Orf1p protein, which has been shown to contain a "nucleic acid chaperone" activity that can facilitate exchange of strands. The 3′ end in the target DNA is then used as a primer for extension by the RT activity along the element RNA.

The steps following first-strand cDNA synthesis are unclear. To complete the reaction, the second strand of the element cDNA must be synthesized, the target DNA must be cleaved on the other strand, and the double-strand break in the target DNA must be sealed. At present, these steps are not reproduced in cell-free reactions, complicating their study. Some evidence suggests that the element-encoded nuclease may cleave the other strand as well, possibly allowing the 3′ end on the other strand to act as a primer for second-strand synthesis. Further studies will likely clarify this pathway.

This replication scheme accounts for the structure of the integrated elements. The frequent 5′ truncations can be understood as failures of RT to traverse the element RNA completely. The poly(A) sequences of the element mRNA are preserved because they pair with poly(T) sequences in the target site DNA. The variable duplications arise from a variable relationship between the point of cleavage and priming on the two strands.

The site-specific non-LTR retrotransposons use a similar strategy but employ a site-specific endonuclease. The R2Bm element, for example, integrates efficiently only in a single sequence found in the ribosomal RNA genes of insects. Specificity comes from the element-encoded endonuclease, which recognizes a specific site in a ribosomal

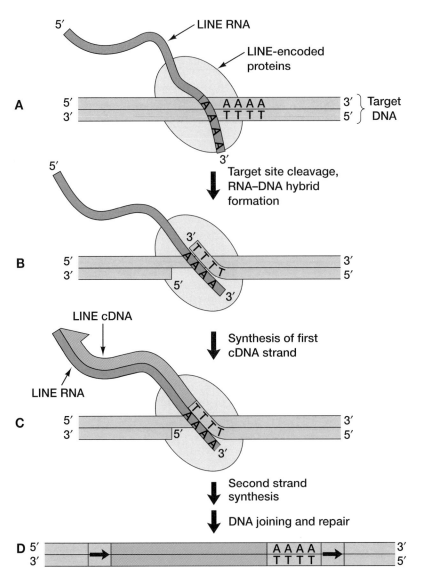

FIGURE 8.5. Target-primed reverse transcription mediating LINE element replication. (*A*) A complex containing the LINE element RNA (*dark green*) and LINE proteins (*light gray*) binds to a chromosomal DNA site rich in A/T base pairs (*gray*). (*B*) The LINE En protein cleaves the target DNA site, and LINE RNA becomes annealed to the target site DNA. (*C*) The 3′ end of the target site DNA acts as a primer for reverse transcription of the LINE RNA. (*D*) The second strand of DNA is synthesized, LINE RNA is removed, and the 5′ end of the LINE DNA becomes joined to host cell DNA (by unknown mechanisms), yielding the integrated LINE copy. (Adapted from Luan et al. 1993.[41])

RNA gene. For these elements, target-primed reverse transcription always takes place from nicks at the same site, leading to a single site of integration. Several site-specific elements are known, and, in each case studied, targeting can be attributed to the specificity of the endonuclease.

Non-LTR Retrotransposons and
Gene Mutation (14, 32, 48, 49, 52)

Active L1 non-LTR retrotransposons were first isolated from mutations in humans caused by L1 insertion. Haig Kazazian and colleagues investigated the molecular basis for human hemophilia, disorders in which the blood clotting system does not function properly. In 2 cases out of 240 studied, they found de novo insertions of L1 elements into the gene for factor VIII, one of the proteins of the clotting cascade. The elements were inserted into an exon of the factor VIII gene, disrupting the coding region. The inserted L1 elements were truncated at the 5´ end, as is usual for these elements. However, the sequences were distinct enough that the inserted regions could be used to isolate the parent copies, providing researchers with an active copy of a human L1 element.

In another case, an L1 element was implicated in causing a human breast carcinoma. Analysis of the *myc* gene in tumor tissue revealed that one allele was altered by insertion of an L1 element. As discussed in Chapter 6, the *myc* gene is also often altered by retroviral integration, thereby promoting tumor formation. The human mutation could be shown to be specific to the tumor, since nearby normal tissue did not show the insertion.

As implied by the above data, non-LTR retrotransposons are still active in the human genome. In fact, different human populations have differences in the locations of newly inserted elements (Chapter 10). The two other types of elements found to cause insertional mutations in humans, Alu elements and SVA (SINE-R) elements, probably use the L1 machinery for their mobility.

In mice, insertion of murine L1 elements is also known to have caused several mutations. The *spa* mutation of mice causes muscular spasticity due to mutation of a glycine receptor gene (*Glrb*). Sequencing revealed a new L1 insertion into an intron of *Glrb*. In the "Orleans reeler" mouse, which displays uncoordinated locomotion, an L1 insertion was found in a gene encoding an extracellular protein of developing neurons. In this case also, the L1 element was inserted into the middle of the gene, causing production of an altered mRNA. Closely related L1 elements were inserted in both cases, and functional tests in cultured cells revealed the elements to be active for transposition. Thus, these studies of mutations in mice also resulted in the isolation of an active murine non-LTR retrotransposon.

Lateral Transfer of Non-LTR
Retrotransposons (7, 15, 36, 42, 43, 47)

Lateral transfer of the non-LTR retrotransposons has been a topic of energetic debate. D. Kordis and F. Gubensek studied a non-LTR retrotransposon called Bov-B LINE, which was found in snakes, some lizards, and ruminants (the group including cattle, sheep, deer, and camels). They compared the sequences of Bov-B LINEs in the different groups, and concluded that the degree of similarity (75–80%) was much greater than that expected if the elements had resided in their respective genomes since the divergence of the progenitor of the different vertebrate lineages. The patchy distribution in lizards also supported the idea that the elements had been transferred laterally.

A similar study proposed that the *jockey* non-LTR retrotransposon of *Drosophila* had been transferred between *Drosophila* species. L. Mizrokhi and A. Mazo studied the distribution of *jockey* in diverse *Drosophila* species and found the element only in *D. melanogaster* and the distantly related *D. funebris*. The similarity between elements was judged to be greater than that expected given the overall divergence between the host species, supporting an interpretation of lateral transfer. Related arguments have been made for a few other cases.

However, a sophisticated analysis of all the non-LTR retrotransposon family sequences by Tom Eickbush and coworkers reached a different conclusion. They began their study by sorting the non-LTR retrotransposons into groups, which they called "clades," based on the similarity of the element sequences. Several types of evidence, such as measurement of rates of sequence divergence and acquisition of different transposon proteins, supported the idea that each clade was very ancient, stretching back to before the divergence of the major animal phyla. For analysis of lateral transfer, they argued that meaningful comparisons could only be made within clades, since comparison between clades would give incorrectly large divergence times. From their analysis of each clade, Eickbush and coworkers concluded that the extent of retrotransposon sequence divergence within each clade does indeed match the divergence expected for organismic sequences generally. Those elements that have persisted, according to this view, have been inherited vertically since before the Cambrian era. Perhaps some aspect of non-LTR retrotransposon biology obstructs their lateral transfer. For example, Eickbush and coworkers speculate that the integration mechanism of non-LTR retrotransposons, probably involving extensive use of host DNA repair enzymes in the late steps, may be unsuitable for hitchhiking on DNA viruses.

Very recently, new data from the human genome project have reopened this debate. Arian Smit has pointed out that if the Bov-B sequence is present in other vertebrates, it really should be detectable in the human genome. There has not been enough time for Bov-B to mutate to such an extent that it could no longer be detectable, but it has not yet been found in the human sequence. Thus, recent lateral transfer into the lineage leading to cows and some reptiles is again a potential explanation.

MOBILE INTRONS (16, 37, 38, 50)

As the name implies, the mobile introns are mobile elements that reside within genes, but the element sequences are spliced out of RNA transcripts after synthesis. Their discovery can be traced back to the late 1970s, when genetic experiments indicated that the mitochondrial genome of baker's yeast (*Saccharomyces cerevisiae*) contains introns. A few years later, it was realized that those introns were not identical to the introns that normally reside within eukaryotic nuclear genes. Additionally, it was proposed that those intron RNAs could adopt specific secondary structures and, on that basis, were classified as group I and group II introns. Numerous other group I introns have since been found in bacteria; in the organelles of fungi, algae, and plants; as well as in some nuclear genes of protozoa and algae. Curiously, the only group I introns known in metazoans are in the mitochondria of some anemones (cnidarians). The distribution of group II introns parallels that of group I (bacteria, the organelles of fungi, algae, and plants), with the notable exception that no group II introns have been discovered in any nuclear gene nor in any animal taxa.

In the 1980s it was discovered that members of both these classes of RNA introns were autocatalytic (ribozymes), i.e., they were able to self-splice out of their precursor mRNAs in vitro in the absence of any proteins. This provided one of the first examples of catalysis by an RNA, strengthening the idea that RNA molecules were central to the early evolution of life ("The RNA World"). For the discovery of RNA catalysis in group I introns, Thomas Cech was awarded the Nobel Prize in 1989.

Numerous studies have since been carried out on both group I and II introns, which we now know are unrelated structurally and mechanistically. An interesting evolutionary connection may exist, however, between group II introns and the machinery that catalyzes pre-mRNA splicing of eukaryotic genes (the spliceosome). Structural similarities between the domains of the group II introns and the RNA components of the spliceosome (the snRNAs) as well as a number of similarities in the chemical mechanism of splicing suggest that the spliceosome may have evolved from group II intron-like molecules. In an additional twist, group I and group II introns have been shown to be mobile genetic elements as well.

Figure 8.6 shows a diagram of the *COX1* gene, which encodes the cytochrome oxidase protein of *S. cerevisiae* mitochondria. The *COX1* gene is interrupted by two group II mobile introns (aI1 and aI2), a third group II intron which is not mobile (aI5γ), and seven group I introns as well (some of them mobile) .

The mobile group II introns contain an open reading frame (ORF), which encodes a "maturase" protein composed of multiple domains with separate functions. Starting from the amino terminus, there is a domain that weakly resembles retroviral proteases, whose function is unclear. Next comes an RT domain, which in sequence most closely resembles the RTs of non-LTR retrotransposons. Next is the so-called X-domain, which binds nucleic acids and assists splicing of the catalytic RNA intron (hence the name maturase for the whole protein, reflecting its role in forming the mature RNA). The carboxy-terminal domain, marked Zn, contains a zinc finger structural motif that has a DNA nuclease activity important in mobility. Many group II intron maturases (especially in bacteria and plant chloroplasts) lack some of these domains, particularly the protease-like domain and the Zn domain.

The synthesis of the intron-encoded maturases differs among mobile group II introns. In the case of the yeast introns, the element proteins are synthesized as fusion proteins attached to cytochrome oxidase. The *Lactobacillus lactis ltrB* intron, however, differs in that the maturase is translated independently of the *ltrB* gene product (this characteristic is common for all known bacterial group II introns).

Transposition of mobile introns often takes place between two copies of the same gene, one of which contains the mobile intron and the other of which does not. This reaction has been called "homing," because the intron inserts in the same DNA sequence from which it was derived (just in the other gene copy). Transposition of the mobile introns can be challenging to understand, because homing can take place by several different pathways, and furthermore, mobile introns can also move to unrelated sites (ectopic insertion) as well. Table 8.2 summarizes six of the known pathways. We begin by describing homing of group II introns involving reverse transcription catalyzed by the maturase RT (retrohoming; Table 8.2, line 3).

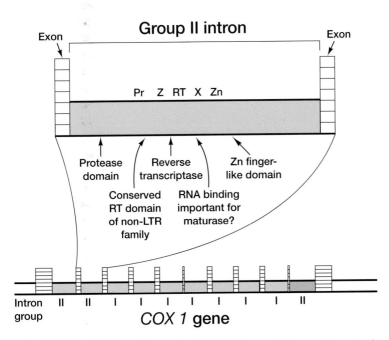

FIGURE 8.6. Self-splicing introns in the *COX1* gene of *S. cerevisiae* mitochondria. An intact Group II mobile intron is shown at the top (*green*), exons are shown with the horizontal hatching. Proteins encoded by the element are indicated. The *COX1* gene contains fully 10 introns, three Group II introns and 7 Group I introns. (Adapted from Lambowitz et al. 1999.[38])

The mechanism of splicing by group II introns closely resembles that of eukaryotic nuclear introns, raising intriguing questions about the evolution of the two. Not only is the chemical mechanism of splicing identical, resulting in production of a lariat RNA, but the specific RNA sequences involved have many similarities. One theory holds that ectopic insertion of group II introns (or their progenitors) gave rise to the modern introns seen in eukaryotic nuclear genomes. According to this idea, intron insertion would have been more active early in the evolution of eukaryotes, since many intron positions are conserved and group II introns have not been found in the nuclear genomes of modern eukaryotes. Although other theories have been advanced to explain the origins of introns (Chapter 14), this idea is quite attractive.

TABLE 8.2. Mobile Intron Transposition Pathways

	Homologous site or many sites?	Substrate (intron)	Substrate (target)	Enzymatic activities involved
Group I				
(1) Homing	homologous	DNA (allele containing intron)	DNA (allele lacking intron)	maturase endonuclease
(2) Ectopic insertion	many sites	excised intron RNP	cellular mRNA	reverse splicing to insert intron into another RNA, then cellular RT
Group II				
(3) Retrohoming	homologous	excised intron RNP	DNA (allele lacking intron)	reverse splicing to insert in DNA, endonuclease for initial DNA cleavage, maturase RT
(4) Retrohoming with upstream conversion	homologous	excised intron RNP, also another pre-mRNA	DNA (allele lacking intron)	reverse splicing to insert in DNA, endonuclease for initial DNA cleavage, maturase RT
(5) RT-independent homing	homologous	excised intron RNP	DNA (two parental mitochondrial genomes involved)	reverse splicing, maturase endonuclease for initial DNA cleavage
(6) Ectopic insertion (retrotransposition)	many sites	excised intron RNP	cellular mRNA	reverse splicing to insert intron into another RNA, then cellular RT

Eskes 2000 [16]; Lambowitz et al. 1999 [38]; Pyle 2000 [50].

Pathways of Mobile Intron Transposition

Transposition of group II introns is carried out by an RNA–protein complex consisting of the excised intronic RNA bound to the maturase protein. Figure 8.7 illustrates the RNA splicing reaction of group II mobile introns that forms the complex active in transposition. Prior to splicing, the intron RNA embedded in the pre-mRNA folds into a higher-order structure that forms the active site for the RNA cleavage and joining reactions. Splicing begins with the recruitment of an RNA 2′ hydroxyl group within the intron to initiate an attack on the 5′ intron–exon junction. The resulting cleavage-joining reaction leaves the A residue joined through a 2′–5′ phosphodiester bond to the 5′ end of the intron, forming a "lariat" structure. The 3′ end of the upstream exon exposed by the first cleavage then attacks the 3′ intron–exon border, joining the two exons and excising the lariat intron RNA. Although the intron RNA donates the active site for self-splicing, the maturase and host proteins also contribute by binding the RNA and promoting the splicing reaction.

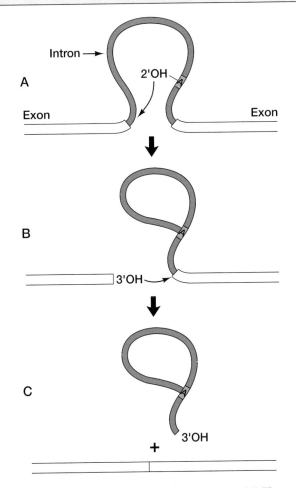

FIGURE 8.7. The self-splicing reaction of Group II introns. (*A*) The reaction begins when a 2′ hydroxyl of an internal RNA nucleotide attacks the 5′ intron–exon junction, resulting in the formation of a lariat intermediate (*B*). The free exon 3′ end then attacks the 3′ intron–exon boundary. A second exchange (transesterification) reaction joins the two exons and frees the lariat RNA. (Adapted from Lambowitz et al. 1999.[38])

Remarkably, this reaction can be run backward to insert the intron RNA into a gene copy lacking the intron (Fig. 8.8). After splicing, the excised intron RNA remains associated with maturase protein, which also assists the reverse splicing reaction. Retrohoming begins with recognition of the target site by interactions between the maturase protein and target DNA. Maturase then unwinds the target site DNA, allowing sequences in the intron to pair with matching sequences in the target DNA. Specificity for the site of insertion comes from these maturase–DNA and RNA–DNA interactions.

The reverse splicing reaction is initiated by the free 3′ end in the intron RNA attacking the DNA target, joining the RNA and breaking one target DNA strand. In some instances, the reverse splicing proceeds to completion and the intron RNA becomes fully integrated into the DNA strand. Meanwhile, the maturase nuclease activity cleaves the other strand of the target DNA at a position 10 bp from the RNA–DNA junction. The target DNA 3′ end freed by this cleavage serves as a primer for reverse transcriptase, which carries out cDNA synthesis across the element RNA. Host DNA repair enzymes are thought to then synthesize the opposite strands and seal up the breaks, integrating the intron, now as DNA, into a new chromosomal site and completing the homing reaction.

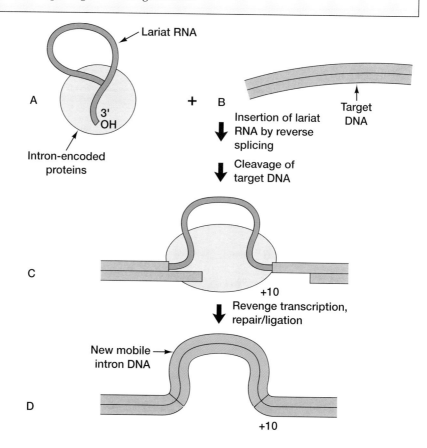

FIGURE 8.8. Insertion of a mobile Group II intron by reverse splicing. (*A*) The excised lariat RNA remains bound to element-encoded proteins. (*B*) Reversal of the steps in splicing (attack of the RNA 3′ end on DNA, thereby joining the RNA to DNA) attaches the mobile intron into one DNA strand. (*C*) Cleavage of the other DNA strand, followed by polymerization along one DNA strand, then removal of the RNA and repair DNA synthesis, yields the fully integrated DNA copy of the mobile intron (*D*). (Adapted from Lambowitz et al. 1999.([8])

Other homing pathways begin with the same steps but involve fewer reactions catalyzed by the intron-encoded maturase. For example, if reverse transcription is incomplete, host cell enzymes can repair the chromosomal double-strand break generated during the initial steps of homing (Table 8.2, line 4). Repair of the double-strand break involves copying sequences matching the new reverse transcript from the other intron-containing allele (recombinational repair), resulting in incorporation of the intron at the site of the break. If reverse transcriptase activity is not present at all—for example, if maturase is mutated—then homing can still take place. The RNA joining and second-strand cleavage can still occur, generating a double-strand DNA break, but host cell enzymes take over even earlier, again capturing the intron by recombinational repair against the other allele (Table 8.2, line 5).

The transposition pathways used by group II introns vary among organisms. In bacteria, retrohoming by complete reverse splicing and independent of the host *recA* system is the predominant mechanism, whereas in yeast several retrohoming pathways as well as RT-independent homing pathways coexist in proportions that are intron and host-dependent.

Group II introns can also be transferred to new sites by inserting into RNA instead of DNA (Fig. 8.10; Table 8.2, lines 6). To move by the "ectopic insertion" pathway, the excised intron carries out reverse splicing, but in this case reverse splicing into a cellular mRNA rather than a DNA site. If the resulting RNA is reverse-transcribed and integrated by homologous recombination, the intron will appear in the gene that generated the mRNA used as substrate for reverse splicing. This reaction is thought to have a reduced sequence specificity, allowing dispersal of mobile introns to new sites.

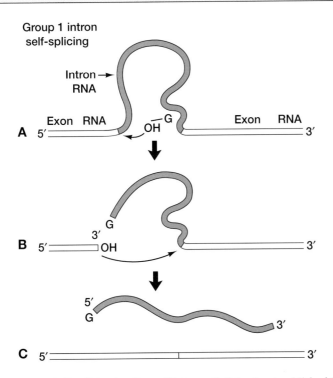

FIGURE 8.9. Self-splicing by Group I introns. Splicing begins (*A*) by binding of a G residue to an RNA site within the intron. The 3′ OH of the G residue then attacks the 5′ intron–exon junction. As result, the G residue becomes attached to the intron 5′ end (*B*). The free 3′ end of the exon then attacks the 3′ intron–exon junction (*C*), joining the two exons and releasing the intron RNA. (Adapted from Lambowitz et al. 1999.[38])

Group II introns can have profound effects on the host organism. One phenomenon is senescence in the fungus *Podospora anserina*, in which a mobile group II intron contributes to the instability of the mitochondrial genome, which, in turn, triggers the degenerative process. In the world of photosynthetic organisms, the chloroplast genome of *Euglena gracilis* contains no fewer than 149 introns! About half of them are group II introns, the others being a related "shortened" version (group III) and complex introns in which introns interrupt other introns ("twintrons"). Splicing of the twintrons and the more complex introns requires ordered, sequential splicing of the internal introns, which demonstrates that they were formed by mobility events in which group II introns transposed into one another.

Group I introns are also both catalytic RNAs and mobile elements. Figure 8.9 presents the group I self-splicing mechanism. These introns also fold into catalytic RNAs that direct their own removal from mRNA, but the mechanism is different from group II introns. The intron RNA binds a free G nucleotide, the 3′ hydroxyl of which is then used to attack the 5′ junction between the intron and exon RNA. This covalently links the G residue to the cleaved RNA. The 3′ end of the exon exposed by this cleavage then attacks the 3′ junction between the

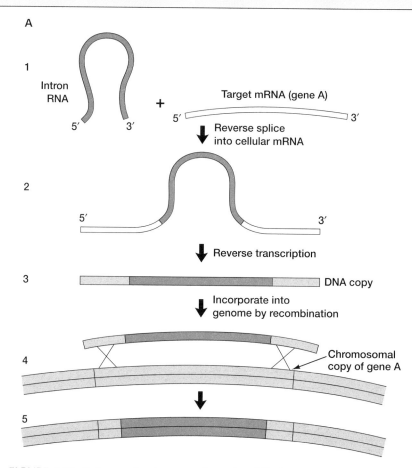

FIGURE 8.10. Pathways for incorporation of mobile introns into cellular DNA involving homologous DNA recombination. (*A*) A free mobile intron (1) becomes incorporated into a cellular mRNA by an exact reversal of normal self-splicing. This intermediate is either a lariat RNA (Group II introns) or the linear intron with the 5′ terminal G (Group I) (2). Reverse transcription of this sequence (3) yields a DNA copy (shown here as single-stranded, but involvement of double-stranded DNA is not ruled out). Recombination of the reverse transcript with the chromosomal gene

intron and the downstream exon, connecting the two exons and excising the intron. In this reaction, the intron is released as a linear RNA rather than as a lariat.

A unique mechanism mediates the pathway for movement of group I introns (Fig. 8.9B; Table 8.2, line 1). The group I maturase differs from that of group II maturases in that it lacks the RT function but contains a site-specific DNA endonuclease activity. This mechanism operates in a situation where there are two gene copies present, one of which contains the intron and the other does not (homing). The element-encoded nuclease cleaves the empty target site DNA, thereby initiating host-cell-mediated repair by homologous recombination. Host nucleases resect DNA in either direction from the break, then the DNA gap is repaired by copying information from the intron-containing gene copy to replace the missing sequence. The net effect is to transfer the intron to the previously intron-free gene copy. In another pathway, group I introns can also transpose to new sites by reverse splicing into other mRNAs followed by reverse transcription and homologous recombination catalyzed by host enzymes (Table 8.2, line 2), as with group II introns.

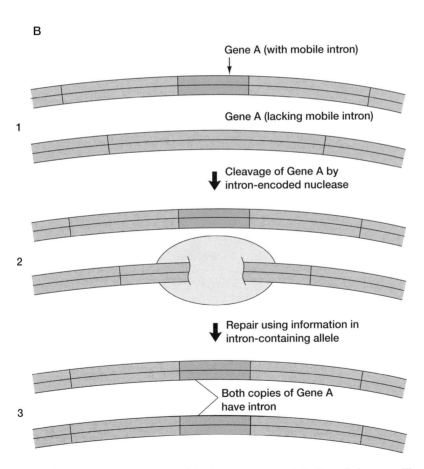

B

Gene A (with mobile intron)

Gene A (lacking mobile intron)

1

Cleavage of Gene A by intron-encoded nuclease

2

Repair using information in intron-containing allele

Both copies of Gene A have intron

3

copy (4) can install a DNA copy of the intron sequence into the cellular gene (5). (*B*) A DNA-mediated pathway for movement of Group I mobile introns. This pathway operates in cases where one sister chromosome contains a copy of the intron and the other does not (homing) (1). An intron-encoded nuclease cleaves the copy of gene A that lacks the intron (2). Repair of the lesion by homologous recombination using the other allele of gene A as template installs the intron in the sister copy of gene A (3). (Adapted from Lambowitz et al. 1999.[38])

Lateral Transfer of Mobile Introns (11, 37, 40, 58)

A dramatic example of probable lateral transfer of group I introns comes from studies of plant mitochondria by Jeffrey Palmer and coworkers. They found that a group I intron was present in the *cox1* mitochondrial gene of 49 out of 335 plant genera studied. The distribution was exceptionally patchy, consistent with acquisition of the introns by lateral transfer. But what argued against the alternative—acquisition by inheritance together with frequent loss? First, the intron sequence similarity was much greater than the similarity of other genes, most easily explained by recent lateral transfer. The sequence-based phylogenies for introns did not match their hosts, again supportive of lateral transfer. Strikingly, an analysis of cotransfer of intron-flanking sequences showed that different short stretches of DNA were cotransferred with the introns in different plant species. This is as expected for the DNA-based homing mechanism, which is expected to transfer flanking sequences frequently as well (termed coconversion), and the differences in the coconversion patches argues strongly for independent events. These findings provide a strong independent argument for lateral transfer and specify the probable mechanism involved.

Mobile Introns for Targeted Gene Manipulation (23)

The mechanism of mobile group II intron homing has proven to be flexible enough to allow reengineered mobile introns to be targeted to new sequences, a trick with applications in biotechnology. Recall that retrohoming begins by the recognition of specific DNA sequences by the maturase protein, and then Watson–Crick base-pairing of the intron RNA with target DNA sequences. Alan Lambowitz and coworkers asked whether it was possible to retarget group II introns to inactivate genes in mammalian cells. They identified sequences that would be expected to bind maturase, then reengineered the intron RNA to pair with the local DNA sequences. The requirements for recognition are few enough that most genes contain suitable target sequences. The modified introns were indeed capable of inserting into the chosen target genes inside cells. This clever strategy may provide a convenient means for targeted insertional mutagenesis in both prokaryotic and eukaryotic cells.

A close study of the transferred segments suggested that the *cox1* intron has been acquired separately no fewer than 32 times among the plants studied. Extrapolating across all the angiosperms (flowering plants), Palmer and coworkers estimate that the *cox1* intron has been transferred over 1000 times to different plant species. Comparison of the plant introns with sequences of other mobile introns suggests that fungal species might have served as the donor that launched the wave of lateral transfer.

In another case, the mitochondria of the protozoa *Acanthamoeba castellanii* was found to have several group I introns inserted into its ribosomal RNA. The green alga *Chlamydomonas reinhardtii*, in contrast, lacked the mitochondrial introns but contained two identically placed introns in its chloroplast ribosomal RNA gene. The introns were found to be missing in intermediate species. These and other observations support the proposal that the introns moved between mitochondria and chloroplasts, explaining the present distribution. A number of other examples of possible transfer have also been proposed on the basis of DNA sequence data, including transfer of group I introns between mtDNA of *C. smithii* and *N. crassa* and between *S. pombe* and *Aspergillus nidulans*. For all of these cases, the mechanism by which mobile introns move between cells is unclear.

PROCESSED PSEUDOGENES, SINES, AND "RETRO-GENES" (3, 13, 17, 39, 51, 52)

The presence of retrotransposons has profound consequences for the genome of the host. Analysis of the eukaryotic genomes has revealed many sequences that are related to known genes but are clearly defective. These sequences share several features reminiscent of mRNA structure, suggesting that they were formed by reverse transcription of mRNAs and integration, hence the name "processed pseudogenes."

Typically, processed pseudogenes contain point mutations in the coding regions, including stop codons or frameshift mutations, that eliminate normal protein production. Pseudogenes are often missing introns present in the functional gene copies. Furthermore, the pseudogenes typically have poly(A) sequences at the 3′ end of the inferred coding region. Many processed pseudogenes are flanked by duplications of host DNA, and these duplications vary in length among pseudogenes as with non-LTR retrotransposons.

All of these features can be accommodated in a model in which cellular mRNAs are occasionally reverse-transcribed and integrated by non-LTR retrotransposon machinery. This accounts for the loss of introns and the presence of poly(A) tails. Laboratory studies have demonstrated the production of processed pseudogenes by overproduction of retrotransposon proteins.

In a few cases, this mechanism may yield a functional copy of the gene that is lacking introns and flanked by a poly(A) tail, but is still competent for directing protein synthesis. The draft human genome sequence reveals several candidates for such "retro-genes."

A pathway resembling processed pseudogene formation probably accounts for the formation of the short interspersed nuclear elements (SINEs). These sequences are short repeats, usually 100–400 bp in length, that have become greatly amplified in some genomes. In humans, for example, a single SINE family, the *Alu* elements, comprises 10% of the genome sequence.

Lateral Transfer of SINEs (24, 56)

One report has suggested that SINEs may also be subject to lateral transfer. A SINE named *Sma*I (for a restriction enzyme that cleaves the

Why Are Some Host RNAs Recruited to Form SINEs but Not Others? (2, 24)

Why are only a few of the cellular small RNAs recruited to form SINEs? One frequent feature of SINE-forming RNAs is the presence of transcriptional promoter sequences within the transcription unit. This allows newly transposed elements to be transcriptionally active in their new chromosomal sites, thereby producing templates for new rounds of retrotransposition and SINE formation. Genes for tRNAs, for example, contain internal promoter regions and are commonly found amplified as SINEs.

At least two other features probably also contribute. Okada and coworkers found that the 3′ ends of some SINE RNAs match the 3′ ends of a particular LINE (non-LTR retrotransposon) element RNA. The 3′ RNA sequence of LINEs probably binds specifically to the element-encoded reverse transcription machinery. SINEs that have captured the 3′ sequence of a particular LINE can thus also bind the LINE-encoded proteins and multiply by retrotransposition.

Jef Boeke has pointed out another intriguing theme connecting many SINEs. RNAs that have formed SINEs tend to be those expected to bind to the ribosome. RNAs that have expanded as SINEs include tRNAs and 7SL RNA, the latter being a component of the cellular protein sorting machinery that binds to ribosomes. Other RNA components of the translational machinery have also been expanded as SINEs. It seems likely that the LINE proteins bind to the LINE mRNA that encodes them during the course of their translation. This allows translation of the LINE genome to promote replication of itself and not that of another competing element. Thus, for a SINE RNA to bind to LINE-encoded proteins, it would help to be tethered near the site of protein synthesis, i.e., on the ribosome. This idea, although speculative, neatly accounts for the nature of the gene families that have become amplified as SINEs.

element) was initially found in genomes of chum and pink salmon. The element was absent in close relatives, but present in all fishes in the subfamily *Coregoninae*, which are primitive salmonoids including the whitefish and cisco. The element sequences were found to be more similar to each other than expected by analysis of the host genomes, supporting a proposal of lateral transfer between the salmonoid groups.

Additional studies hint at possible mechanisms of lateral transfer. Another study found the fish SINE *Hpa*I to be inserted into a *Tc1*-like transposon. As discussed in the next chapter, there is extensive evidence for lateral transfer of the *Tc1/mariner* family of DNA transposons. These and other data raise the interesting possibility that SINE elements may move between cells as parasites on other transposons.

Summary

Retroelements, mobile elements replicating by means of reverse transcription, are a heterogeneous class with members displaying several quite different life-styles. The LTR retrotransposons are similar to retroviruses, differing only by the lack of an extracellular phase in the life cycle. Indeed, for some elements in the Ty3-*gypsy* group, it can be hard to tell whether they are retroviruses, retrotransposons, or both. Like retroviruses, the LTR transposons prime reverse transcription of the RNA genome by binding RNA primers at the edges of the LTRs, and use integrase to integrate a preformed linear cDNA into the genome. The non-LTR retrotransposons replicate by a different means, starting reverse transcription by nicking the target DNA and using the DNA 3′ end thus exposed as a primer for RT. The group I and group II mobile introns are distinctive for encoding splicing signals at either end of the element and a ribozyme that catalyzes the splicing internally. This allows the element to be spliced out after transcription, minimizing disruption of the gene in which it is embedded. Mobile introns can move within genomes by a number of different mechanisms. For the group II mobile introns, some of the pathways employ an element to encode a reverse transcriptase protein. Other types of elements, such as retroplasmids and retrons, are also synthesized by reverse transcription, but are not known to be major vehicles for transporting sequences between species.

Several studies document lateral transfer of retroelements. For the LTR elements, there is strong evidence that the *copia* element has been transferred from D. *melanogaster* to D. *willistoni*, probably since the arrival of D. *melanogaster* in the New World. In other cases, possible transfer of the LTR retrotransposons among plants has been inferred by the finding of closely related elements in distantly related host species. Data on transfer of mobile introns is also compelling, since, in addition to the above arguments based on homology, the finding of sequences flanking introns likely subject to cotransfer also supports models of recent introduction. Proposals for lateral transfer of non-LTR retrotransposons, however, are presently equivocal. Several possible examples have been reported, but recent thorough studies by Eickbush and colleagues suggest that multiple independent lineages have been inherited mainly by descent.

Several plausible scenarios can be imagined for transfer of retrotransposons between organisms. There is abundant laboratory evidence that eukaryotic viruses can be engineered to transport DNAs of any kind between cells, so one group of models involves natural transfer by hitchhiking on viruses. Several cases are known of transposon sequences residing in the genomes of naturally isolated viruses, bolstering this idea. Retroviral "pseudotyping" provides another mechanism, the packaging of retroelement genomes by proteins encoded by a retrovirus resident in the same cell. Genomes so packaged could then be transferred, reverse-transcribed, and integrated by fully viral mechanisms. Transport as pseudotypes is particularly likely for the LTR retrotransposons, the group most closely related to retroviruses. Other mechanisms involve the transient exposure of the cellular genome to a retroelement, as in wide crosses in plants. Still further possibilities include uptake of cellular DNA by bacteria or fungi, followed by introduction of the transported DNA into a cell of an infected host.

Our genes are floating on a sea of retroelements, and studies of this ocean are central to understanding our genetic heritage. In Chapter 10 we return to the role of retroelements in eukaryotic genomes, but first we review the other major class of eukaryotic mobile elements, the DNA transposons.

REFERENCES

1. Bennetzen J.L. and A. Kumar 1999. Plant retrotransposons. *Annu. Rev. Genet.* **33:** 479–532.

2. Boeke J.D. 1997. LINEs and Alus—the polyA connection. *Nat. Genet.* **16:** 6–7.

3. Boeke J.D. and Stoye J.P. 1997. Retrotransposons, endogenous retroviruses, and the evolution of retroelements. In *Retroviruses* (ed. J.M. Coffin et al.), pp. 343–435. Cold Spring Harbor Laboratory Press, Cold Spring Harbor, New York.

4. Boeke J.D., Garfinkel D.J., Styles C.A., and Fink G.R. 1985. Ty elements transpose through an RNA intermediate. *Cell* **40:** 491–500.

5. Boissinot S., Chevret P., and Furano A. 2000. L1 (LINE 1) retrotransposon evolution and amplification in recent human history. *Mol. Biol. Evol.* **17:** 915–928.

6. Bucheton A. 1995. The relationship between flamenco gene and gypsy in *Drosophila:* How to tame a retrovirus. *Trends Genet.* **11:** 349–353.

7. Burke W.D., Malik H.S., Lathe W.C., and Eickbush T.H. 1998. Are retrotransposons long-term hitchhikers? *Nature* **392:** 141–142.

8. Bushman F.D. 1995. Targeting retroviral integration. *Science* **267:** 1443–1444.

9. Chaboissier M.-C., Busseau I., Prosser J., Finnegan D.J., and Bucheton A. 1990. Identification of a potential RNA intermediate for transposition of the LINE-like element 1 factor in *Drosophila melanogaster. EMBO J.* **9:** 3557–3563.

10. Chalker D.L. and Sandmeyer S.B. 1992. Ty3 integrates within the region of RNA polymerase III transcription initiation. *Genes Dev.* **6:** 117–128.

11. Cho Y., Qiu Y.-L., Kuhlman P., and Palmer J.D. 1998. Explosive invasion of plant mitochondria by a group I intron. *Proc. Natl. Acad. Sci.* **95:** 14244–14249.

12. Coffin J.M., Hughes S.H., and Varmus H.E. 1997. *Retroviruses.* Cold Spring Harbor Laboratory Press, Cold Spring Harbor, New York.

13. Derr L.K., Strathern J.N., and Garfinkel D.J. 1991. RNA-mediated recombination in *S. cerevisiae. Cell* **67:** 355–364.

14. Dombroski B., Mathias S., Nanthakumar E., Scott A., and Kazazian Jr., H.H. 1991. Isolation of an active human transposable element. *Science* **254:** 1805–1808.

15. Drew A.C. and Brindley P.J. 1997. A retrotransposon of the non-long terminal repeat class from the human blood fluke *Schistosoma mansoni.* Similarities to the chicken-repeat-1-like elements of vertebrates. *Mol. Biol. Evol.* **14:** 602–610.

16. Eskes R., Liu L., Ma H., Chao M.Y., Dickson L., Lambowitz A.M., and Perlman P.S. 2000. Multiple homing pathways used by yeast mitochondrial group II introns. *Mol. Cell. Biol.* **20:** 8432–8446.

17. Esnault C., Maestre J., and Heidmann T. 2000. Human LINE retrotransposons generate processed pseudogenes. *Nat. Genet.* **24:** 363–367.

18. Feng Q., Moran J.V., Kazazian H.H., and Boeke J.D. 1996. Human L1 retrotransposon encodes a conserved endonuclease required for retrotransposition. *Cell* **87:** 905–916.

19. Flavell A.J. 1999. Long terminal repeat retrotransposons jump between species. *Proc. Natl. Acad. Sci.* **96:** 12211–12212.

20. Flavell A.J., Smith D.B., and Kumar A. 1992. Extreme heterogeneity of Ty1-copia group retrotransposons in plants. *Mol. Gen. Genet.* **231:** 233–242.

21. Friesen P.D. and Nissen M.S. 1990. Gene organization and transcription of TED, a lepidopteran retrotransposon integrated within the baculovirus genome. *Mol. Biol. Evol.* **10:** 3067–3077.

22. Gonzalez P. and Lessios H.A. 1999. Evolution of sea urchin retroviral-like (SURL) elements: Evidence from 40 echinoid species. *Mol. Biol. Evol.* **16:** 938–952.

23. Guo H., Karberg M., Long M., Jones J.P., Sullenger B., and Lambowitz A.M. 2000. Group II introns designed to insert into therapeutically relevant DNA target sites in human cells. *Science* **289:** 452–457.

24. Hamada M., Kido Y., Himberg M., Reist J.D., Ying C., Hasegawa M., and Okada N. 1997. A newly isolated family of short interspersed repetitive elements (SINEs) in coregonid fishes (whitefish) with sequences that are almost identical to those of the SmaI family of repeats: Possible evidence

for the horizontal transfer of SINEs. *Genetics* **146:** 355–367.

25. Hofmann J., Schumann G., Borschet G., Gosseringer R., Bach M., Bertling W., Marschalek R., and Dingermann T. 1991. Transfer RNA genes from *Dictyoselium discoideum* are frequently associated with repetitive elements and contain consensus boxes in their 5′ and 3′-flanking regions. *J. Mol. Biol.* **222:** 537–552.

26. Hollinger, F. B. 1996. Hepatitis B. In *Field's Virology,* 3rd edition (ed. B.N. Fields et al.), pp. 2739–2807. Lippincott-Raven, Philadelphia, Pennsylvania.

27. Houck M.A., Clark J.B., Peterson K.R., and Kidwell M.G. 1991. Possible horizontal transfer of *Drosophila* genes by the mite *Proctolaelaps regalis*. *Science* **253:** 1125–1128.

28. Isfort R., Jones D., Kost R., Witter R., and Kung H.-J. 1992. Retrovirus insertion into herpesvirus in vitro and in vivo. *Proc. Natl. Acad. Sci.* **89:** 991–995.

29. Ji H., Moore D.P., Blomberg M.A., Braiterman L.T., Voytas D.F., Natsoulis G., and Boeke J.D. 1993. Hotspots for unselected Ty1 transposition events on yeast chromosome III are near tRNA genes and LTR sequences. *Cell* **73:** 1–20.

30. Jordan I.K., Matyunina L.V., and McDonald J.F. 1999. Evidence for the recent horizontal transfer of long terminal repeat retrotransposon. *Proc. Natl. Acad. Sci.* **96:** 12621–12625.

31. Kazazian H.H. 2000. L1 retrotransposons shape the mammalian genome. *Science* **289:** 1152–1153.

32. Kazazian H.H., Wong C., Youssoufian H., Scott A.F., Phillips D., and Antonarakis S.E. 1988. Haemophilia A resulting from de novo insertion of L1 sequences represents a novel mechanism for mutation in man. *Nature* **332:** 164–166.

33. Kim J.M., Vanguri S., Boeke J.D., Gabriel A., and Voytas D.F. 1998. Transposable elements and genome organization: A comprehensive survey of retrotransposons revealed by the complete *Saccharomyes cerevisiae* genome sequence. *Genome Res.* **8:** 464–478.

34. Kinsey J.A., Garrett-Engele P.W., Cambareri E.B., and Selker E.U. 1994. The Neurospora transposon *Tad* is sensitive to repeat-induced point mutation (RIP). *Genetics* **138:** 657–664.

35. Kirchner J., Connolly C.M., and Sandmeyer S.B. 1995. In vitro position-specific integration of a retroviruslike element requires Pol III transcription factors. *Science* **267:** 1488–1491.

36. Kordis D. and Gubensek F. 1998. Unusual horizontal transfer of a long interspersed nuclear element between distant vertebrate classes. *Proc. Natl. Acad. Sci.* **95:** 10704–10709.

37. Lambowitz A. and Belfort M. 1993. Introns as mobile genetic elements. *Annu. Rev. Biochem.* **62:** 587–622.

38. Lambowitz A., Caprara M.G., Zimmerly S., and Perlman P.S. 1999. Group I and Group II ribozymes as RNPs: Clues to the past and guides to the future. In *The RNA world*, 2nd edition (ed. R.F. Gesteland et al.), pp. 451–485. Cold Spring Harbor Laboratory Press, Cold Spring Harbor, New York.

39. Lander E., Linton L.M., Birren B., Nusbaum C., Zody M.C., Baldwin J., Devon K., Dewar K., Doyle M., FitzHugh W., et al. 2001. Initial sequencing and analysis of the human genome. *Nature* **409:** 860–921.

40. Lonergan K.M. and Gray M.W. 1994. The ribosomal RNA gene region in

Acanthamoeba castellanii mitochondrial DNA. A case of evolutionary transfer of introns between mitochondria and plastids? *J. Mol. Biol.* **17:** 476–499.

41. Luan D.D., Korman M.H., Jakubczak J.L., and Eickbush T.H. 1993. Reverse transcription of R2Bm RNA is primed by a nick at the chromosomal target site: A mechanism for non-LTR retrotransposition. *Cell* **72:** 595– 605.

42. Malik H.S. and Eickbush T. 2000. NeSL-1, an ancient lineage of site-specific non-LTR retrotransposons from *Caenorhabditis elegans. Genetics* **154:** 193–203.

43. Malik H.S., Burke W.D., and Eickbush T.H. 1999. The age and evolution of non-LTR retrotransposable elements. *Mol. Biol. Evol.* **16:** 793–805.

44. Marschalek R., Brechner T., Amon-Bohm E., and Dingermann T. 1989. Transfer RNA genes: Landmarks for integration of mobile genetic elements in *Dictyostelium discoideum. Science* **244:** 1493–1496.

45. Martin S.L. and Bushman F.D. 2001. Nucleic acid chaperone activity of the ORF1 protein from the mouse LINE-1 retrotransposon. *Mol. Cell. Biol.* **21:** 467–475.

46. Miller D.W. and Miller L.K. 1982. A virus mutant with an insertion of a copia-like transposable element. *Nature* **299:** 562–564.

47. Mizrokhi L.J. and Mazo A.M. 1990. Evidence for horizontal transmission of the mobile element *jockey* between distant *Drosophila* species. *Proc. Natl. Acad. Sci.* **87:** 9216–9220.

48. Morse B., Rotherg P., South V., Spandorfer J., and Astin S. 1988. Insertional mutagenesis of the myc locus by a LINE-1 sequence in a human breast carcinoma. *Nature* **333:** 87–90.

49. Naas T., DeBerardinis R.J., Moran J.V., Ostertag E.M., Kingsmore S.F., Seldin M.F., Hayashizaki Y., Martin S.L., and Kazazian H.H. 1998. An actively retrotransposing, novel subfamily of mouse L1 elements. *EMBO J.* **17:** 590–597.

50. Pyle A.M. 2000. New tricks from an itinerant intron. *Nat. Struct. Biol.* **7:** 352–354.

51. Roeder G.S. and Fink G.R. 1982. Movement of yeast transposable elements by gene conversion. *Proc. Natl. Acad. Sci.* **79:** 5621–5625.

52. Smit A.F. 1999. Interspersed repeats and other mementos of transposable elements in mammalian genomes. *Curr. Opin. Genet. Dev.* **9:** 657–663.

53. Song S.U., Kurkulos M., Boeke J.D., and Corces V.G. 1997. Infection of the germ line by retroviral particles produced in the follicle cells: A possible mechanism for the mobilization of the *gypsy* retroelement of *Drosophila. Development* **124:** 2789–2798.

54. Song S.U., Gerasimova T., Kurkulos M., Boeke J.D., and Corces V.G. 1994. An *env*-like protein encoded by a *Drosophila* retroelement: Evidence that gypsy is an infectious retrovirus. *Genes Dev.* **8:** 2046–2057.

55. Stewart S.A. and Weinberg R.A. 2000. Telomerase and human tumorigenesis. *Semin. Cancer Biol.* **10:** 399–406.

56. Takasaki N., Park L., Kaeriyama M., Gharrrett A.J., and Okada N. 1996. Characterization of species-specifically amplified SINEs in three salmonid species–chum salmon, pink salmon, and kokanee: The local enviornment of the genome may be important for the generation of a dominant source gene at a newly retroposed locus. *J. Mol. Evol.* **42:** 103–116.

57. Travisano M. and Inouye M. 1995. Retrons: Retroelements of no known function. *Trends Microbiol.* **3:** 209–211.

58. Turmel M., Cote V., Otis C., Mercier J.P., Gray M.W., Lonergan K.M., and Lemieux C. 1995. Evolutionary transfer of ORF-containing group I introns between different subcellular compartments (chloroplast and mitochondrion). *Mol. Biol. Evol.* **12:** 533–545.
59. Venter J.C. et al. 2001. The sequence of the human genome. *Science* **291:** 1304– 1351.
60. Wang H. and Lambowitz A.M. 1993. The Mauriceville plasmid reverse transcriptase can initiate cDNA synthesis de novo and may be related to reverse transcriptase and DNA polymerase progenitor. *Cell* **75:** 1071–1081.
61. Wright D.A. and Voytas D.F. 1998. Potential retroviruses in plants: Tat1 is related to a group of *Arabidopsis thaliana* Ty3/gypsy retrotransposons that encode envelope-like proteins. *Genetics* **149:** 703–715.
62. Xiong Y. and Eickbush T. 1990. Origin and evolution of retroelements based upon their reverse transcriptase sequences. *EMBO J.* **9:** 3353–3362.
63. Zhu Y., Zou S., Wright D.A., and Voytas D.F. 1999. Tagging chromatin with retrotransposons: Target specificity of the *Saccharomyces* Ty5 retrotransposon with the chromosomal localization of Sir3p and Sir4p. *Genes Dev.* **13:** 2738–2749.
64. Zou S. and Voytas D.F. 1997. Silent chromatin determines target preferences of the *Saccharomyces* retrotransposon Ty5. *Proc. Natl. Acad. Sci.* **94:** 7412–7416.

9

The DNA Transposons of Eukaryotes: Mariners Sailing to Survive?

EUKARYOTIC GENOMES CONTAIN A DIVERSE ARRAY of transposons that replicate through DNA intermediates in addition to the retrotransposons of the previous chapter. Many of the DNA transposons have a quite simple design, consisting of a transposase gene flanked by DNA sites that bind transposase and thereby potentiate transposition, much like the prokaryotic cut-and-paste transposons (Chapter 3). The transposition reaction involves DNA cleavage to liberate the transposon from the former flanking host DNA, then insertion into a new site elsewhere in the genome. The apparent simplicity notwithstanding, close study reveals great diversity in the mechanisms and consequences of transposition by DNA transposons.

Some of the best-documented examples of lateral transfer of eukaryotic transposons involve the DNA transposons. As discussed in Chapter 8, arguments for lateral transfer are often based on the observation of (1) unexpected similarity between elements in distantly related genomes and (2) lack of related elements in intermediate species. This argument can go awry, however, if the data on which the expectations are based are not exhaustive or carefully analyzed. However, the evidence for lateral transfer of DNA transposons is overwhelming, both in the amount of sequence available and in the degree to which elements in distant species are unexpectedly similar.

This chapter begins with the initial discovery of DNA transposons in maize by Barbara McClintock. The chapter next reviews the types of transposons found in eukaryotes, and then the detailed biology of the most prominent eukaryotic DNA transposons, including their mechanisms of replication and evidence for lateral transfer. The chapter concludes with evidence that lateral transfer may actually be required for the evolutionary persistence of some DNA transposons. Chapter 10 puts the DNA transposons in the larger context of the complete eukaryotic genome sequences.

Barbara McClintock and the Discovery of Transposons (17, 25, 31–34)

Transposons were discovered in the late 1940s by Barbara McClintock, and for this discovery she received the Nobel Prize in 1983 (Fig. 9.1). For most of her career, McClintock studied the genetics and development of maize plants. Before achieving her breakthrough on studies of transposons, she made several key contributions to understanding genetic mechanisms. One of her best-known early studies provided crucial support for the idea that chromosomes were in fact the material embodiment of the genes. McClintock succeeded in correlating the segregation of a morphological feature, a knob on one chromosome in a maize strain, with segregation of genetic markers on the same chromosome. Together with other studies, this established beyond a reasonable doubt the chromosomal theory of inheritance.

Another line of study set the stage for the discovery of transposons. McClintock found that chromosomes with broken ends tended to fuse. In cases where two chromosomes both contain broken ends, they tend to fuse at the tips, forming an abnormal chromosome with two centromeres. During cell division, such a dicentric chromosome will be pulled toward both poles of the cell at the same time, accumulating tension until the chromosome breaks. Daughter cells inherit chromosomes with broken ends, which can again fuse. Thus, the "breakage-fusion-bridge" cycle can be repeated again and again during cell division.

While studying plants undergoing the breakage-fusion-bridge cycle, McClintock noticed a surprisingly high level of variegation; for example, streaks of white in a green leaf, or spots of pigment in white corn kernels. These sectors were genetically distinct from the surrounding normal tissue, arising as a result of mutation during the growth of the plant. It was by focusing on these gene rearrangements during growth that McClintock began to recognize the existence of mobile genetic elements.

An important insight came with the recognition that the genetic changes often correlated with the appearance of chromosomal breaks. These points of breakage were found in the same locations in different maize plants of the same strain, but close study revealed that there was more to the story. Sites of breakage could disappear or move to a new chromosomal location. After such a "transposition" event, chromosomal breaks would occur selectively at the new location. McClintock named these mobile hot spots for breakage "dissociator" (*Ds*) elements. Further genetic crosses revealed that *Ds* elements did not act alone, but required another element present at a different chromosomal location. These elements were named

FIGURE 9.1. Barbara McClintock. (Courtesy of Cold
Spring Harbor Laboratory Archives.)

"activator" (*Ac*) elements for their effect on *Ds*. Not only was McClintock one of the first to
focus on regulation of growth in any system, but she also found that the *Ac* control elements
were mobile.

Over the years, working entirely alone, McClintock found that *Ds* elements could not only
break chromosomes, but could also induce new mutations in maize genes. These mutations
could be unstable if an *Ac* element was also present in the strain, but stable in the absence of
Ac. Furthermore, *Ac* elements could be converted into *Ds* elements. These findings are readily explained, as was foreseen by McClintock, if *Ac* is an intact mobile element and *Ds* a defective element requiring functions supplied by unlinked active elements for mobility. Many
more findings also emerged from her work, such as the discovery of several additional mobile
element families in maize and further features of the regulation of transposition.

McClintock's studies were far ahead of her time. Key to her discoveries was her remarkable ability to make observations and then formulate functional descriptions of what she saw.
As she once told geneticist Marcus Rhoades, "...when I look at a cell, I get down in that cell
and look around."

Although many scientists were interested in her work, for several decades there was little connection to the science of her contemporaries. It was not until the late 1960s that mobile
elements were also discovered in bacteria, flies, and, later, other systems. With the recognition of transposons in almost all organisms studied came a full appreciation of the magni-
tude of McClintock's discovery.

Even today, her discoveries are affecting contemporary biology in new ways. McClintock
believed very strongly that the maize genome, and probably the genomes of all organisms,
respond to stress by undergoing programmed genetic rearrangements. When she induced
the breakage-fusion-bridge cycle in maize, characteristic secondary rearrangements took
place. Today it is clear that many other transposons are activated for transposition in
response to stress. For example, in Chapter 10 we examine the remarkable case of the BARE
transposon of barley, which becomes activated by drought, resulting in BARE elements accumulating to add another 3% to the size of the barley genome. Many laboratories are investigating the presently contentious question of whether environmental stresses can induce a
"hypermutable state" that promotes the appearance of favorable mutations. Barbara
McClintock's experimental work of decades ago continues to inform current debates.

FAMILIES OF EUKARYOTIC TRANSPOSONS (4, 29)

Figure 9.2 summarizes the major types of transposons found in eukaryotes, together with some of the most common remnants of transposon action found in eukaryotic DNA. As discussed in Chapter 8, the retrotransposons are divided into two main types, the long terminal repeat (LTR) and non-LTR retrotransposons. Each of these has also given rise to defective nonautonomous elements distinctive to each class. For LTR retrotransposons, mobilization with proteins supplied in *trans* requires that the *cis* sites (LTRs, flanking sequences for priming reverse transcription, the Ψ packaging site) be intact. For non-LTR retrotransposons, small RNAs can associate with LINE element machinery and become reverse-transcribed, forming the SINEs. Also prominent are truncated LINE elements, probably produced as partially completed replication products.

The mobile introns form another family. As the name implies, these elements are mobile, and the element RNA can be spliced out of pre-mRNAs as introns. The Group II introns encode reverse transcriptase activity within element-encoded maturase protein. The Group I introns encode maturase lacking reverse transcriptase (RT), but do encode a double-stranded DNA endonuclease.

All of the DNA transposons are composed of a probable transposase gene flanked by inverted repeats in the DNA that act as transposase-binding sites. The details differ among the different families, such as the host site sequences used as transposition targets, the splicing patterns in the transposase gene, the length of the inverted repeat sites, the involvement of other proteins, and the DNA sequence relationships of the elements. The DNA transposons are parasitized by a distinctive class of nonautonomous elements that do not encode transposase but consist of a pair of inverted repeats that function as transposase-binding sites. These defective elements have in some cases expanded to great numbers, potentially contributing to the demise of whole element families. Each of the transposon families also contains other members not described in Figure 9.2, in some cases diverged in distinctive and intriguing ways from the other family members.

TRANSPOSITION OF *Tc1/mariner* ELEMENTS (11, 19, 22, 38, 42, 43)

We begin our survey of eukaryotic DNA transposons with the *Tc1/mariner* elements (named for the most extensively studied family members). *Tc1* elements were discovered in worms by David Hirsh and coworkers, and *mariners* in *Drosophila mauritiana* by Daniel Hartl and colleagues. Family members have since been identified in numerous insects, *Caenorhabditis elegans*, flatworms, humans, and many other species. The *Tc1/mariner* group has multiple representatives in the genomes of most metazoans (multicellular organisms) studied.

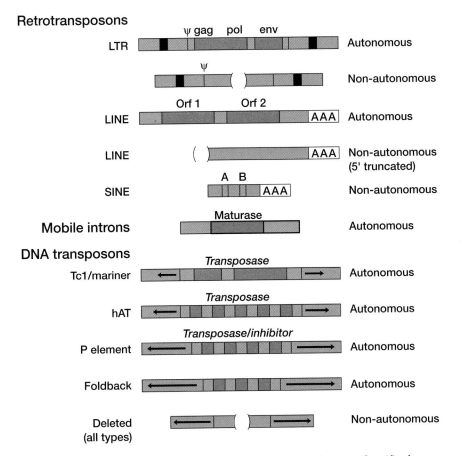

FIGURE 9.2. Eukaryotic transposons and common defective elements. Specific element types are described in Chapters 8 and 9.

Curiously, for unknown reasons, no *Tc1/mariner* family members have been identified in the yeasts. We first review the structure and function of *Tc1/mariner* elements, then the dramatic evidence for extensive lateral transfer.

Tc1/mariner Structure

The *Tc1/mariner* family of transposons contains some of the simplest mobile elements known. These elements are structurally related to the insertion sequence (IS) elements of prokaryotes and consist of a transposase gene flanked by inverted repeat DNA sequences that contain transposase-binding sites. The *Tc1/mariner* family elements are relatively small, on the order of 1.6–2.3 kb, and thus have limited coding capacity. Figure 9.3 illustrates two of the family members, *Tc1* and *Tc3*, found in *C. elegans*.

Most of the *Tc1/mariner* element DNA is dedicated to encoding the transposase protein. The *Tc1* transposase is 282 amino acids in length

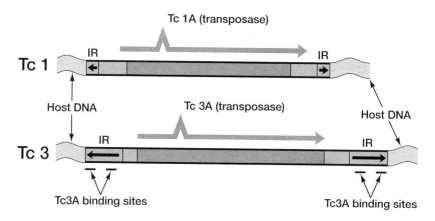

FIGURE 9.3. The *Tc*1 and *Tc*3 transposons of *C. elegans*, representatives of the *Tc1/mariner* family. (Modified, with permission, from Plasterk 1995 [© Springer-Verlag] [37]; see also Pasterk et al. 1999. [38])

(Fig. 9.4). Analysis of protein sequences of members of this family places the transposase protein in the large DDE group of enzymes. "IR" indicates the inverted repeat sequences in the DNA (discussed below).

The terminal inverted repeats of the *Tc1/mariner* elements bind transposase. In the *mariner* element, for example, the repeats are about 500 bp in length and are composed of blocks of about 30-bp sequences that bind transposase. For the case of Tc3 (Fig. 9.3), there are two transposase-binding sites at each end. Deleted elements lacking most of the transposon DNA but retaining these sites are still capable of transposition when supplied with transposase in *trans*, emphasizing the functional importance of these sites.

Tc1/mariner Transposition (20, 38, 43, 51, 54)

The transposition cycle begins with a double-stranded cleavage generated by transposase at each end of the element DNA, here illustrated with *Tc*3 (Fig. 9.5). The location of these cleavage sites defines the ends

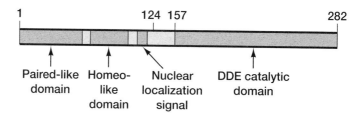

FIGURE 9.4. The Tc1A transposase protein, showing the DDE motif and functional regions. (Adapted from Plasterk 1995 [37]; see also Plasterk et al. 1999. [38])

The DDE Enzyme Family (3, 11, 15, 28, 44)

A prominent feature in many transposase amino acid sequences is the conserved "DDE motif." The three acidic amino acids comprising the motif are known to come close together in the folded three-dimensional structure of well-studied family members, such as the retroviral integrase enzymes and bacterial transposases such as MuA. The three acidic residues bind two metal atoms (Mg^{++} or Mn$^{++)}$, which are central to the chemistry of DNA breaking and joining. Probably most prokaryotic transposons, the eukaryotic DNA transposons, and the retrovirus and retroelement integrases all carry out catalysis using the DDE domain. In a few cases, such as the *mariner* element transposase, the third acidic residue is a D instead of E, but the chemistry of DNA cleavage and joining is likely similar. The DDE catalytic domains are found widely in other polynucleotide phosphotransfer enzymes such as RNase H nucleases and the RuvC Holiday junction-resolving enzyme.

In each of these enzymes, additional protein domains bind substrates and products, thereby selecting the polynucleotide reactants and dictating the course of the reaction. In the *Tc1* transposase, for example, the amino-terminal part of the protein contains two DNA-binding domains, one related to the fruit fly *paired* DNA-binding domain and the other related to the "homeodomain" DNA-binding motif. *mariner* transposase, in contrast, contains a domain related to the bacterial "helix-turn-helix" DNA-binding proteins instead of the *paired* domain in Tc1. Retroviral integrases, in contrast, contain instead an amino-terminal zinc-binding domain that does not bind DNA sequence specifically.

of the transposon DNA. The positions of cleavage are staggered, so that the 3′ DNA end of the element forms a single-stranded protrusion. The excised DNA remains associated with the transposase, forming the protein–DNA complex that carries out the DNA insertion steps. Circular *Tc1/mariner* DNA molecules have been found, but these are thought to be dead-end forms rather than replication intermediates.

One DNA strand at each end of the *Tc3* element DNA is then attached to sites in the host target DNA. All known *Tc1/mariner* family members favor integration into target sites with the sequence 5′-TA-3′, and, as with most other transposons, the sites of joining are staggered in the target DNA. Thus, after repair of the integration intermediate, presumably by host cell DNA repair enzymes, a duplication of the TA sequence is found at each end of the *Tc1/mariner* element DNA.

What becomes of the original donor DNA site, formerly containing the transposon but now burdened with a double-stranded break? It appears that the target site break can be rejoined in more than one way (Fig. 9.6). In some cases, the two broken ends at the donor site are ligated together, usually accompanied by the accumulation of a few extra nucleotides not present in the original chromosomal site. In the example in Figure 9.5, polymerization from each 3′ end at the double-stranded break makes each end suitable for blunt-end ligation. After ligation, the TA sequence is spaced by four nucleotides derived from the Tc3 element, termed an excision "footprint." Such sequences can be quite variable from footprint to footprint, depending on the combination of nuclease and polymerase activities that prepare the double-stranded breaks for ligation.

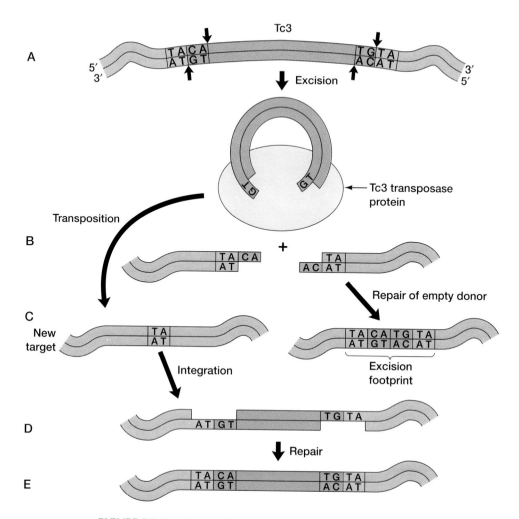

FIGURE 9.5. Excision and integration by Tc3. (*A*) A cycle of excision and integration begins with cleavage at the ends of the element by transposase (*arrows*) to liberate the transposon from flanking host DNA. The transposon DNA remains associated with transposase. (*B*) Because the 5´ cleavage on each strand is internal to the transposon DNA, two bases derived from the 5´ end of the transposon remain at the empty transposon donor site. (*C*) The transposon DNA complex can then carry out integration at a new TA target site. Meanwhile, in the example shown, the empty donor site is repaired, leaving an excision footprint. The diagram shows a footprint in which both single-stranded regions were "filled in" by a DNA polymerase prior to end joining. (*D*) The transposon 3´ ends are then joined to 5´ protruding ends in the target DNA. Repair of the resulting gaps (*E*) completes integration. (Adapted from Plasterk 1995; [37]; van Luenen et al. 1994. [50])

In another repair pathway, homologous sequences elsewhere in the genome can be used for recombinational repair (Fig. 9.6, A–C). Here similar sequences are copied into the gap to fill in missing sequences. If the homologous chromosome is used as a template, this results in copying *Tc1/mariner* element sequences into the gap. If the excised

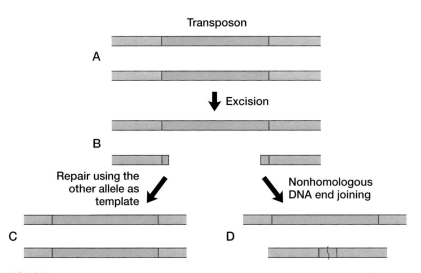

FIGURE 9.6. Two fates for the empty transposon donor site. (*A*) A pair of sister chromatids is shown, each bearing a type II transposon. Excision of one (*B*) leaves a DNA gap with short sequences derived from the transposon. Repair can take place by gene conversion, using the other sister chromatid as template (*C*). Alternatively, repair can take place by nonhomologous DNA end-joining (*D*). (Adapted from Plasterk 1995. [37])

copy then integrates elsewhere in the genome, this results in a net increase in element copy number. The gap resulting from excision can also be repaired by copying in homologous sequences from elsewhere in the genome, which results in related sequences changing position in the genome because of excision.

An interesting feature of *Tc1/mariner* transposition is that it does not appear to be strongly dependent on host-specific functions. As discussed below, there is now an extensive body of work documenting the ability of *Tc1/mariner* elements from a particular species to transpose in cells of distantly related organisms. The host-encoded DNA repair functions needed for transposition appear to be ubiquitous. Hugh Robertson and coworkers have taken this to an extreme by developing modified *mariner* elements that can even transpose in Archaea! Evidently any required host factors are shared among all the organisms tested.

Lateral Transfer of *mariner* Elements (20, 38, 42, 43)

Extensive phylogenetic studies have provided strong evidence for lateral transfer of *Tc1/mariner* elements. Elements found in many organisms can be grouped into subfamilies with at least 40% predicted amino acid identity in the transposase genes. Several of the groups contain elements from animals that diverged up to 200–600 million years ago. The expected rates of sequence change in *Tc1/mariner* elements are not

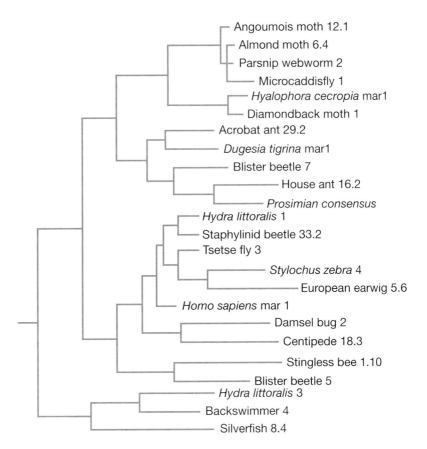

FIGURE 9.7. Sequence relationship among *mariner* elements of the cecropia sub-family. Note that a human element clusters with several insects, whereas a prosimi-an sequence clusters in a different part of the tree with other insects. The discor-dance between transposon phylogeny and that of their hosts is strongly suggestive of lateral transfer. (Modified, with permission, from Robertson et al. 1998. [43])

easy to judge, but the reasonable assumption of neutral drift would predict 1% DNA sequence change per million years. It appears that *Tc1/mariner* element transposase sequences are not under any unusual selective pressure, judging by the rates and types of substitutions observed in transposase genes determined to have diverged by con-ventional vertical transmission. *Tc1/mariner* elements from different phyla are much more closely related, strongly supporting the idea that lateral transfer accounts for the observed similarity among elements.

Consider the cecropia family of *Tc1/mariner* elements, named for the giant silk moth *Hyalophora cecropia*, in which the first family member was found (Fig. 9.7). Members of this family have been found in insects, flatworms, a hydra, and humans. One subgroup of *mariners* from humans, insects, and hydra share 85% amino acid identity, an amazing degree of similarity for species believed to have shared a common ancestor 600 million years ago. In another subgroup, *mariners* from ants and beetles cluster with those of flatworms and prosimians.

There is evi-dence for mariner-like elements in a wide range of eukaryotic genomes, including plants and humans.

In the *Drosophila irritans mariner* subfamily, sequences from green lacewing, horn fly, stable fly, and *Drosophila ananassae* show 88% identity. Again the remarkable sequence similarity strongly supports recent lateral transfer, in this case among different orders of insects. The second human *mariner* sequence discovered may also be a distantly related member of this group. For the lacewing and hornfly *mariners*, the encoded transposase sequences differ by only one amino acid substitution, in this case, in insect suborders believed to have diverged 200 million years ago. These dramatic similarities between elements strongly support the idea that the elements spread by lateral transfer.

In one case, *mariner* proliferation has been implicated in actually expanding the total size of an insect genome. The horn fly has an enormous genome of 2.2×10^9 bp, ten times the size of *D. melanogaster*. At least 18 Tc/*mariner* family members are present. One of these elements has been found to account for fully 1% of the genome. The full contribution of *mariner* elements has not been determined, but expansion of *mariner* sequences may well account for the increased genome size.

A series of dramatic laboratory experiments has documented the ability of *Tc1/mariner* transposons to hop in diverse cell types, including cells form organisms of different kingdoms. We return to these studies below in considering the possible link between lateral transfer and evolutionary persistence of these elements.

P-ELEMENT TRANSPOSITION (14, 26, 27)

Hybrid Dysgenesis

The P-element transposons of *Drosophila* were first discovered unexpectedly in genetic studies of natural fly populations. Crosses between different strains displayed a puzzling nonreciprocal character. When females of strains that had been kept in the laboratory for many years were mated with males from recently caught strains, sickly offspring were produced. Abnormal traits included high mutation rates, sterility, and recombination in males, which does not normally take place in flies. Mutations caused by these crosses were found to be genetically unstable. The two strain types were named M (lack P elements) and P (have P elements). Reciprocal crosses between P-strain females and M-strain males differed, yielding mostly normal progeny. This effect came to be known as as hybrid dysgenesis.

Genetic studies revealed that the factor responsible for P-M hybrid dysgenesis could be mapped to a specific locus in P strains, but that the locus had different locations from strain to strain. The locations of the P elements were found to coincide with hot spots for chromosomal breakage. These observations strongly suggested that the P elements were DNA transposons, and that dysgenic crosses somehow activated transposition.

Today, hybrid dysgensis is understood in some detail. In the dysgenic cross, the P-strain male donates a P-element-bearing chromosome to the egg cell of an M-strain female. M strains lack P elements. P elements also encode inhibitors of their own activity, but upon introduction into the M strain, the inhibitor does not accumulate immediately. The intact P element from the male produces transposase that allows the elements to transpose, creating mutations that result in defective progeny. In the reciprocal (nondysgenic) cross, the P-strain female contains P elements, which synthesize the transposition inhibitor. Thus, transposition is not stimulated and no mutations accumulate in the progeny. It turns out that flies with P elements build up a supply of the transposase inhibitor over time, which largely immobilizes the resident P elements.

Further studies revealed the hybrid dysgenesis phenomena in several different transposon systems. The I-R system, a non-LTR retrotransposon of flies, was found to display hybrid dysgenesis as well as the P-M system, and other examples have been found in flies as well.

P-element Organization (1, 2, 9, 24)

The organization of a P element is shown in Figure 9.8. The transposon is only about 3 kb long and encodes a single gene for transposase. The transposase gene is flanked by inverted repeats that are 31 bp long. Overall, these features are typical of small DNA transposons, but this apparent simplicity conceals a number of surprises.

First, the inverted repeats are not binding sites for transposase, but rather a cellular protein, named IRBP for *inverted repeat binding protein*. IRBP is the *Drosophila* homolog of the vertebrate Ku proteins, which have been closely studied for their role in repair of double-

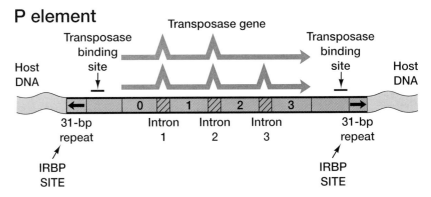

FIGURE 9.8. Diagram of the *Drosophila* P element. Two transcripts are produced from the transposase gene, the triply spliced transposase message and the doubly spliced transposition inhibitor. Transposase protein binds to the transposase-binding site; the fly Ku protein binds to the IRBP sites. (Adapted from Clark and Kidwell 1998. [9])

stranded DNA breaks. The transposase actually binds more internally, at sites 40–60 bp from each end.

The transposase coding region is not colinear in the DNA but broken up with three introns. Transposase expression requires removal of the introns by RNA splicing. Here splicing itself is tightly regulated, and alternative splicing provides a means of regulating transposase activity.

P-element Transposition (2, 24, 49)

P-element transposition begins with transposase binding to the internal DNA sites together with the IRBP. The two ends of the transposon are brought together to form a synaptic complex. Transposase then cleaves both DNA strands at each end, liberating the transposon DNA.

The free transposase–DNA complex then captures a target DNA site and joins the 3′ ends of the element DNA to 5′ ends in the target DNA. The points of joining are spaced by 8 bp in the target DNA. The intermediate is then unfolded and the resulting DNA gaps repaired, yielding 8-bp repeats of the target DNA sequence at the site of joining.

Meanwhile, the empty transposon donor DNA site, now containing a double- stranded break, is repaired. As was discussed for *Tc1/mariner* elements, the target site can have several fates. The single-stranded regions at the ends can be made double-stranded by a polymerase, or trimmed by a nuclease, in either case creating blunt ends that can be ligated together. For the case of P elements, the IRBP may potentially facilitate the rejoining of target site ends, as IRBP resembles proteins known to participate in rejoining of DNA double-stranded breaks. The DNA break can also be repaired by copying homologous sequences elsewhere in the genome into the gap.

Today P elements are crucial tools in genetic research. *Drosophila* geneticists seeking to isolate a gene responsible for a particular characteristic routinely use P elements as mutagens, because genes altered by P-element insertion can be readily isolated using the P-element sequence as a "tag." P-element transposition also provides a means of stably incorporating new genes into the fly germ line. The gene of interest is cloned between P-element termini, and P-element transposase is supplied in *trans*, allowing transposition of the gene into a chromosome.

Regulation of P-element Transposition (1, 2, 24)

P-element transposition is regulated so that elements hop efficiently only in the germ line, not in somatic tissue. This is accomplished by controlling the splicing of the transposase gene. The first two introns are removed efficiently in both somatic and germ cells, but the third intron is spliced efficiently in germ cells only. Selective splicing has been traced to the presence in somatic cells of a specific suppressor of intron 3 splicing, named PSI. PSI binds to the transposase RNA and suppresses intron 3 splicing, blocking transposase production.

Inhibition also has another component. The transposase mRNA containing intron 3 still retains the ability to encode a protein termed KP, which is inactive for transposition but acts instead as an inhibitor. Thus, the production of KP protein provides a second mechanism of negative regulation. The presence of KP inhibitor in P-type female eggs in large measure explains the nonreciprocal character of hybrid dysgenesis.

It is an open question as to whether other mechanisms may also suppress transposition. Chapter 13 reviews several regulatory systems, such as the RNA interference (RNAi) system, that could also potentially contribute.

P Elements Mobilized by Mites? (9, 21)

The pattern of P-element distribution in natural populations of flies suggests that P elements have been newly introduced by horizontal transfer. Strains of *Drosophila melanogaster* isolated from wild populations prior to the late 1970s and maintained in isolation lack P elements. Strains caught more recently contain active P elements. These observations suggest that P elements have spread in *D. melanogaster* within the last 50 years.

Two other observations provide further support for recent horizontal transfer: (1) P elements from geographically distant strains are virtually identical. (2) P elements in *D. melanogaster* and in members of the *D. willistoni* species group are very closely related, more so than would be expected, given that the species groups are thought to have diverged 50 million years ago. It has been proposed that *D. melanogaster* recently acquired P elements from *D. willistoni* concomitant with the introduction of *D. melanogaster* into the New World and the first mixing of the two species. Similar reasoning identifies at least one other probable example of P-element transfer between fly species.

A strong case has been made that the mite *Proctolaelaps regalis* (Fig. 9.9) was responsible for transporting P elements between fly species. *P. regalis* and other mites frequently cause infestations of laboratory fly cultures and are a nightmare for fly geneticists. *P. regalis* is semiparasitic on flies, feeding by sucking fluid from eggs and immature larvae. *P. regalis* thrusts its mouth parts into its prey, withdraws bodily fluids, then moves quickly to new victims. Some eggs and larvae are killed, but others survive the feeding bout.

Intriguingly, *P. regalis* could be shown to pick up fly DNA, including *P* element DNA, by feeding. Other mites infesting flies but not feeding in the same fashion did not acquire host DNA detectably. This suggested that the *P. regalis* mouthparts and feeding habits are particularly suited to trapping host material and transferring DNA. In fact, the feeding style of *P. regalis*—stabbing into eggs and embryos with a feeding shaft and quickly moving between animals—mimics the methods used by experimentalists to introduce DNA deliberately. Thus, a sim-

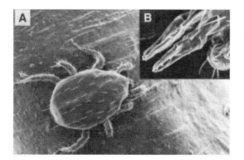

FIGURE 9.9. (*A*) The mite *Proctolaelaps regalis,* which is implicated in horizontal transfer of P elements. (*B*) A close-up view of the chelicerae (mouth parts). (Reprinted, with permission, from Houck et al. 1991 [© American Association for the Advancement of Science]. [21])

ple explanation for P-element transfer between fly species would be transfer during predation by *P. regalis* on mixed fly populations.

hobo, Ac AND *Tam:* THE *hAT* FAMILY (6, 7, 10, 16)

The second major family of DNA transposons is the *hAT* family, which comprises the *hobo* element of flies, the *Ac* element of maize, the *Tam* elements of snapdragons, and many others. The name is taken from the first letter of each transposon. The P-element family does not appear to be as widely distributed as the *hAT* family, possibly reflecting a greater reliance on species-specific host factors for mobility. Below we review the three family members and their lateral transfer.

hobo

The *hobo* elements are 2959 bp long and encode a single open reading frame. The element termini are flanked by 12-bp inverted repeats. *hobos,* like P elements, are found in flies recently captured from natural settings. Fly strains that have been long kept in laboratories tend to lack *hobo* elements. Like P elements, *hobo* displays hybrid dysgenesis. Matings of *hobo*-containing males (H strains) with *hobo*-free females (E or empty strains) result in activation of transposition. This hybrid dysgenic system is found in *D. melanogaster,* and in the relatives *D. simulans* and *D. mauritiana.*

The transposition mechanism of animal *hAT* elements has not yet been studied in detail. Particularly lacking is a test-tube system recapitulating transposition, a crucial resource for mechanistic studies. From an analysis of *hAT* transposon structure, it seems likely that these are typical cut-and-paste transposons, resembling *Tc1/mariner* and P elements in broad outline. However, as discussed below, transposition is inferred to proceed through an intermediate involving a hairpin structure in the target DNA.

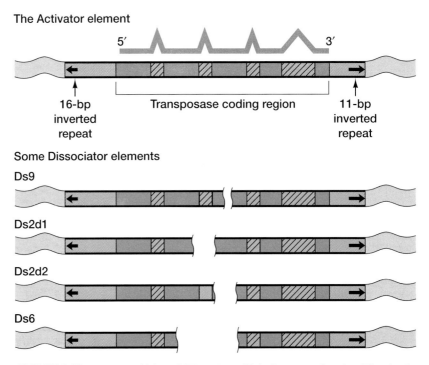

FIGURE 9.10. Activator (*Ac*) and Dissociator (*Ds*) elements of maize. The *Ac* element (*top*) encodes transposase protein from a transcript containing four introns (*hatched*). *Ds* elements are derived from *Ac* elements by internal deletions. *Ds* elements contain transposase-binding sites at the ends of the element but do not encode intact transposase protein. (Adapted from Fedoroff 1989. [16])

Ac

The *Ac* transposon, discovered and characterized by McClintock, is shown diagrammatically in Figure 9.10. The element is 4.6 kb in length and encodes a single gene for transposase that is interrupted by four introns. At each end are 11-bp perfect inverted repeat sequences that probably comprise part of the binding sites for transposase.

Ds elements were found by molecular studies to be deleted *Ac* elements. The *Ds* elements cannot direct the synthesis of transposase, but they do contain the necessary *cis* sites for transposase action. Studies of deleted *Ac* elements indicate that sequences internal to the terminal inverted repeats are also important for transposition. The relationship of *Ds* to *Ac* elements largely explains McClintock's observation of *Ac* as a mobile control region, since *Ac* can supply transposase to *Ds* in *trans*.

McClintock and others observed that *Ac* elements could be reversibly inactivated, a finding that today can be explained by reversible methylation of the element DNA. As discussed in detail in Chapter 13, many eukaryotes encode methylase enzymes that can attach methyl groups to C residues at CpG or CpNpG sequences. Extensive methylation of the DNA correlates with transcriptional inactivation. The reversible inactivation of *Ac* (and many other transposable elements) probably arises from methylation of element DNA and

FIGURE 9.11. An 1838 illustration of snapdragon (*Antirrhinum majus*) flowers showing variegation due to excision of *Tam* elements. (Reprinted from *Paxton's Magazine of Botany.*)

shutoff of transposase gene transcription. How methyl-directed gene inactivation is targeted and controlled is an intriguing mystery.

Tam

The snapdragon *Tam* elements, which resemble *Ac* in sequence and function, provide one of the most beautiful examples of mobile element activity, because they dictate the coloration in the variegated lobes of snapdragon flowers (Fig. 9.11). Today the molecular mechanism of variegation is well understood, and analysis of this system, together with parallel studies in maize, has helped elucidate the molecular mechanism of transposition.

Variegation in snapdragon flowers arises when transposons insert into one of the genes for pigment biosynthesis, forming an unstable mutant. Flower tissue is consequently white, because pigment synthesis is blocked. On occasion, however, the transposon excises, restoring activity to the gene and permitting pigment synthesis. This results in a pigmented sector in the flower against a background of white mutant tissue. Many of the snapdragon flowers grown commercially actually contain such unstable mutants, resulting in complex and beautiful patterns of color in flowers. The variegated patterns seen in McClintock's corn ears have a similar origin.

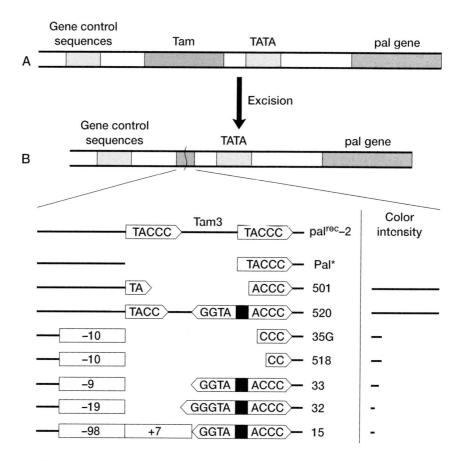

FIGURE 9.12. A *pal* gene allele with a *Tam* element inserted in the 5′ region (*A*) and excision footprints found in partial revertants (*B*). The structures of the locus and the product of excision are shown schematically in *A*. The detailed sequence changes in excision footprints are shown in *B* with the color intensity, which corresponds to the level of gene activity, shown at the right. Note that many of the excision footprints show partial inverted repeat sequences. (*B*, modified, with permission, from Coen et al. 1989. [10])

The ability to recognize plants containing excision products has allowed the molecular nature of the excision reaction to be characterized in detail. Excision proceeds by transposase-catalyzed cleavage of both DNA strands, freeing the element for integration elsewhere in the genome. The two DNA ends in the flanking chromsomal DNA are then rejoined to restore continuity to the chromosome. Analysis of the sequence of excision products suggests that the cleavage reaction does not proceed by a simple cut on each DNA strand, but rather through a hairpin intermediate. As shown in Figure 9.12, excision does not restore the starting sequence in the host DNA, but leaves a "footprint" of altered sequence, probably a consequence of hairpin formation (Fig. 9.13). First, one strand is cleaved by transposase, then the free end attacks the other DNA strand, breaking the chromosome and forming the hairpin in a single-step exchange (transesterification) reaction. To rejoin the chromosome ends after excision, the DNA ends are opened

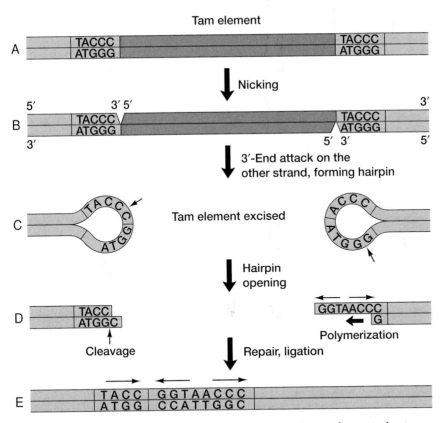

FIGURE 9.13. Generation of excision footprints containing inverted repeats due to DNA hairpin formation. Excision of a *Tam* element (*A*) begins with nicking at each 5′ edge of the element DNA (*B*). Attack of the 3′ end in the target DNA on the other DNA strand yields a hairpin in the target DNA and releases the transposon DNA (*C*). Opening of the hairpin (*D*), followed by repair of the ends, in the example, by cleavage of the right sequence and filling in on the left, yields the excision footprint in *E*. (Adapted from Coen et al. 1989. [10])

by cleavage with a nuclease (possibly transposase), the ragged ends are filled in with a DNA polymerase, and the two ends joined. This pathway then generates the characteristic inverted repeats seen with excision footprints of *Tam* elements and also *Ac*.

As discussed in Chapter 11, the RAG recombination system that constructs the antigen-binding genes of vertebrates also generates hairpins in DNA. This observation provided key support to the idea that the RAG system evolved from a progenitor transposon.

Lateral Transfer of *hAT* Elements (47, 52)

DNA sequencing studies suggest that the *hobo* transposon has proliferated among insects by lateral transfer. The *hobo* element has been found in several species of the *Drosophila* group, including *D. melanogaster*, *D. simulans*, and *D. mauritiana*. The elements in these species differ by only 0.1%, whereas the nuclear genes differ by typically 5%, suggesting that recent transfer of *hobo* elements explains their similarities. A survey of *hobo* "fossils," mutant inactive elements of presumably ancient pedigree, indicated that *D. simulans* may have been the species of origin of modern *hobo*s. The data suggest that inactive *hobo* elements are inherited vertically (by descent), whereas the active elements move laterally among fly species.

A recent laboratory study of the *Ac* element reveals that this transposon class is at least potentially very widely transmissible. C. Weil and R. Kunze transferred a maize *Ac* element to the yeast *Saccharomyces cerevisiae* and found the elements were capable of transposition in the new host. Yeasts normally do not have DNA transposons, but this study showed that yeast cells are not intrinsically unsuitable hosts. Possibly yeasts are very efficient at killing off DNA transposons when they attempt to invade the cell. The study also indicates that any host factors required for *Ac* multiplication are common to plants and yeast.

THE *FOLDBACK* FAMILY (5, 30, 40, 46, 53)

The ends of foldback elements are long palindromic sequences.

The *foldback* (*F,* or long inverted repeat*)* elements were first identified in *D. melanogaster* but are now known to be present in many species (Fig. 9.14A). *foldback* transposons got their name from their unusually long terminal inverted sequences, which, when present in purified DNA, can fold back to form branched DNA structures (Fig. 9.14B). The terminal inverted repeats are composed of two domains, an outer tandemly arranged subrepeat and an inner nonrepetitive region. Presumably, some of these bind proteins important for transposition, but this has not yet been studied in detail. *foldback* elements can potentially encode three proteins, none of which as yet has known functions, although for some family members a probable transposase sequence can be discerned.

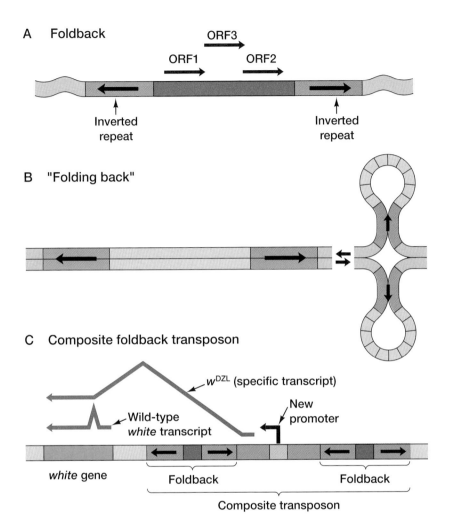

FIGURE 9.14. The *Drosophila foldback* element. (*A*) Organization of the foldback element. (*B*) The long inverted repeat sequences at either edge of the foldback element allow the DNA to "fold back" in vitro, yielding pairs of hairpin structures, and giving the element family its name. (*C*) A composite transposon composed of two foldback elements and an internal region derived from the cellular chromosome integrated near the *white* locus, giving rise to the w^{DZL} mutation. (Adapted from Bingham and Zachar 1989. [5])

foldback elements are widely distributed, with elements now described in many *Drosophila* species, in plants (FARE of *Arabidopsis*, SoFT of the Solanaceae group), worms, and sea squirts. The last instance documents the presence of these elements in chordates, a group defined by development involving a notochord and including vertebrates. Curiously, there do not seem to be representatives of this group in vertebrates proper.

As discussed below, *foldback* elements are resposible for diverse chromosomal rearrangements in *Drosophila*. The *foldback* elements have also been found to form composite structures like the *Tn* elements of bacteria, resulting in the translocation of blocks of chromosome DNA (Fig. 9.14C). *foldback* elements are also frequently associated with sites of DNA breakage, all reactions serving to diversify the genomes in which active *foldback* elements are present.

Analysis of sequence data led Silber and coworkers to propose that *foldback* elements are not subject to lateral DNA transfer. That is, the extent of sequence variation, at least among the *Drosophila* species studied, was about as expected for the timing of radiation of the various species. As discussed in more detail below, many of the DNA transposons may need to switch host species occasionally by lateral transfer to persist evolutionarily. If *foldback* elements are not subject to transfer, it may help explain their absence in some lineages, such as the vertebrates.

LATERAL TRANSFER TO AVOID EXTINCTION? (18, 20, 21, 23, 42, 45)

An intriguing argument can be made that the DNA transposons of animals must actually undergo lateral transfer to avoid extinction. Many studies reveal mechanisms that repress transposon activity (Chapter 13). Some are encoded by the transposons themselves, such as the KP inhibitor encoded by P elements. Other inhibitory functions are encoded by the host, such as DNA methylases and the RNAi system (Chapter 13).

Another type of inhibitor may be transposon end sequences. In eukaryotes, the transposase message must be translated in the cytoplasm, and then the transposase protein returned to the nucleus after synthesis. Once in the nucleus, transposase can bind to the ends of any element—the active element from which it was expressed, or inactive elements elsewhere in the genome. Thus, defective elements can engage in transposition if they just retain correctly positioned transposase-binding sites. Defective elements can thus proliferate, providing an expanding sink for transposase protein. If one further hypothesizes that active elements resident in the genome are occasionally inactivated by chance mutation, active element families will, with time, be extinguished, leaving only inactive fossil elements. Many examples are known of element families that are presently immobile but must once have been active, such as the *mariner* elements of vertebrates, all of which appear to be inactive, or the P elements in some M strains of flies. In some cases, clearly defective elements have expanded, such as the Tc7 defective transposon of worms, documenting that defective elements act as expanding sinks for transposase.

The only way for DNA transposons to dodge extinction may be to switch hosts periodically. Presumably, the naive host could not inactivate the invading transposon immediately, allowing time for the element to proliferate. A new host would also be free of inactive elements with intact transposase-binding sites to sequester newly synthesized transposase. Once the new host cell began to squelch transposition, the element would need to move on again. Such a life-style would explain the widespread examples of lateral transfer for this class of elements. The most successful classes, *Tc1/mariner* and *hAT*, would be expected to have minimal reliance on host factors, so that the elements could replicate efficiently in each new environment.

A number of dramatic laboratory studies confirm that *mariner* elements have particularly broad host ranges. The *mariner* element of *D. mauritiana* was shown to transpose in the parasitic protozoa *Leishmania major*. The worm element *Tc1* was able to transpose in human cells. A mutant *Tc1*-like element from a fish was resurrected by repairing the mutations, and the reactivated element, *Sleeping Beauty*, was shown to transpose in a wide range of vertebrate cells, from fish to humans. In fact, *Sleeping Beauty* is the only known active DNA element from vertebrates, testifying to the efficiency with which they are inactivated. All of these studies document the wide host ranges of individual *Tc1/mariner* elements, supporting the speculation that they have evolved to switch hosts to promote their evolutionary persistence.

Transposase Cleavage Reactions Dictate the Structure of DNA Products (10, 12, 13, 35, 36, 50)

One of the main themes of this book is that seemingly complicated DNA transactions can be explained by relatively simple mechanisms. A particularly striking example is provided by a comparison of the DNA cleavage reactions carried out by transposase and integrase proteins. Each transposase protein cleaves at the ends of the transposon DNA to permit subsequent reaction steps, but the detailed nature of the cleavages differs (Fig. 9.15), resulting in different reaction products.

The position of the cleavage at the 3′ end defines the boundary of the element because the 3′ end then becomes attached to target DNA. The 5′ cut, on the other hand, can take place at many different locations, and the placement of this second cut specifies the DNA products generated.

The Ty1 cDNA generated by reverse transcription is blunt-ended and requires no DNA cleavage prior to integration. The 3′ ends become connected to the target DNA directly. Retroviruses such as HIV remove two nucleotides from the cDNA ends prior to integration. The terminal cleavage reaction may serve two purposes: (1) to remove messy extra bases occasionally added by the reverse transcriptase enzyme during synthesis and (2) to stabilize the binding of integrase to the viral cDNA.

Phage Mu transposase, in contrast, does not cleave off the 5′ DNA at all. Joining of the 3′ end generated by cleavage yields the Shapiro intermediate, and replication through this structure yields a cointegrate, as discussed in Chapter 3. Tn7, Tn5, and Tn10 cleave both strands in the flanking target DNA, freeing the transposon DNA from the target DNA site. Integration of the free transposon yields a simple insertion. Here we see another nuance, in that Tn5 and Tn10, like most transposons, make both cleavages with a single protein, but Tn7 uses two different enzymes to cleave each of the two strands. Tn5 and Tn10 form hairpins in the element DNA during cleavage, whereas Tn7 does not.

The *mariner* and P-element transposases adopt yet another cleavage strategy. The 3′ cleavage defines the edge of the element, but the 5′ cleavage is actually within the element DNA. The 5′ cleavage is 2 bases internal in the *Tc1/mariner* family member Tc3 and fully 17 bases internal in P elements. Consequently, the 3′ ends of the elements are single-stranded after excision over the 2 or 17 bases at the ends. Upon joining to target DNA, the single-stranded regions are filled in to complete transposition. These internal cuts have no consequence for the transposed element, but they can affect the repair of the transposon donor site. One pathway for repair of the donor involves polymerization to make the single-stranded regions double-stranded, followed by ligation of the two DNA ends. As a result, short transposon DNA sequences are often found at the empty donor site after transposition.

For *hAT* transposons, hairpins are inferred to be produced on the target DNA side. This accounts for the inverted repeat sequences often seen with excision footprints of these elements.

Probably the chemistry of DNA joining is identical for all the elements illustrated, involving a direct attack of the element 3´ end on the target DNA (single-step transesterification). Host DNA repair enzymes then join the remaining DNA ends, completing transposition.

In summary, the different placements of 5´ end cleavages in Figure 9.15 dictate much of the structures of transposition products. Complicated DNA structures can be understood as resulting from simple variations in the placement of the initial DNA cleavages.

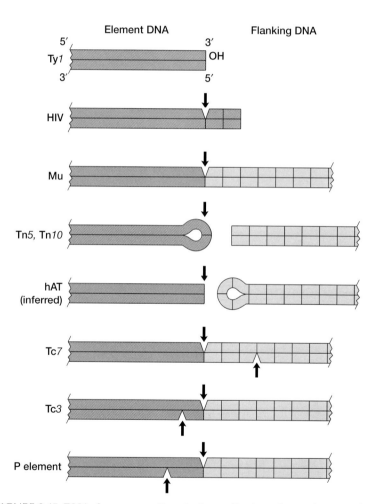

FIGURE 9.15. DNA cleavage reactions in the replication of retroviruses and transposons. Green indicates DNA from the transposon or virus; gray indicates flanking DNA. Vertical black arrows indicate locations of cleavage sites. Cleavage takes place by hydrolysis except in the cases of the Tn5/Tn10 and hAT families, where cleavage takes place by nicking on one strand, followed by attack of the free 3´ end on the other strand, so that a DNA transesterification reaction mediates both cleavage and joining to form a hairpin. The different locations of DNA cleavage dictate the nature of the DNA integration and excision products formed. (Modified, with permission, from van Luenen et al. 1994 [© Elsevier]. [50])

DNA TRANSPOSONS AND GENOME STRUCTURE (8, 38, 39, 41, 48)

In this section, we review some of the mechanisms by which DNA transposons can rearrange the host genome. In the next chapter, we investigate in detail the transposon complement of different organisms as revealed in the genome sequencing projects, and the consequences of transposon activity. Transposons can alter the host genome by providing portable regions of homology that act as substrates for the host general recombination apparatus. Transposons can also alter the genome by mechanisms involving catalysis by transposase protein instead of cellular enzymes. These pathways are illustrated below with a few well-documented examples.

The first pathway is illustrated by two human neurological diseases, Charcot-Marie-Tooth disease type 1A (CMT1A) and hereditary neuropathy with liability to pressure palsies (HNPP). Most cases of these diseases correlate with rearrangements of chromosome 17, a small duplication in the case of CMT1A and a small deletion of the same region for HNPP. Molecular analysis revealed that normal chromosome 17 contains a short duplicated sequence in this region, derived in part from insertion of human *mariner*-like elements. The rearrangements represent the products of unequal crossing-over between the repeated *mariner* sequences. Thus, the insertion of two *mariner* elements at nearby sites created a region of genetic instability in humans. This is almost certainly mediated by cellular enzymes and not transposase, since there are no known active *mariner* elements in human cells.

An example of the second, transposase-mediated, pathway of chromosomal rearrangements is provided by the male recombination observed during P-element hybrid dysgenesis. Normal males in *D. melanogaster* do not carry out DNA recombination in the germ line. However, during dysgenic crosses, male recombination is seen. A variety of observations support a model in which male recombination is carried out by transposase protein produced during the dysgenic cross. Normal P-element transposition involves recruitment of a left end and a right end from a single element into the active transposase–DNA complex. However, if transposon ends on different chromosomes pair and undergo transposition, then some of the products can be recombination events. As expected from this model, P-element sequences are present at the recombination junctions, and recombination produces short duplications and deletions. This mechanism, recruitment of abnormal pairs of ends followed by DNA cutting and rejoining, can create a diverse array of genomic rearrangements in any organism containing active DNA transposons.

The *foldback* elements of flies are of special interest for their effects on chromosome structure. In flies, chromosomal deletions, inversions, and reciprocal translocations frequently contain a *foldback* element at one or both break points. Rearrangements involving *foldback* elements take place in about 1 in 1000 fly chromosomes, a very high rate. Studies in flies have revealed that *foldback* elements can also form composite transposons that mobilize cellular genes, much as with composite transposons of bacteria. In one case, insertion of *foldback* elements on either side of the *white* and *roughest* genes allowed the full unit, *white* and *roughest* and flanking transposons, to be transferred to new locations in the genome. In another strain, a fragment of the *white* locus was found to be transferred between *foldback* elements, in this case resulting in mobilization of transcriptional controls that affected nearby genes (Fig. 9.14c).

Tc1/mariner family transposons may also alter genes by another means. The most common excision footprint leaves a sequence (5′-CATG-3′) with a reasonable resemblance to a 5′ splice donor site. Thus, integration and excision of a mariner element can potentially create new splicing patterns in the host genome. To what extent this mechanism leads to the formation of new genes is unknown, although many genes display organism-specific splicing patterns potentially arising from this mechanism.

To what degree transposon sequences themselves actually become incorporated into new host genes is an intriguing question. Chapter 11 deals with a likely example of this in the evolution of the vertebrate immune system. In another example, the human CENP-B protein, one of the proteins that bind the centromeres of chromosomes, is related in sequence to transposase of the DNA transposon *pogo*. This suggests that a *pogo*-related transposon invaded the germ line of a human progenitor and was subsequently inactivated, as can be seen from the dead *pogo* elements in human DNA, but a transposase remnant was recruited to form the CENP-B gene. Analysis of the draft of the human genome sequence so far reveals 49 probable genes that incorporated transposon sequences, emphasizing the opportunistic nature of new gene creation and evolution more broadly.

Summary

Work of Barbara McClintock provided the first example of transposons, the DNA transposons of maize. Today the eukaryotic DNA transposons provide some of the clearest examples of lateral transfer in animals.

The main element families—*Tc1/mariner*, *hAT*, P elements, and *foldback*—are relatively simple in structure. Each contains a gene encoding the transposase protein, and flanking inverted repeats that serve as transposase binding sites. Close inspection, however, reveals that this simple organization conceals interesting twists. For example, P elements encode shortened fragments of transposase that serve as inhibitors of transposition, probably helping to minimize the burden of the transposons on cells. A few elements, such as the *foldback* elements, encode additional proteins of unknown function.

The DNA transposons move by a cut-and-paste mechanism, in which the element DNA is initially cleaved on both DNA strands at each end of the element. The excised linear transposon sequences are then reintegrated at new locations in the genome. The empty target site can be repaired by joining of the two ends by host DNA repair proteins or it can be repaired by copying new sequences into the DNA gap (recombinational repair).

The DNA elements are widespread, having been found in all multicellular eukaryotes in which they have been sought. They are also abundant; for example, in humans some 300,000 transposon-related sequences comprise about 2.8% of the human genome. Genomes of many organisms contain defective elements that consist of little more than inverted repeats, parasites on transposons that can bind transposase and move without encoding their own enzyme.

The evidence for lateral transfer is particularly clear for some of the DNA transposons. For *Tc1/mariner*, *hAT*, and P elements, there are cases of almost identical elements in distantly related organisms, very likely reflecting recent lateral transfer. For P elements, there is even a credible proposal for a mechanism of transfer. Parasitic mites rove between eggs and larvae, piercing their prey with needle-like mouth parts, feeding, then moving on. The feeding style is reminiscent of exactly what experimentalists do when seeking to transfer genes deliberately!

For the DNA transposons, lateral transfer may be more than just happenstance, instead representing a requirement for evolutionary persistence. Several mechanisms down-regulate transposition, both host-encoded and element-encoded (Chapter 13). Hybrid dysgenesis provides an example of derepression of transposons upon transfer to a naive cellular environment. Defective elements arise and proliferate, providing sinks for newly synthesized transposase. It has been argued that these forces necessarily direct DNA transposons on a path to extinction. The only escape, according to this idea, is to transfer laterally and reset the clock. This argument may be more pertinent for some transposons than others. For *Tc1/mariner* in particular, the element shows extremely broad host range and widespread transfer, so for *Tc1/mariners*, sailing to new homes may be required for survival.

REFERENCES

1. Adams M.D., Tarng R.S., and Rio D.C. 1997. The alternative splicing factor PSI regulates P-element third intron splicing in vivo. *Genes Dev.* **11:** 129–138.
2. Beall E.L. and Rio D.C. 1996. *Drosophila* IRBP/Ku p70 corresponds to the mutagen-sensitive mus309 gene and is involved in P-element excision in vivo. *Genes Dev.* **10:** 921–933.
3. Beese L.S. and Steitz T.A. 1991. Structural basis for the 3′-5′ exonuclease activity of *Escherichia coli* DNA polymerase I: A two metal ion mechanism. *EMBO J.* **10:** 25–33.
4. Berg D.E. and Howe M.M. , Eds. 1989. *Mobile DNA.* American Society for Microbiology, Washington, D.C.
5. Bingham P.M. and Zachar Z. 1989. Retrotransposons and the FB transposon from *Drosophila melanogaster.* In *Mobile DNA* (ed. D.E. Berg and M.M. Howe), pp. 485–502. American Society for Microbiology, Washington, D.C.
6. Blackman R., Koehler M., Grimaila R., and Gelbart W. 1989. Identification of a fully-functional *hobo* transposable element and its use for germ-line transformation of *Drosophila. EMBO J.* **8:** 211–217.
7. Calvi B., Hong T., Findley S., and Gelbart W. 1991. Evidence for a common evolutionary origin of inverted repeat transposons in *Drosophila* and plants: *hobo, Activator* and *Tam3. Cell* **66:** 465–471.
8. Carr B. and Anderson P. 1994. Imprecise excision of the *Caenorhabditis elegans* transposon *Tc1* creates functional 5′ splice sites. *Mol. Cell. Biol.* **14:** 3426–3433.
9. Clark J.B. and Kidwell M.G. 1998. Evidence for horizontal transfer of the

P transposable element between *Drosophilia* species. In *Horizontal gene transfer* (ed. M. Syvanen and C.I. Kado), pp. 253–266. Chapman & Hall, London.

10. Coen E.S., Robbins T.P., Almeida J., Hudson A., and Carpenter R. 1989. Consequences and mechanisms of transposition in *Antirrhinum majus*. In *Mobile DNA* (ed. D.E. Berg and M.M. Howe), pp. 413–436. American Society for Microbiology, Washington, D.C.

11. Doak T.G., Doerder F.P., Jahn C.L., and Herrick G. 1994. A proposed super-family of transposase genes: Transposon-like elements in ciliated proto-zoa and a common "D35E" motif. *Proc. Natl. Acad. Sci.* **91:** 942–946.

12. Ellison V. and Brown P.O. 1994. A stable complex between integrase and viral DNA ends mediates human immunodeficiency virus integration *in vitro. Proc. Natl. Acad. Sci.* **91:** 7316–7320.

13. Engelman A., Mizuuchi K., and Craigie R. 1991. HIV-1 DNA integration: Mechanism of viral DNA cleavage and DNA strand transfer. *Cell* **67:** 1211–1221.

14. Engels W.R. 1989. P elements in *Drosophila melanogaster*. In *Mobile DNA* (ed. D.E. Berg and M.M. Howe), pp. 437–484. American Society for Microbiology, Washington, D.C.

15. Fayet O., Ramond P., Polard P., Prere M.F., and Chandler M. 1990. Functional similarities between the IS3 family of bacterial insertion ele-ments? *Mol. Microbiol.* **4:** 1771–1777.

16. Fedoroff N. 1989. Maize transposable elements. In *Mobile DNA* (ed. D.E. Berg and M.M. Howe), pp. 375–411. American Society for Microbiology, Washington, D.C.

17. Fedoroff N. and Botstein D. 1993. *The dynamic genome.* Cold Spring Harbor Laboratory Press, Cold Spring Harbor, New York.

18. Gueiros-Filho F.J. and Beverley S.M. 1997. Trans-kingdom transposition of the *Drosophila* element *mariner* within the protozoan *Leishmania. Science* **276:** 1716–1719.

19. Hartl D.L. 1989. Transposable element *mariner* in *Drosophila* species. In *Mobile DNA* (ed. D.E. Berg and M.M. Howe), pp. 531–536. American Society for Microbiology, Washington, D.C.

20. Hartl D.L. 2001. Discovery of the transposable element *Mariner. Genetics* **157:** 471–476.

21. Houck M.A., Clark J.B., Peterson K.R., and Kidwell M.G. 1991. Possible horizontal transfer of *Drosophila* genes by the mite *Proctolaelaps regalis. Science* **253:** 1125–1128.

22. Ivics Z., Hackett P.B., Plasterk R., and Izsvak Z. 1997. Molecular recon-struction of *Sleeping Beauty*, a *Tc1*-like transposon from fish, and its trans-position in human cells. *Cell* **91:** 501–510.

23. Izsvak Z., Ivics Z., and Plasterk R. 2000. *Sleeping Beauty*, a wide host-range transposon vector for genetic transformation in vertebrates. *J. Mol. Biol.* **302:** 93–102.

24. Kaufman P.D. and Rio D.C. 1992. P element transposition in vitro pro-ceeds by a cut-and-paste mechanism and uses GTP as a cofactor. *Cell* **69:** 27–39.

25. Keller E.F. 1983. *A feeling for the organism.* W.H. Freeman, San Francisco, California.

26. Kidwell M.G. and Kidwell J.F. 1975. Cytoplasm-chromosome interactions in *Drosphila melanogaster. Nature* **253:** 755–756.

27. Kidwell M.G. and Lisch D. 1997. Transposable elements as sources of variation in animals and plants. *Proc. Natl. Acad. Sci.* **94:** 7704–7711.

28. Kulkosky J., Jones K.S., Katz R.A., Mack J.P.G., and Skalka A.M. 1992. Residues critical for retroviral integrative recombination in a region that is highly conserved among retroviral/retrotransposon integrases and bacterial insertion sequence transposases. *Mol. Cell. Biol.* **12:** 2331–2338.

29. Lander E., Linton L.M., Birren B., Nusbaum C., Zody M.C., Baldwin J., Devon K., Dewar K., Doyle M., FitzHugh W., et al. 2001. Initial sequencing and analysis of the human genome. *Nature* **409:** 860–921.

30. Lovering R., Harden N., and Ashburner M. 1991. The molecular structure of TE146 and its derivatives in *Drosophila melanogaster. Genetics* **128:** 357–372.

31. McClintock B. 1957. Controlling elements and the gene. *Cold Spring Harbor Symp. Quant. Biol.* **21:** 215.

32. McClintock B. 1948. Mutable loci in maize. *Carnegie Inst. Wash. Year Book* **47:** 159.

33. McClintock B. 1950. The origin and behavior of mutable loci in maize. *Proc. Natl. Acad. Sci.* **36:** 347.

34. McClintock B. 1978. Mechanisms that rapidly reorganize the genome. *Stadler Genet. Symp.* **10:** 25–48.

35. Miller M.D., Farnet C.M., and Bushman F.D. 1997. Human immunodeficiency virus type 1 preintegration complexes: Studies of organization and composition. *J. Virol.* **71:** 5382–5390.

36. Mizuuchi K. and Adzuma K. 1991. Inversion of the phosphate chirality at the target site of Mu DNA strand transfer: Evidence for a one-step transesterification mechanism. *Cell* **66:** 129–140.

37. Plasterk R. 1995. The *Tc1/mariner* transposon family. *Curr. Top. Microbiol. Immunol.* **204:** 125–143.

38. Plasterk R., Izsvák Z., and Ivics Z. 1999. Resident aliens: The *Tc1/mariner* superfamily of transposable elements. *Trends Genet.* **15:** 326–332.

39. Preston C.R. and Engels W.R. 1996. P-element-induced male recombination and gene conversion in *Drosophila. Genetics* **144:** 1611–1622.

40. Rebatchouk D. and Narita J.O. 1997. Foldback transposable elements in plants. *Plant Mol. Biol.* **34:** 831–835.

41. Reiter L.T., Murakami T., Koeuth T., Pentao L., Muzny D.M., Gibbs R.A., and Lupski J.R. 1996. A recombinational hotspot responsible for two inherited peripheral neuropathies is located near a mariner transposon-like element. *Nat. Genet.* **12:** 288–297.

42. Robertson H.M. 1993. The *mariner* transposable element is widespread in insects. *Nature* **362:** 241–245.

43. Robertson H.M., Soto-Adames F.N., Walden K.K.O., Avancini R.M.P., and Lampe D.J. 1998. The *mariner* transposons of animals: Horizontally jumping genes. In *Horizontal gene transfer* (ed. M. Syvanen and C.I. Kado), p. 268-283. Chapman & Hall, London.

44. Rowland S.J., Sherratt D.J., Stark W.M., and Boocock M.R. 1995. Tn552 transposase purification and in vitro activities. *EMBO J.* **14:** 196–205.

45. Schouten G.J., van Luenen H.G., Verra N.C., Valerio D., and Plasterk R. 1998. Transposon Tc1 of the nematode *Caenorhabditis elegans* jumps in human cells. *Nucleic Acids Res.* **26:** 3013–3017.

46. Simmen M.W. and Bird A. 2000. Sequence analysis of transposable elements in sea squirt, *Ciona intestinalis. Mol. Biol. Evol.* **17:** 1685–1694.

47. Simmons G.M., Plummer D., Simon A., Boussy I.A., Frantsve J., and Itoh M. 1998. Horizontal and vertical transmission of *hobo*-related sequences between *Drosophila melanogaster* and *Drosophila simulans*. In *Horizontal gene transfer* (ed. M. Syvanen and C.I. Kado), pp. 285–293. Chapman & Hall, London.

48. Smit A.F.A. and Riggs A.D. 1996. Tiggers and other DNA transposon fossils in the human genome. *Proc. Natl. Acad. Sci.* **93:** 1443–1448.

49. Spradling A.C. and Rubin G.M. 1982. Transposition of cloned P elements into *Drosophila* germ line chromosomes. *Science* **218:** 341–347.

50. van Luenen H.G., Colloms S.D., and Plasterk R.H. 1994. The mechanism of transposition of Tc3 in *C. elegans*. *Cell* **79:** 293–301.

51. Vos J.C., De Baere I., and Plasterk R. 1996. Transposase is the only nematode protein required for in vitro transposition of Tc1. *Genes Dev.* **10:** 755–761.

52. Weil C.F. and Kunze R. 2000. Transposition of maize Ac/Ds transposable elements in the yeast *Saccharomyces cerevisiae*. *Nat. Genet.* **2:** 187–190.

53. Windsor A.J. and Waddell C.S. 2000. *FARE*, a new family of foldback transposons in *Arabidopsis*. *Genetics* **156:** 1983–1995.

54. Zhang J.K., Pritchett M.A., Lampe D.J., Robertson H.M., and Metcalf W.W. 2000. *In vivo* transposon mutagenesis of the methanogenic archaeon *Methanosarcina acetivorans* C2A using a modified version of insect *mariner*-family transposable element *Himar1*. *Proc. Natl. Acad. Sci.* **97:** 9665–9670.

Lateral Transfer in Eukaryotic Genomes: Fluidity in the Human Blueprint

AT THE TIME OF THIS WRITING, COMPLETE SEQUENCES have been reported for the yeast *Saccharomyces cerevisiae* and the roundworm *Caenorhabditis elegans*, and nearly complete sequences for the fruit fly *Drosophila melanogaster*, the mustard weed *Arabidopsis thaliana*, and, quite recently, human as well (Table 10.1). As with the prokaryotic genome sequences, the eukaryotic genomes provide a rich record of fluidity in the eukaryotic blueprint.

DNA mobility results in continuous change in eukaryotic genomes. Retrotransposons have expanded to comprise a large fraction of the DNA in humans and some plants. Human individuals differ in the locations of some elements, indicating contemporary mobility. Formation of processed pseudogenes by retroelement reverse transcriptase provides a flow of new genetic information back into the chromosomes. Transposons can also act as portable regions of homology, providing the substrate for altering chromosomes by homologous recombination. Capture of new sequences by lateral transfer clearly affects eukaryotic genomic evolution, although to a lesser degree than in prokaryotes.

Much of the DNA between genes is made up of mobile elements, providing a dynamic substrate on which the genes reside. Indeed, in the human genome only 1.5% of the DNA consists of gene-coding sequences (exons) whereas 45% consists of mobile elements or their remnants.

Chapters 6–9 surveyed integrating eukaryotic viruses and transposons. For different transposon families, the evidence for lateral transfer ranges from strong to suggestive at best, and the long interspersed nuclear elements (LINEs) may not be subject to lateral transfer at all. Other types of sequences are also captured by lateral transfer,

TABLE 10.1. Extensively Sequenced Eukaryotic Genomes

Organism	Genome size (Mb)	Chromosome number	Estimated gene number	Reference
Yeast *Saccharomyces cerevisiae*	12	16	6,340	Goffeau et al. (1996) (18)
Worm *Caenorhabditis elegans*	97	6	19,000	The *C. elegans* Sequencing Consortium (1998) (6)
Fruit fly *Drosophila melanogaster*	180	5	13,600	Adams et al. (2000) (1)
Human *Homo sapiens*	3,400	46	37,000	Lander et al. (2001); Venter et al. (2001) (29, 48)
Mustard plant *Arabidopsis thaliana*	125	5	25,500	Arabidopsis Genome Initiative (2000) (2)

notably as imports from the organelles. This chapter provides an overview of eukaryotic genome sequences, emphasizing the consequences of DNA mobility in constructing modern genomes.

YEAST (18)

The sequence of the *S. cerevisiae* genome was reported in 1997 by a consortium of 100 research laboratories, providing the first complete sequence of a eukaryote. The sequence of 12,068 kilobases, arrayed on 16 chromosomes, reveals about 5,885 genes encoding proteins, 140 genes for ribosomal RNA, 275 genes for transfer RNA, and 40 genes for other small nuclear RNAs. The gene numbers may be subject to slight corrections, as with all the complete sequences, because it can be surprisingly tricky to identify genes in the primary sequence. The gene density in *S. cerevisiae* is high, with about 70% of the genome consisting of open reading frames (orfs).

Transposons in the Yeast Genome (4, 15, 20, 26)

The *S. cerevisiae* genome contains two broad classes of retrotransposons, the Ty elements in the nuclear genome and mobile introns in the mitochondrial DNA (Chapter 8). The mobile introns appear to have relatively modest long-term effects on the mitochondrial genome. The Ty elements, on the other hand, strongly influence the yeast nuclear genome over time.

The complete *S. cerevisiae* sequence reveals 52 complete Ty elements as well as 264 solo long terminal repeats (LTRs) or other element fragments, about 3.3% of the total sequence. As discussed in Chapter 6, the solo LTRs are formed by integration followed by recombination between the directly repeated LTRs, thereby excising the Ty genome but leaving a single LTR. These elements are not randomly distributed but are clustered with each other and near tRNA genes (Fig. 10.1). If a cluster is taken as any site with one or more elements or fragments, there are 193 clusters. Of the five types of Ty elements, Ty1–4 elements are found near tRNA genes about 90% of the time. Ty5 is different, favoring integration into the ends of chromosomes (telomeres) and silent genes at the yeast mating type cassettes. Both of these types of Ty5 target sequence are known to bind common proteins that probably direct Ty5 target site selection.

This precise targeting of Ty integration is probably crucial for survival of both the element and the host cell. Yeasts spend some of their time growing as haploids, containing only a single copy of the genome. If Ty integration disrupted a required cellular gene, as would often happen by random integration, this could be lethal for both the element and the host. Thus, the fastidious targeting seen with the Ty elements is probably strongly favored by natural selection.

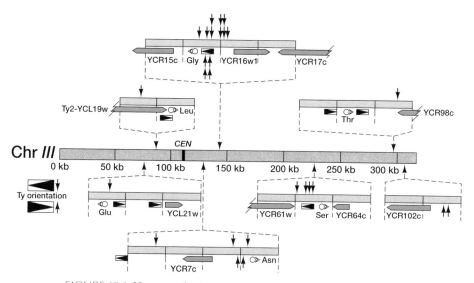

FIGURE 10.1. Hot spots for insertion of Ty elements in *S. cerevisiae* chromosome III. Locations of hot spots are shown expanded (*dotted brackets*). Protein-coding genes are shown as dark gray arrows; the direction of the arrow indicates the transcriptional orientation. The tRNA genes are shown as circles with arrows also indicating the direction of transcription. Ty1 integration sites are shown by the vertical arrows. LTR sequences are shown as boxes enclosing arrows, which indicate the transcriptional orientation. (Modified, with permission, from Ji et al. 1993 [© Elsevier]. [20])

Are Ty elements subject to lateral transfer? A comparison of variations among Ty elements suggests that they are. Comparison of LTR sequences showed a broad range of sequence similarity, ranging from 70% to 99% identity. However, the Ty3 and Ty4 LTRs are very similar within each family, displaying >96% identity. LTR sequences are more diverged within each of the other families. This suggests that Ty3 and Ty4 may be relatively recent imports into the *S. cerevisiae* genome that have been proliferating over a relatively short (evolutionary) time. The donor of the Ty3 and Ty4 sequences and the mechanism of transfer are unknown.

Ty elements are well documented as forces for reorganizing the yeast genome. Studies of close relatives of *S. cerevisiae* reveal that their genomes contain the same number of chromosomes but differ by chromosomal translocations. In most cases, segments of chromosomes that were colinear in *S. cerevisiae* were exchanged between two chromosomes in the other species. Analysis of sequences at the chromosomal break points by E. Fischer, E. Louis, and their colleagues revealed Ty elements were present at 9 out of 15 junctions. The other break points contained other types of repeated sequences, providing regions of homology for recombination. This study documents that Ty elements can facilitate chromosomal rearrangements associated with speciation.

A Genome of Pseudogenes? (13)

One interesting speculation from Gerry Fink connects the surprising observation that *S. cerevisiae* genes generally lack introns with the action of Ty elements. Metazoan genomes contain many processed pseudogenes, usually nonfunctional sequences lacking introns and containing poly(A) sequences, apparently formed by reverse transcription of mRNAs. The *S. cerevisiae* genome lacks such processed pseudogenes, but Fink has proposed that most or all of the chromosomal genes may in fact contain sequences resulting from reverse transcription and recombination back into the genome.

Laboratory studies show that mRNA transcripts of cellular genes can occasionally be reverse-transcribed by Ty1-encoded machinery. *S. cerevisiae* has highly active homologous recombination systems but less prominent illegitimate recombination. Thus, cDNAs produced by Ty-mediated reverse transcription may re-enter the genome by recombination with the original gene copy resident in a yeast chromosome. The recombination product would match the original gene, except that any introns formerly present would now be deleted by the cycle of transcription, mRNA splicing, reverse transcription, and recombination. Those few introns that are present tend to be near the 5′ ends of genes, a finding that supports the model because reverse transcription primed internally on an mRNA might often fail to extend all the way to the 5′ end of the message, yielding the observed bias. By this reasoning, most of the *S. cerevisiae* genome may in fact qualify as functional pseudogenes.

Organelle Imports (14, 16, 33, 40)

One of the most prominent means of lateral gene transfer in eukaryotes is transfer from organelle genomes, the mitochondria and chloroplasts, to the nuclear genome. This topic is treated in detail in Chapter 12, which covers lateral gene transfer between the domains of life. Gene imports noticed so far in each genome are summarized in this chapter along with organism-specific features.

To date, detailed genome-wide surveys of bacterial-derived sequences have not been reported for yeast. A few reports have identified potential bacterial-derived genes in the yeast nuclear genome, such as the gene for chymotrypsin.

Particularly striking is a recent report from Bernard Dujon and coworkers that yeast mitochondrial DNA can appear at sites of repaired DNA double-strand breaks. Double-strand breaks in yeast are repaired 90% of the time by recombination. Many of the remaining events involve direct ligation of the broken ends, often at short regions of sequence homology near the two ends. About 1% of the time, repair is accompanied by insertion of short sequences between the initially broken ends. Evidently repair in these rare cases involves polymerization primed at each broken end accross a free DNA fragment. Ty sequences, for example, are commonly found at repaired sites with insertions, probably reflecting polymerization along unintegrated Ty DNA genomes as a step in the repair reaction. Dujon and coworkers created experimental double-strand breaks in a haploid yeast strain that was impaired in repairing double-strand breaks by the homologous recombination pathway. Ninety-nine percent of cells died. Rare survivors typically showed conventional end joining, but a few survivors were found to have short insertions (40–200 bp) of mitochondrial DNA. The implication is that free sequences from mitochondrial DNA are also sometimes present in the nucleus and used as repair templates, providing a mechanism for incorporation of mitochondrial sequences. In fact, analysis of the yeast genome sequence revealed 30 short homologies with mitochondrial DNA in the nuclear chromosomes. Other studies have shown that a genetic marker incorporated into the mitochondrial DNA can be transferred to the nuclear genome at a substantial rate. These data clearly document the flow of sequences from the mitochondrial to the nuclear genome and establish a mechanism for their incorporation.

Missing Transposon Families in Yeast (50)

Curiously, no non-LTR retrotransposons or DNA-based transposons have been reported in yeast, an intriguing but unexplained observation. There is apparently no reason that DNA transposons cannot exist in this organism, as documented by the remarkable demonstration by Clifford Weil and Reinhard Kunze that Ac elements of maize can transpose when introduced into yeast. It will be interesting to see whether experiments with Ac can help explain why DNA transposons are normally absent.

THE WORM *CAENORHABDITIS ELEGANS* (6, 32, 37, 39)

The genome sequence of the worm *C. elegans* was completed in 1998 by the Washington University Genome Sequence Center and the Sanger Center, providing the first full sequence of a multicellular eukaryote. The *C. elegans* genome contains about 19,000 genes distributed over 97 Mb of DNA. The genome is divided among six chromosomes: five autosomes (I–V) and the X sex chromosome.

Mobile Element Content (32, 37, 39)

Some 38 probable mobile element types were recognized in the sequence, some of which were known from previous work, but many of which were newly discovered. Active elements are present together with many apparently inactive element fossils. The *Tc* transposons, which are DNA-based transposons of the *Tc1/mariner* family discussed in Chapter 9, are well represented. *C. elegans* strains differ in the activity of the *Tc* elements. The Bristol strain contains about 30 *Tc1* elements, all inactive for transposition in the worm germ line, whereas the Bergerac strain also contains active elements. This was once thought to indicate that the Bristol elements were defective, but it turns out that this is not the case. Instead, the Bristol elements are potentially active, but the worm strain encodes effective inhibitors of their activity. The Bergerac strain lacks the inhibitory system. The worm regulatory system probably involved—the co-supression/RNAi pathway—is treated in Chapter 13. At this writing, the factor that differs between the two strains has not been identified.

Also present in the worm genome are defective *Tc7* elements, which are 921 bp long with inverted repeats similar to those of *Tc1*. The flanking sequences also contain the duplicated TA sequence characteristic of the *Tc1/mariner* family, but the *Tc7* elements do not encode their own transposase proteins. Ron Plasterk and coworkers have shown that *Tc7* can transpose in the presence of *Tc1* transposase, documenting that these defective elements are parasites on full-length *Tc1* elements.

One prominent aspect of mobile element ecology seen with larger eukaryotes is absent in *S. cerevisiae* and worms. In flies, plants, and vertebrates, the centromeres are sites of giant congregations of mobile elements, but this is not the case in worms and yeast.

Yeast have short specific sequences that act as centromeres with no enrichment of mobile sequences nearby. Worms have dispersed centromeres, which lack any special location on the linear sequence, precluding possible congregation of mobile elements.

C. elegans contains several classes of non-LTR retrotransposons, including RTE, CR1, and NeSL-1. These are conventional in many respects, displaying, for example, frequent truncations at the 5′ ends of the elements. However, the last of these, NeSL-1, displays some intriguing idiosyncrasies.

The strategy adopted by NeSL-1 for gene expression is unusual and highly adapted to the *C. elegans* host (Fig. 10.2). These worms, trypanosomes, and a few other organisms form the majority of their mRNAs by an unusual pathway in which a 5′ leader region is spliced onto mRNAs after synthesis. The mRNAs are thus composites of sequences found at two separate locations in the genome. The NeSL-1 element inserts directly into DNA encoding the spliced leader sequences, thereby picking up the needed leader RNA and also signals for efficient gene expression. The name of the element derives from *Ne*matode *S*pliced *L*eader. The leader sequences are present in many copies, so integration of NeSL-1 does not harm the nematode host. A similar target-site specificity is seen with another class of non-LTR retrotransposons in trypanosomes, which also employ *trans* splicing to attach leaders to their mRNAs, emphasizing the effectiveness of this targeting strategy.

The centromeres serve as the binding sites for microtubules and associated machinery important in segregating one copy of each chromosome to daughter cells during cell division. Their organization varies dramatically among eukaryotes.

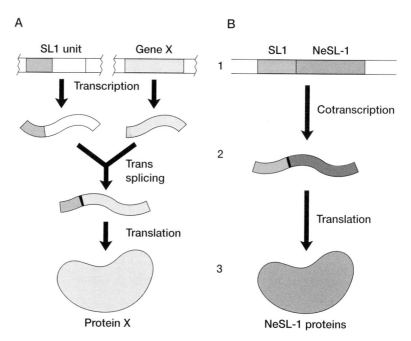

FIGURE 10.2. Nematode *trans* splicing and NeSL-1 replication. (*A*) Synthesis of nematode proteins involving *trans* splicing. (1) Messenger RNAs (gene X in the example) are transcribed separately from the spliced leader RNA (SL1). Genes for the SL1 RNA are present in multiple copies in the nematode genome. (2) The leader RNA is then covalently transferred to the 5′ end of the message for gene X, yielding the mature mRNA (3). Translation (4) yields the protein produced by gene X. (*B*) Replication strategy of the NeSL-1 transposon. The NeSL-1 integrates into genes for the spliced leader sequence (1). Transcription initiating at the 5′ end of the leader RNA (2) yields a fusion of the leader RNA and the NeSL-1 RNA. The fused mRNA can be translated directly without a need for *trans* splicing. (Modified, with permission, from Malik and Eickbush 2000. [32])

Bacterial-derived Genes in Worms (52)

Potential genes imported from bacteria have recently been cataloged in the *C. elegans* sequence by Eugene Koonin and coworkers. Fully 185 genes could be classified as recent bacterial imports by comparison to the sequence database. Many of these were probably acquired from mitochondria, but analysis of the sources of these imports is just beginning.

DROSOPHILA MELANOGASTER (1, 10, 17, 30)

The fruit fly *D. melanogaster* has been the topic of intensive genetic research since the early 1900s. In March 2000, a nearly complete sequence of the fly genome was reported by Celera Genomics in collaboration with the Berkeley Drosophila Genome Project and others, allowing a detailed look at the chromosomes of this very well-studied insect. The genetic and sequence data together provide a rich record of DNA fluidity.

Genome Structure (1)

The *Drosophila* genome is about 180 Mb, distributed over three autosomes and two sex chromosomes, X and Y. The chromosome sizes are quite heterogeneous, with chromosomes 2, 3, and X being much larger than 4 and Y. About 120 Mb is "euchromatin," active gene-rich regions. The rest is "heterochromatin," less active gene-poor regions comprising in part the centromeres and telomeres. The *Drosophila* genome contains about 13,600 genes, spaced on average one per 9 kb.

Mobile Element Content (1)

The fruit fly genome contains over 50 families of transposable elements, many of which are known to be active for transposition. The retrotransposon group is most abundant, including probable retroviruses such as *gypsy* (e.g., LTR retrotransposons containing *env* genes), conventional LTR retrotransposons (*copia*), and non-LTR retrotransposons (*I* elements and *jockey*). Also present are several classes of DNA transposons, including *foldback* and, in some strains, P elements.

These elements are found in many sites over the fly chromosomes, with the distribution of integration sites varying from element to element (Fig. 10.3). The detailed pattern has been found in some cases to vary among fly strains, indicating contemporary mobility of transposons.

Centromere Structure (22, 46)

Transposon sequences are particularly enriched in the regions flanking the centromeres. Some centromeres are built up of short tandem repeated sequences such as AATAT and AAGAG, spaced by rare integrated transposons. Others appear to be composed entirely of mobile element remnants, although this finding remains uncertain because many of the highly repeated regions of the genome are still not fully sequenced. Analysis of the centromeric sequences revealed numerous copies not only of the roughly 50 known transposon types, but also of a few new ones as well. Genes, in contrast, are rare or absent. Why the centromeric regions are particularly enriched in transposons is unclear. Perhaps these regions are particularly tolerant of insertions, or perhaps insertions are not removed efficiently. More speculatively, possibly the insertions serve a purpose; for example, acting as DNA spacers between the centromeric regions and flanking gene-rich euchromatin. Similar centromeric clusters of transposons have been seen in the human and plant genomes, and may be characteristic of higher eukaryotic chromosomes generally.

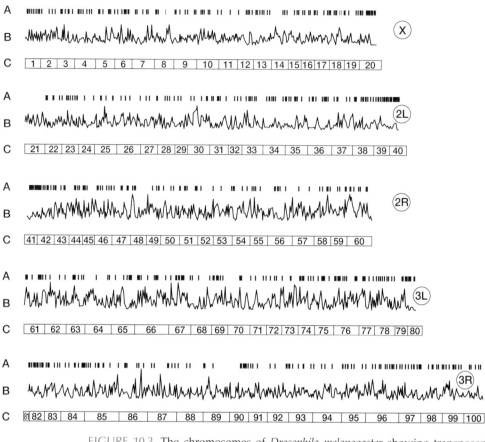

FIGURE 10.3. The chromosomes of *Drosophila melanogaster* showing transposon insertions (row *A*). Also shown are the relative gene densities (*B*) and the polytene chromosome divisions (*C*). (Modified, with permission, from Adams et al. 2000 [© American Association for the Advancement of Science]. [1])

Retrotransposons as Fly Telomeres (30, 36)

In flies we see one of the most dramatic examples of recruitment of transposon machinery by the host cell. Typically, transposons are selfish DNA, existing for their own benefit. Their effects on the host cell are usually neutral at best, and often deleterious. However, *Drosophila* and other dipteran insects appear to have recruited retrotransposons to stabilize the ends of chromosomes, replacing the telomeres found in all other eukaryotes studied, an example of mutualism between transposon and host.

The DNA replication machinery cannot efficiently copy the DNA at the tips of linear molecules such as eukaryotic chromosomes. Thus, chromosomes risk getting shorter with each round of DNA replication. In most cells, this is counteracted by the action of the telomerase enzyme, which attaches short repeated sequences to the chromosome ends and thereby maintains a consistent length. As discussed in Chapter 8, telomerase is a divergent reverse transcriptase, which uses

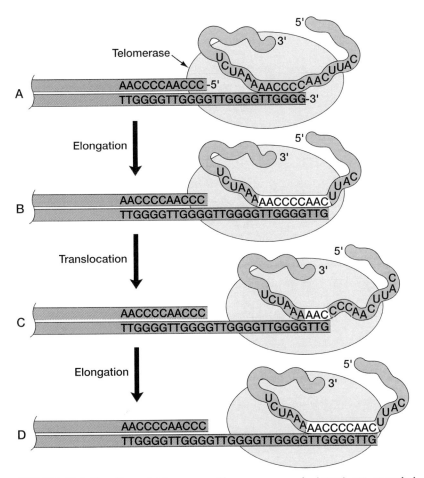

FIGURE 10.4. Function of telomerase. Chromosome ends (*green*) are extended using an internal guide RNA (*dark gray*) within the telomerase enzyme (*light gray*) as template. (*A*) The 3′ end of the chromosome can be extended (*B*) using the internal guide RNA as template. The internally repeated structure of the guide RNA allows the 5′-TTG-3′ sequence at the chromosome terminus (*C*) to switch binding partners to a more 3′ AAC in the internal guide RNA, permitting further extension of the growing chain (*D*). Repeated cycles of extension and translocation yield the repeated sequences found at chromosome termini. (Redrawn, with permission, from Kornberg and Baker 1991. [28])

an "internal guide" RNA to direct synthesis of the DNA repeats that become affixed to chromosome ends. The terminal 3′ end of the chromosomal DNA serves as the primer for addition of the terminal repeats (Fig. 10.4).

In *Drosophila*, the characteristic telomeric repeats have never been detected, but instead, retrotransposon sequences are associated with the chromosome ends (Fig. 10.5). Of the 50 or so element families in flies, only two non-LTR retrotransposons, TART and HeT-A, substitute for the telomerase system.

A simple modification of the usual non-LTR retrotransposition mechanism explains how these elements stabilize chromosome ends

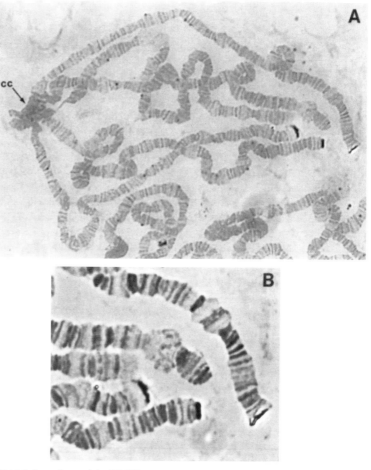

FIGURE 10.5. Locations of the TART transposon in *Drosophila* chromosomes shown by in situ hybridization. (*A*) A full set of chromosomes stained to reveal TART sequences (*black*). (*B*) Close-up of some chromosome tips, highlighting the telomeric location of TART elements (*black*). (Reprinted, with permission, from Levis et al. 1993 [© Elsevier]. [30])

(Fig. 10.6). Non-LTR retrotranspositions normally use a free 3′ end in the target DNA for priming reverse transcription on element RNA. TART and HeT-A appear to replicate by a similar means, but using chromosomal 3′ ends as a primer, thereby attaching element sequences to the end of the chromosome. How the other strand of element DNA is synthesized is unknown, but it may involve cellular DNA repair enzymes. Thus, as long as retrotransposon sequences are added to chromosome ends at a rate similar to that of DNA loss, the ends are stable during replication indefinitely.

Hybrid Dysgenesis (12, 24, 25)

Drosophila has played a central role in understanding the regulation of transposition through studies of hybrid dysgenesis. Hybrid dysgene-

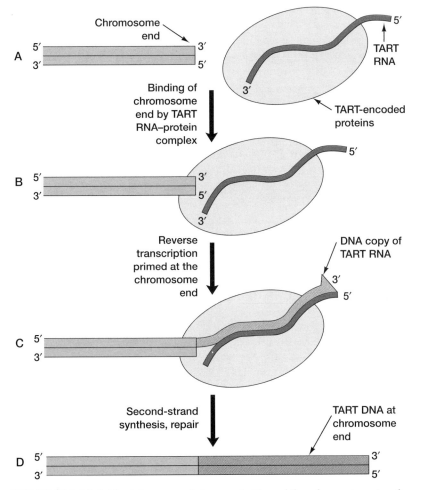

FIGURE 10.6. Model for telomere maintenance in *Drosophila melanogaster* using the non-LTR (LINE-like) retrotransposon TART. The chromosome end is shown dark gray, TART element RNA is green, TART element proteins are light gray. RNA is shown thinner than DNA. Extension of the chromosome tip begins when the TART element ribonucleoprotein complex (*A*) binds to the chromosome tip (*B*), allowing extension of the chromosome 3´ end by reverse transcription (*C*). Synthesis of the second DNA strand and ligation of remaining free ends completes the attachment of a TART DNA copy to the chromosome tip. (Adapted from Levis et al. 1993. [30])

sis was introduced in Chapter 9 in our discussion of P elements and comes up again in Chapter 13 (regulation of mobile element activity). For several types of fly transposons, there are strains known that either contain or lack the elements. Crosses between a male containing active transposons and a female lacking them can lead to hyperactive transposition in their progeny. The resulting hybrid dysgenesis syndrome includes high rates of mutation and sterility. If the female parent in the initial cross contains active copies of the transposon in question, the hybrid dysgenesis syndrome is suppressed. The presence of active transposons apparently programs the egg to suppress further transposition. To date, at least five different *Drosophila* transposons have displayed hybrid dysgenesis in appropriate crosses, disclosing a potent force for reorganizing the fly genome.

HUMANS (29, 48)

In one of the most dramatic achievements of biological science, a provisional draft of the human genome sequence was reported in February of 2001, determined jointly after a fierce competition by publicly funded efforts and Celera Genomics. The human genome comprises about 3,400 Mb of DNA, distributed over 46 chromosomes in the diploid complement. Full completion of the sequence, involving filling in gaps and sequencing highly repetitive regions, will still take a while longer. Analyzing the data will go on for many years.

The Number of Human Genes (19a, 29, 48)

A relatively small fraction of the human genome turns out to be made up of genes. Early estimates ranged from 40,000 to 120,000 human genes, but the initial analysis sequence suggested that the number is even smaller, some 30,000–38,000. It can be surprisingly difficult to recognize genes in the primary DNA sequence, particularly when the noise level is high due to large amounts of noncoding DNA. Exons have a mean length of only 150 bp, whereas introns average 3,365 bp. The mean number of exons per gene is 9, for a size (together with 5′ and 3′ untranslated regions) of 27 kb for the mean transcription unit. A few genes are enormous. The dystrophin gene is 2.4 Mb, and the titin gene, which encodes the largest protein, comprises 80,780 bp of coding sequence distributed over 178 exons. About 1.5% of the human genome is estimated to consist of exons, and about one-third is transcribed.

However, the analysis is at an early stage and the number is likely to increase. John Hogensch, Michael Cooke, and coworkers carried out a comparison of the gene predictions made by Celera Genomics and the public sequencing consortium and found remarkably little overlap between the two. When 11,000 well-characterized genes are removed from both data sets (the "Refseq" set), leaving only the theoretically predicted genes, collectively nearly 80% of genes were predicted by one group and not the other. The gene sets predicted by each group seem to be largely real, since 80% of the predicted genes sampled were found to be expressed in at least one of 13 human tissues. These findings suggest that there may be substantially more genes than have been identified—Hogensch and Cooke have speculated that the number may reach 42,000–50,000.

The more provisional estimate of around 37,000 is also deceptive for another reason. Many genes are composed of multiple exons, and it is already clear that individual mRNAs can contain different collections of exons. Thus, individual genes often direct the combinatorial synthesis of many different proteins. A lower estimate of the fraction of genes subject to alternative mRNA splicing is 35%. Whether the proteins encoded by splice variants carry out different functions is generally unclear, although it is known that they do in some cases. Nevertheless, the provisional number of human genes is surprisingly small, only about twice as large as *C. elegans*.

LINEs and SINEs in the Human Genome (5, 23, 27, 44, 45, 48)

Interspersed repeats, mostly transposon fragments and elements that rely on them for replication, comprise much more of the human genome than coding regions, fully 45% of the sequence so far. Many of the repeated sequences can be recognized as retrotransposons. However, using conventional search routines, around half of the human DNA is not identified as anything in particular. Although the nature of these sequences is unclear, it seems likely that much of this DNA is also composed of transposon remnants that just have accumulated mutations to the point that they are no longer recognizable.

Table 10.2 presents a summary of the deduced copy numbers of dispersed repeated sequences. As discussed in Chapter 8, repeated sequences in vertebrates are often referred to as SINEs (short interspersed nuclear elements) and LINEs (long interspersed nuclear elements). For the SINEs, there are about a million *Alu* elements in human DNA, comprising about 10.6% of the genome. In addition, another family of SINEs, called MIRs (mammalian-wide interspersed repeats) is found in about 400,000 copies. This SINE is apparently evolutionarily ancient, as it is found in all mammals, including marsupials and monotremes. MIRs appear to be derived from tRNA-like sequences. RNAs enriched in the vicinity of ribosomes, such as tRNAs, may form SINE families by hijacking LINE proteins as they are synthesized and replicating by reverse transcription (Chapter 8).

LINEs (non-LTR retrotransposons) comprise a whopping 20% of the genome. The overwhelming majority of these are 5′ truncated fragments rather than active elements. The LINEs can be subdivided into three families based on sequence resemblance, and each of these can be further subdivided from there.

Anthony Furano and coworkers have described a striking correlation between the age of LINEs in the human genome, as inferred by the divergence of sequences within families, and proficiency for replication. The youngest elements, the Ta-0 and Ta-1 members of the LINE-1 family, both contain active elements, whereas the older subfamilies do not. Remarkably, the younger Ta-1 elements are polymorphic in different human populations, with 69% of insertions polymorphic for the presence or absence of the element in different human groups. For the slightly older Ta-0 elements, the number is 29%. These dramatic find-

TABLE 10.2. Number of Copies and Fraction of Human Genome for Classes of Interspersed Repeat

	Number of copies (x1,000)	Total number of bases in the draft genome sequence (Mb)	Fraction of the draft genome sequence (%)	Number of families (subfamilies)
SINEs	1,558	359.6	13.14	
Alu	1,090	290.1	10.60	1(~20)
Mir	393	60.1	2.20	1(1)
Mir3	75	9.3	0.34	1(1)
LINEs	868	558.8	20.42	
LINE1	516	462.1	16.89	1(~55)
LINE2	315	88.2	3.22	1(2)
LINE3	37	8.4	0.31	1(2)
LTR elements (proviruses)	443	227.0	8.29	
HERV-class I	112	79.2	2.89	72(132)
HERV (K)-class II	8	8.5	0.31	10(20)
HERV (L)-class III	83	39.5	1.44	21(42)
MaLR	240	99.8	3.65	1(31)
DNA elements	294	77.6	2.84	
hAT group				
MER1-Charlie	182	38.1	1.39	25(50)
Zaphod	13	4.3	0.16	4(10)
Tc-1 group				
MER2-Tigger	57	28.0	1.02	12(28)
Tc2	4	0.9	0.03	1(5)
Mariner	14	2.6	0.10	4(5)
PiggyBac-like	2	0.5	0.02	10(20)
Unclassified	22	3.2	0.12	7(7)
Unclassified	3	3.8	0.14	3(4)
Total interspersed repeats		1,226.8	44.83	

(Reprinted, with permission, from Lander et al. 2001 [© Macmillan Magazines Ltd.]. [29])

ings clearly establish that new LINE insertions have been accumulating in the genome in recent human history. Older families of LINE elements are inactive. These findings suggest a process of element expansion followed by suppression and eventual death. The mechanism for this process can be specified with some confidence for the DNA transposons (discussed below), but it is less easy to understand for the LINE elements.

The LINE replication mechanism provides another means for reshaping the genome. LINE transcription extends through the poly(A) addition signal at the end of the element quite frequently, thereby transcribing the cellular 3′ flanking sequences. Retrotransposition of the extended LINE RNA carries with it a copy of the former 3′ flanking sequence, installing the old flanking sequence at the new integration site along with the LINE sequence. Remarkably, recent studies suggest that such 3′ sequence transduction may have formed fully 0.5–1% of the human genome.

The analysis of the human genome has also reopened the question of whether LINE elements can be subject to lateral transfer. Arian Smit has pointed out that the Bov-B elements, LINEs found in cows, are clearly absent in humans. The expectation is that if these elements were present in the common ancestor of humans and cows, they should be detectable as fossils in the human genome, but they do not appear to be present. Thus, it seems possible that Bov-B elements are instead recent imports in the bovine lineage. The debate about lateral transfer of LINE elements goes on!

HERV Elements (29, 38, 48)

The draft of the human genome contains some 50,000 copies of human endogenous retroviruses (HERVs), comprising 4.6% of the total sequence. The HERVs can be subdivided into three classes, including the HERV(K) and HERV(L) classes discussed in Chapter 6. A total of 85% of these are present as solo LTRs, probably relics of retroviruses that integrated into the germ line but then were deleted by recombination between LTRs. The HERV elements are not obviously clustered in the genome, suggesting that forces favoring integration or persistence of integrated copies are similar over much of the chromosomes.

Analysis of the complete sequence indicates that the majority of HERV elements are dead or nearly so. Most sequences appear to have mutations that would inactivate replication. Only a single copy of a HERVK10 element is known to have integrated since the divergence of humans and chimpanzees, and only one copy is known that is divergent among human groups. Inspection of the sequence revealed three copies with three open reading frames intact out of four, leaving open the possibility that there may be a few active HERV elements, or that elements could be reactivated by experimentalists for study. Retrovirus-like particles containing HERV proteins are found in some germ-cell-derived tumor cell lines. These might still be infectious, keeping open the question of whether there are active HERV elements.

Are there LTR retrotransposons in the human genome? It is difficult to answer this question, because none of the elements is active, precluding direct study, and a retrovirus fossil deleted for the *env* region can be mistaken for an LTR retrotransposon. The MaLR elements (mammalian LTR repeats) represent a possible class of LTR retrotransposons, but these are also (distantly) similar in sequence to the HERV-L class, leaving them probable retroviral derivatives. Adding together the HERVs and MaLR elements suggests that about 8% of the human genome is derived from probable endogenous retroviruses.

DNA Transposons (29, 44, 45, 48)

Fossils of DNA-based transposons are also present in the human sequence, comprising roughly 2.8% of the draft sequence. This group was only recognized recently, although similar DNA-based transposons have long been studied in invertebrates and plants. The human genome contains seven families, including members of the *Tc1/mariner* and *hAT* families. Several more of these families appear to be specific to humans.

Are there active *mariner* or *hAT* transposons in the human genome? The answer is still uncertain, because it is not always straightforward to recognize active elements, but it seems unlikely that any remain functional. As discussed in Chapter 9, families of DNA transposons are prone to death due to their style of replication. A transposase-encoding mRNA synthesized in the nucleus must migrate to the cytoplasm to be translated. The tranposase protein, upon return to the nucleus, has no way to distinguish active elements from inactive ones. Inactive elements with proper transposase-binding sites can bind transposase and form new copies in the genome, providing an expanding sink for transposase. Coupled with random inactivation of elements by mutation, this provides a "ratchet" mechanism leading to death of the transposon family. Further mechanisms also squelch transposition (Chapter 13), although to what degree they operate in human cells is still being worked out. The human genome sequence provides a detailed record of the invasion of DNA transposons, their expansion, and ultimate demise.

Consistent with this mechanism, humans also contain more than 150,000 copies of what appear to be internally deleted DNA transposons. That is, they contain the inverted repeats necessary for binding transposase, but do not contain an intact transposase gene (like McClintock's *Ds* elements; see Chapter 9). The nature of the host DNA sequences flanking these sequences reveals the element types likely involved in their creation. Some are flanked by the TA sequence duplication characteristic of *Tc1/mariner* family elements. Others are flanked by 8-bp repeats of target DNA characteristic of *hAT* family elements. As described above, it appears that transposase hijacking by inactive elements has allowed these parasites to proliferate in the human genome.

Patchy Distribution of Mobile Elements (7, 27, 29, 48)

The human chromosomes are divided into morphologically distinct regions that differ in their content of retroelement DNA. Staining of human chromosomes with the Giemsa stain reveals a pattern of Giemsa-bright bands (G/Q bands) and Giemsa-dark bands (R or reverse bands). Analysis of early sequence data suggested that the G/Q bands were rich in guanine plus cytosine base pair (G/C-rich), rich in genes, and rich in *Alu* repeats; R bands were A/T-rich, gene-poor, and rich in LINEs but not *Alu* elements.

Analysis of the human genome has confirmed the differential distribution of *Alu* elements and LINEs in the G/Q and R bands and raised an intriguing mystery. If the *Alu* elements and LINEs use the same LINE machinery for replication, why are they found integrated

FIGURE 10.7. Two regions of the human genome, illustrating the startling lack of repeated sequences in the HoxD gene cluster. In the two regions, the bars above the line indicate exons and the bars below the line interspersed repeats. The upper diagram shows a fairly average region of human chromosome 22, the lower diagram a region of chromosome 2 containing the HoxD gene cluster. (Reprinted, with permission, from Lander et al. 2001 [© Macmillan Magazines Ltd.]. [29])

into different parts of the genome? The presence of LINE elements in R bands can be explained by the finding that LINE insertion is favored in A/T-rich DNA characteristic of R bands. But if *Alu* elements become integrated by LINE element machinery, why are they found in G/C-rich DNA? An analysis of the age and distribution of *Alu* elements indicates that the youngest family members are indeed integrated in A/T-rich DNA. Thus, the differential distribution must be a result of subsequent selection, either for *Alu* elements in G/C-rich DNA, or against *Alu* repeats in A/T-rich DNA. At present the explanation is elusive, but one speculation has been advanced based on the idea that *Alu* elements are indeed mutualists. Various forms of genotoxic stress can induce the cellular PKR kinase, an enzyme that acts by closing down translation in the stressed cell (PKR and the interferon response of which it is a part are treated in Chapter 13). *Alu* RNA inhibits PKR, possibly modulating the effects of stress by avoiding hyperactivation of PKR.

Other inhomogeneities in mobile element distribution provide fascinating hints about genetic function. For example, the homeobox gene clusters are exceptionally poor in integrated mobile elements of all kinds (Fig. 10.7). Studies in flies and other organisms have shown that these loci are key regulators of development, and that the multigene complexes function as single units to direct the differentiation of the animal body during development. The sequencing data suggest that either the homeotic gene cluster DNAs are poor integration targets or, perhaps more likely, that insertions into these loci are counterselected, either way emphasizing the unitary nature and unusual function of the homeo-gene clusters.

Genetic Diseases due to Recombination among Mobile Elements (29, 38, 48)

Recombination among mobile elements can lead to the rearrangements of required sequences and so cause human disorders. Many examples are known of rearrangements due to recombination between *Alu* elements. Recombination between HERV15 elements on the Y chromosome can cause a deletion resulting in male infertility (Chapter 6). Recombination between *mariner* elements on chromosome 17 can yield either a deletion or duplication of a short region. These result in Charcot-Marie-Tooth disease type 1A or hereditary neuropathy with liability to pressure palsies, depending on the genetic change (Chapter 9). Transposition of *Alu* elements or LINE elements can also disrupt required genes, as in the LINE element insertion in hemophilia, or activate oncogenes, as with LINE insertions in the *myc* locus in breast cancer (Chapter 8).

Gene Imports from Prokaryotes [29, 41]

The initial analysis of the human genome suggested that at least 223 genes from prokaryotes had been imported into the human genome. As discussed in detail in Chapter 12, eukaryotic genomes typically contain a few genes that appear by sequence comparison to be derived from bacteria, potentially by transmission from the organelle genome to the nuclear genome. According to this idea, on occasion the new organelle form has replaced a nuclear gene for the same function, a process called gene displacement. Comparison of the human genes to those in the other four complete eukaryotic sequences initially suggested that 223 had best matches to bacterial genes and not genes of the other eukaryotes. On this basis, it was proposed that the 223 genes were probably acquired by lateral transfer. Surprisingly, the genes did not appear to be derived disproportionately from the α-proteobacteria group, the expected progenitor of the mitochondria.

Reanalysis by several groups, however, has challenged this idea. Jonathan Eisen and coworkers have pointed out that addition of more sequence information from eukaryotes (from unfinished genomes) changes the outcome of the analysis. Many of the proposed human bacterial imports in fact do have eukaryotic close relatives in the enlarged sequence data set. Reanalysis taking this and other factors into account leaves only about 40 genes as candidates instead of 223. Eisen and coworkers point out that further expansion of the eukaryotic sequences available for comparison will probably further reduce the number of human genes with bacterial closest relatives. Two other groups have published similar conclusions. It is unclear how many genes will remain as candidates for bacteria-to-human transfer, but the number now appears to be much lower than initially estimated.

Nevertheless, there are still a variety of candidates for recent imports of mitochondrial sequences into the human nuclear genome. Among the roughly 40 genes left after the analysis by Eisen are two of probable mitochondrial origin that are likely to be derived from lateral transfer. Arian Smit has pointed out that many short mitochondrial DNA fragments can also be detected in the human genome, although they have not yet been cataloged in detail. Thus, there do seem to be bacterial-derived sequences in the human genome, but their nature remains to be fully clarified. We return to the topic of bacterial imports in Chapter 12.

New Genes Created from Transposon Fragments (8, 9, 11, 29, 34, 35, 44)

In the draft human sequence, 49 human genes were recognized as probably derived from transposable elements (20 were known previously), emphasizing the generative capacity of mobile DNA (Table 10.3). For example, the major centromere-binding protein CENPB is apparently derived from a DNA transposon. Another DNA-based transposon, present in only one known copy, appears to have been recruited to found the vertebrate adaptive immune system. This remarkable story is the topic of Chapter 11. The human telomerase enzyme also resembles other reverse transcriptases including the LINE family elements. Curiously, of the 49 genes, only 5 are derived from retrotransposons, the rest arising from the numerically less abundant DNA transposons. Why the transposase genes of DNA transposons are favored as new gene components is unclear. It will be of great interest to learn more about the functions of these newly created genes and their contribution to evolution in the human lineage.

Another positive effect of retroelement replication is the production of pseudogenes, most of which are inactive but a few of which probably are functional as new gene duplications. Eight genes were identified in the human sequence as probable functional pseudogenes, and many more intronless genes may also be in this category.

Interpreting the Human Genome

Biomedical science has been changed greatly by the draft sequence of the human genome. Any gene isolation strategy that can be engineered to yield a fragment of human sequence will immediately yield the complete gene, its chromosomal location, and possible association with genetic diseases (which are also mapped onto the sequence). Any new assignment of a particular gene to a biological function will be immediately followed by the question of whether related genes are present and performing related functions. Any number of new genome-based technologies have become possible. For example, DNA chips (Chapter 5) can now be made that interrogate the activity of all the genes in the human genome at the same time from a single sample.

The sequence also poses unanticipated new questions. For example, the developmentally crucial homeobox gene clusters are almost devoid of transposable elements, emphasizing the unitary nature of these gene clusters and refocusing questions about their function (Fig. 10.7).

Some basic tenets of biology have been shaken by the new sequence. For example, Stephen Jay Gould pronounced the "one-gene, one-enzyme" dogma dead as a consequence of so much alternative splicing. Many would call this a bit extreme—alternative splicing represents an elaboration of the previous idea rather than its destruction—but this illustrates the debates that arise. The unsettling paucity of total human genes will be something to wrestle with indefinitely—we are left rather more similar to flies and worms than might have been altogether comfortable.

Mobile element families have expanded in the genome in dramatic waves. Families of LINE elements appear to have proliferated and died out repeatedly, and the most recent wave is still going on. Families of DNA transposons appear to have arrived in the human genome sporadically over evolutionary time, proliferated, and died. Our genetic heritage is largely a story of genomic parasites. Although it is perhaps a disturbing thought, this activity of mobile DNA elements generates variation that provides a substrate for natural selection and evolutionary change, as revealed, for example, by the many mobile element fragments incorporated into new genes.

As important as the scientific findings, but harder to evaluate, are the moral and ethical issues posed by the human genome sequence. What information should be determined about each individual, and what is better left unknown? One technical consequence of the sequencing projects is the identification of over 2 million single nucleotide polymorphisms (SNPs), sites of genetic variation between individuals. This vast panel allows mapping of disease genes, of great medical value, but also potentially mapping of other types of inherited characteristics. Are there genes that predispose individuals to violence, substance abuse, or psychosis? If such genes are found, it is essentially certain that the effects of different alleles will not be fully penetrant; that is, genes will only affect the statistical likelihood of the outcome, not guarantee it. If a gene is found that increases a person's vulnerability to alcoholism by 10%, should an employer be allowed to assay for that gene and make an employment decision on that basis? In a related set of issues, if a disease gene is segregating in a family, what do scientists, who isolate and patent the gene, owe to the afflicted family? Human genetic information has now acquired economic value, a development not strictly due to the complete genome sequence but certainly accelerated by it. Individuals and organizations that can profit using this information will clearly press to maximize returns. It is crucial that an educated public form their own views of ethical policies and work to see them enacted, so that a fair balance can be reached.

These data emphasize that Selfish DNA, while having no necessary benefit to the host, can nevertheless provide a substrate for natural selection. The human genomic parasites exist for their own benefit, but their action can lead secondarily to the birth of new genes. Looking back, the early theoretical paper from Leslie Orgel and Francis Crick on Selfish DNA made these points and had it right.

MOUSE (19b, 29, 38, 48)

The genome of the mouse is presently being completed, but at this writing the results have not been published. Celera Genomics reports that they have sequenced 15.8 billion base pairs, allowing 5.5-fold coverage of the genome (somewhat higher levels of coverage are needed for reliable determination and closure of sequence gaps).

Parts of the mouse genome resemble the human genome over large regions. In these so-called syntenic regions, the order and orientation of genes are largely conserved. The mouse genome is related to the human genome by a series of inversions and translocations of large gene segments that have occurred since the divergence from the last common ancestor. The human genome itself contains myriad segmental duplications—chromosomal segments have been duplicated and reintegrated at a new location—providing evidence for chromosomal rearrangements during mammalian evolution.

TABLE 10.3. Human Genes Potentially Derived from Transposable Elements

GenBank ID	Gene name	Related transposon family
nID 3150436	BC200	FLAM Alu
pID 2330017	Telomerase	non-LTR retrotransposons
pID 1196425	HERV-3env	Retrovirdae/HERV-R
pID 4773880	Syncytin	Retrovirdae/HERV-W
pID 131827	RAG 1 and 2	Tc1-like
pID 29863	CENP-B	Tc1/Pogo
EST 2529718		Tc1/Pogo
PID 10047247		Tc1/Pogo/Pogo
EST 4524463		Tc1/Pogo/Pogo
pID 4504807	Jerky	Tc1/Pogo/Tigger
pID 7513096	JRKL	Tc1/Pogo/Tigger
EST 5112721		Tc1/Pogo/Tigger
EST 11097233		Tc1/Pogo/Tigger
EST 6986275	Sancho	Tc1/Pogo/Tigger
EST 8616450		Tc1/Pogo/Tigger
EST 8750408		Tc1/Pogo/Tigger
EST 5177004		Tc1/Pogo/Tigger
PID 3413884	KIAA0461	Tc1/Pogo/Tc2
PID 7959287	KIAA1513	Tc1/Pogo/Tc2
PID 2231380		Tc1/Mariner/Hsmar1
EST 10219887		hAT/Hobo
PID 6581095	Buster 1	hAT/Charlie
PID 7243087	Buster 2	hAT/Charlie
PID 6581097	Buster 3	hAT/Charlie
PID 7662294	KIAA0766	hAT/Charlie
PID 10439678		hAT/Charlie
PID 7243087	KIAA1353	hAT/Charlie
PID 7021900		hAT/Charlie/Charlie 3
PID 4263748		hAT/Charlie/Charlie 8
EST 8161741		hAT/Charlie/Charlie 9
pID 4758872	DAP4, pP52^{rtPK}	hAT/Tip100/Zaphod
EST 10990063		hAT/Tip100/Zaphod
EST 10101591		hAT/Tip100/Zaphod
pID 7513011	KIAA0543	hAT/Tip100/Tip100
pID 10439744		hAT/Tip100/Tip100
pID 10047247	KIAA1586	hAT/Tip100/Tip100
pID 10439762		hAT/Tip100
EST 10459804		hAT/Tip100
pID 4160548	Tramp	hAT/Tam3
BAC 3522927		hAT/Tam3
pID 3327088	KIAA0637	hAT/Tam3
EST 1928552		hAT/Tam3
pID 6453533		piggyBac/MER85
EST 3594004		piggyBac/MER85
BAC 4309921		piggyBac/MER85
EST 4073914		piggyBac/MER75
EST 1963278		piggyBac

(Reprinted, with permission, from Lander et al. 2001 [©Macmillan Magazines Ltd.]. [29])

The repeated sequences present in mouse differ considerably from those in human. The LINE elements are present, but the types of RNAs expanded as SINEs differ. It appears that the infestation of *Alu* elements found in humans did not take place in the lineage leading to mice. However, unlike humans, mice do have active endogenous retroviruses. Some are very recent imports, since they are present in some, but not all, mouse strains, whereas others are quite ancient (i.e., present in all strains). Some of the endogenous retroviruses are xenotropic, meaning that they are no longer able to infect mouse cells, probably indicative of ancient infection. The xenotropic character could have arisen by mutation of the viral envelope, so that it no longer binds to the murine receptor, or the receptor may have mutated, perhaps under the pressure of viral infection.

The paper on the human genome sequence from the public consortium suggested that mice differ from humans in containing more active mobile element families, but caution is warranted in drawing this conclusion from the available data. The suggestion arises from the observation that mice contain active LTR-retrotransposons (IAP elements) and more active LINE elements. Some 3,000 LINE elements are estimated to be active in mice, whereas only 40–60 are thought to be active in humans. In mice, LINE element insertions comprise about 10% of spontaneous mutations but only about 0.1% in humans. The numbers of active elements may be greater in mice; however, this may be a transient condition, the happenstance of timing in waves of activity or a sampling artifact. Neither species appears to harbor active DNA transposons. Humans do appear to lack active endogenous retroviruses, although active HERVs could still be discovered.

A potentially important variable is inbreeding in laboratory mice. Lab strains have been repeatedly back-crossed to generated lines that are genetically uniform. Humans, of course, are outbred and genetically heterogeneous. Any effects on mobile element activity due to inbreeding and prolonged maintenance in captivity could score as differences between species. Although farther off, it will be important to see whether wild mice differ from captive mice and more resemble humans.

Completion of the mouse genome (and completion of the human genome) will allow the genetic differences between the two to be specified. What makes humans different from mice? Early accounts suggest that there are about 300 human genes that have no obvious counterpart in the mouse, hinting at the genes that distinguish humans among mammals. The completion of the mouse genome should greatly illuminate our genetic heritage, all in the context of an organism highly suitable for biological study.

THE MUSTARD PLANT *ARABIDOPSIS THALIANA* (2)

The sequence of the *A. thaliana* genome was mostly completed at the end of 2000 by a consortium of laboratories, the Arabidopsis Genome

Initiative, providing the first comprehensive look at a plant genetic blueprint (Plate 8). The rice genome has also been mostly sequenced, but the study was carried out by biotechnology companies and the data are not publicly available. Comparing the avaliable data on genomes of diverse plants, both fully and partially sequenced, provides a dramatic example of genomic change directed by mobile DNA, particularly the retroelements.

A. thaliana has a relatively small genome for a plant, only 125 million base pairs distributed among five chromosomes. Indeed, the small size of the genome was one of the features that made this plant attractive to experimentalists. In 2000 a nearly complete sequence of the *A. thaliana* genome was reported, lacking only a few highly repetitive regions from the rDNA and centromeric DNA.

About 39% of the DNA is made up of genes, numbering about 25,498 total, less compact than the *S. cerevisiae* genome but much more gene-dense than human. Plate 8 presents an annotated map of the *A. thaliana* chromosomes. The top two lines depict the density of genes and known expressed sequences (expressed sequence tags or ESTs), both color-coded for density. Genes are present at high density, up to 200 per 100 kb, in the chromosome arms. An exception is the regions of the rDNA repeats on chromosomes 2 and 4 (magenta color on the chromosome diagram), which are composed of tandem repeats of the rDNA sequences. The centromeric regions are less gene-dense.

The distribution of transposons is highly asymmetric and opposite to that of the genes (line marked TE in Plate 8). Transposable elements are enriched in the centromeres and relatively sparse in the chromosome arms. The centromeres are also rich in simple repeated sequences, notably a family of 180-bp tandem repeats.

Table 10.4 summarizes the retroelement types found in the *Arabidopsis* genome, and their abundance in chromosome 2. DNA elements from the *mariner* and *hAT* family are well represented. The major classes of retroelements are also found, including non-LTR retrotransposons, LTR retrotransposons of the *Ty1/copia* and *Ty3/gypsy* classes, and retrovirus-like elements. The retrotransposons comprise the most abundant group of elements, accounting for roughly 5.3% of the draft sequence. Thus, we see preserved in plants most of the major transposon families also found in animals. As discussed below, *A. thaliana*, although infested with retrotransposons, still has fewer than do many plants and consequently a higher gene density.

As discussed in detail in Chapter 12, genes are frequently transferred from endosymbiotic organelle genomes, the mitochondria and chloroplasts, to the eukaryotic nuclear genome. In Plate 8, the line marked MT/CP indicates the location of mitochondrial or chloroplast-derived sequences in the *A. thaliana* nuclear genome. Remarkably, a large block of mitochondrial DNA is present within the centromeric DNA of chromosome 2. Initially, this was estimated at 270 kb, but reanalysis revealed that the sequencing strategy used had yielded a falsely simple picture, and that the insert is really about 620 kb in length

TABLE 10.4. Transposable Elements on *Arabidopsis* Chromosome 2

Transposon class	Subclass	Family	Number of open reading frames
DNA elements			
		Mutator	73
		En–Spm/Tam 1/Psi	71
		Ac-Ds	12
		mariner	2
		hAT	1
Retroelements			
	Non-LTR LINE-like retrotransposons		
		Ta11-like	152
		Tscl	1
	LTR retrotransposons		
		Athila	75
		Other	141
		Ty1-copia-like (Ta1)	
		Ty3-gypsy-like (Tat 1)	
	Retroviral-like	Replication protein	28
		Helicase	7
		Total	563

(Reprinted, with permission, from Lin et al. 1999 [© Macmillan Magazines Litd.]. [31])

and multiply rearranged. This sequence appears to have been introduced recently, since it is 99% identical to the mitochondrial sequence. Other chromosomal genes also show evidence of transfer from organelle to nuclear genomes. As discussed in Chapter 12, since the organelle genomes are probably of prokaryotic origin, this represents a mechanism for lateral gene transfer between the domains of life.

Comparison of *Arabidopsis* to Other Plants (3, 19, 21, 43, 51)

So far, *Arabidopsis* is the only plant genome for which a nearly complete genome sequence is available. In this section, we contrast *Arabidopsis* with other plants based on the available information, emphasizing the effect of mobile DNA on genome size.

For the case of maize, the total copy number of retrotransposons is estimated to be 300,000, comprising nearly two dozen element types. This vast amount of retrotransposon DNA accounts for a remarkable 50–80% of the maize genome. Figure 10.8 shows a region of the maize chromosome studied by Bennetzen and colleagues, drawn to illustrate the myriad retrotransposons present. Two genes are present in this 280-kb region, *Adh-1* and *u22*. Surrounding these genes are some 10 types of integrated retrotransposons, accounting for about 60% of the determined sequence. Most of the transposons are interrupted by

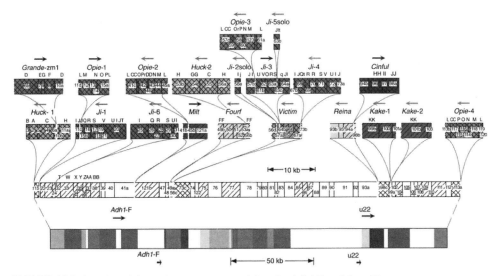

FIGURE 10.8. A region of the maize genome containing the *Adh1-F* and the *u22* genes, illustrating myriad nested insertions of retroelements. Cross-hatching indicates highly repetitive elements (thousands of copies), hatching indicates repetitive elements (hundreds of copies), and open bars indicate low copy elements (<100 copies). See original report for full details. (Reprinted, with permission, from SanMiguel et al. 1996 [© American Association for the Advancement of Science]. [43])

other transposon sequences, indicating a complex history of integrated retrotransposons themselves serving as integration sites for additional elements. Further studies support the idea that the retrotransposon content of this region is typical of much of the maize genome. Analysis of the sequences involved indicates that the maize transposons either arrived or were amplified in the maize genome within the last 2–6 million years.

Studies in plants suggest that the expansion of retrotransposons largely accounts for the remarkable differences in genome size seen among relatively closely related organisms.

The C Value Paradox (19)

Table 10.5 presents the sizes of some genomes, illustrating a puzzle known as the "C-value paradox." In some cases, genome size parallels organismic complexity. Yeast, for example, has a much smaller genome than humans (12 million bp versus 3,400 million). However, many other pair-wise comparisons show a radical discordance. Why should lily (36,000 million bp) have a genome that is ten times larger than human? And why should fern be four times larger than that? Topping the list, *Amoeba dubia* has a genome of 670 billion base pairs, and *Amoeba proteus* of 290 billion base pairs. Several theories have been proposed to explain the C-value paradox, but no consensus has been reached. For example, there is a rough correlation between genome size, nuclear size, and cell size. Possibly "spacer" DNA is needed to fill up a nucleus that needs to be large for other reasons. Perhaps the added DNA is positively selected for some other reason. Still another alternative is that different genomes just happen to have become differentially infested with parasites for no reason other than happenstance. We return to the C-value paradox in Chapter 14.

TABLE 10.5. Estimated Genome Sizes: The C Value Paradox

Organism	Genome size (in base pairs)
Amoeba dubia	670,000,000,000
Amoeba proteus	290,000,000,000
Nematode	97,000,000
Carp	1,700,000,000
Cow	3,651,500,000
Chicken	1,200,000,000
Fern	160,000,000,000
Frog	3,000,000,000
Fruit fly	180,000,000
Gorilla	3,523,200,000
HIV-1	9,750
Housefly	900,000,000
Human	3,400,000,000
Lily	36,000,000,000
Maize	3,200,000,000
Mouse	3,454,200,000
Mustard plant	125,000,000
Pig	3,108,700,000
Pine	68,000,000,000
Rat	3,093,900,000
Yeast	12,067,280

Although these differences in genome size remain puzzling, differences in genome size in several well-studied plants can now be attributed to differences in retroelement content (Table 10.6). Maize and sorghum are closely related grasses, having diverged roughly 15–20 million years ago, but the maize genome is three to four times larger. Sequencing of the *Adh-1* gene region of sorghum reveals that gene content and order are conserved, but the flanking retrotransposon sequences found in maize were absent. Probing of the sorghum genome failed to reveal copies of most of the retrotransposons of maize, suggesting that retrotransposon content was the major determinant of genome-size differences.

Studies of retrotransposon content and genome size in other plants reinforce this conclusion. Rice, the smallest grass genome studied, is estimated to contain only 14% retrotransposon sequences. The barley genome, about ten times larger, is composed of >70% retrotransposon sequences. It will be of interest to study the genomes of lily and fern to see whether they are infested with still greater densities of retrotransposons.

Recent studies suggest that plant genomes are still in a state of flux due to retrotransposon activity. One study investigated individual barley plants growing wild in a valley in Israel. The BARE-1 LTR-retrotransposon comprises about 3% of the barley genome, but the value ranged from 1.8 to 4.7% for the individual plants studied. Plants from

TABLE 10.6. Estimated Retrotransposon Content of Several Plant Genomes

Plant species	Estimated genome size (Mbp)	Estimated retroelement content	Reference
Mustard weed	125	5.3%	Arabidopsis Genome Initiative (2000) (2)
Rice	430	14%	Tarchini et al. (2000) (47)
Maize	3,200	50–80%	SanMiguel and Bennetzen (1998) (42)
Barley	4,800	>70%	Vicient et al. (1999) (49)
Lily	36,000	unknown	
Fern	160,000	unknown	

(Adapted from Wendel and Wessler 2000. [51])

higher up the valley slopes, which were exposed to drier and more stressful conditions, had higher contents of BARE-1 elements. The BARE-1 promoter contains binding sites for a transcription factor known to be induced by drought conditions, suggesting that stress may have triggered BARE-1 expansion. Evidently the BARE-1 element has recently proliferated in the barley genome in response to local environmental conditions. Whether this increase in retroelement content benefited the plants is an intriguing but unanswered question. Individual plants of other species have also been found to vary considerably in their retroelement content, suggesting that the genomic plasticity seen with barley may also hold in other plant species as well.

Whether lateral transfer has played a role in the expansion of retroelements in plant genomes is unknown. The high numbers of retroelement copies complicate the analysis of possible lateral transfer events, since it is hard to generate the exhaustive retroelement phylogenies necessary to identify anomalously similar elements in different organisms. However, the sequence comparisons of LTR-retrotransposons suggest lateral transfer in some cases. As discussed above, studies of mobile introns provide compelling data for lateral transfer of these elements in plant organelles. Whether the wide variation in plant genome sizes is connected to lateral transfer is unclear, but a fascinating possibility.

MOBILE ELEMENT CONTENT IN EUKARYOTIC GENOMES (29)

Table 10.7 summarizes the transposon content of the nuclear genomes of yeast, worm, fly, human, and mustard plant. The smallest eukaryote, yeast, is quite streamlined, with only 3.3% transposon sequences in its DNA, and only one of the major element types (LTR retrotransposons). Flies are also streamlined, with only 3.1% mobile elements, but a much larger complement of different types. Worms have somewhat higher fractions of mobile elements, 6.5%, with all major types present,

TABLE 10.7. Transposon Content of Nuclear Genomes of Several Species

	Human (Homo sapiens)		Fly (Drosophila melanogaster)		Worm (Caenorhabditis elegans)		Mustard weed (Arabidopsis thaliana)		Yeast (Saccharomyces cerevisiae)	
	% of bases	approx. no. of families	% of bases	approx. no. of families	% of bases	approx. no. of families	% of bases	approx. no. of families	% of bases	approx. no. of families
LINE/SINE	33.40%	6	0.70%	20	0.40%	10	0.50%	10	0	0
LTR	8.10%	100	1.50%	50	0.00%	4	4.80%	70	3.3%	5
DNA	2.80%	60	0.70%	20	5.30%	80	5.10%	80	0	0
TOTAL	44.30%	166	2.90%	90	5.70%	94	10.40%	160	3.3%	5

(Modified, with permission, from Lander et al. 2001 [© Macmillan Magazines Ltd.]. [29])

although LTR retrotransposons are scarce. Humans by comparison are infested with parasites, with fully 44% of the genome contributed by transposons, mostly retrotransposons. *Arabidopsis* is only 10% transposons, predominantly DNA transposons. In other plants with larger genomes, the relative retroelement content can be much larger. For all of the metazoan genomes, much of the DNA looks like nothing identifiable, a category that includes about 50% of the human DNA. Much of this was likely contributed by mobile elements, although it is no longer recognizable as such.

The diversity in mobile element content is a fascinating mystery. Why are different element classes differentially abundant in the eukaryotic genomes? Why are some genomes relatively streamlined (yeast and flies), while others—like ours—are greatly expanded by retroelement proliferation? Is there an explanation beyond happenstance and historical accident? Perhaps the life-styles of yeast and flies require relatively streamlined genomes (rapid cell division in yeast and flight in flies?), creating a pressure to eliminate unnecessary sequences. Perhaps humans and many plants need more spacer DNA, or perhaps we are just less efficient at ridding ourselves of parasites. In possible support of the latter idea is the observation that the repeat copies in humans appear to be older on average than repeats in the other organisms, suggesting that humans may be poor at eliminating transposon copies. It will be fascinating to see to what extent further studies can explain the organismic differences in transposon content.

Summary

The first eukaryotic sequence determined was that of the yeast *S. cerevisiae*. This genome stands as the most compact eukaryote studied, with 70% of the DNA encoding some 6340 genes in 12 Mb of DNA. The transposon complement is rather sparse, with only the Ty LTR retrotransposons present in the nuclear genome and mobile introns in mitochondrial DNA. Perhaps because of the high gene density, the Ty retrotransposons show fastidious targeting, probably evolved to minimize the damage of de novo transposon insertions. The full sequence reveals 193 clusters of Ty elements, 90% of them upstream of tRNA genes, benign sites of integration. Analysis of the Ty sequences supports the speculation that Ty3 and Ty4 have been subject to recent lateral transfer. Sequences have been detected in the nuclear genome derived from the mitochondrial genome, another means by which the nuclear genome can acquire new sequences.

The genome of the worm *C. elegans* was the first sequenced for a multicellular eukaryote. The *C. elegans* genome is about eight times larger than that of yeast, containing about 19,000 genes. This genome is rich in identifiable mobile elements, comprising at least 38 types. Worms, like the larger animals and plants studied to date, contain members of all of the main classes of eukaryotic transposons, including DNA transposons and retrotransposons of the LTR and non-LTR varieties. A unique type of retrotransposon was also discovered in the sequence, which integrates into genes for leader mRNAs found in nematodes and so pirates these unusual sequences to promote their own expression.

The genome of the fruit fly *Drosophila melanogaster* was sequenced in 2000, providing a new window on this classic system for genetic research. Flies have 13,600 genes spread over five chromosomes, and more than 50 families of mobile elements. As with other organisms, the mobile elements are not sprinkled evenly over the chromosomes, but rather clustered.

Many cluster within the centromeres in complex arrays of unknown origin and function. This organization is seen in vertebrates and plants also, and so may be a characteristic of higher eukaryotes generally. Dramatically, in flies we see one of the clearest examples of host cell recruitment of mobile element machinery. Flies lack conventional telomerase enzymes to maintain chromosome ends, but instead use non-LTR retrotransposons.

A first draft of the human genome was published in two papers in February 2001. The human genome, roughly 3,400 Mb of DNA, is only sparsely populated with genes, initially estimated at around 37,000. Exons comprise only 1.5% of the total DNA. Much more is made up of LINEs (20%), present in about a million copies, and the SINEs such as *Alu* repeats, which are present in around 1.5 million copies (13%). Other element types include the human endogenous retroviruses (HERVs) present as 8% of the genome, and DNA transposons, comprising 2.8%. Recombination among these repeated sequences is known to underlie a variety of human disorders. Viral or transposon sequences have also been recruited to form 47 known human genes. Fascinatingly, a single-copy DNA transposon appears to have been recruited by vertebrates to carry out the DNA rearrangements that generate diversity in the adaptive immune system (the topic of Chapter 11). Lateral DNA transfer from the mitochondrial genome to the human nuclear genome can also be detected.

The first complete genome sequence of a plant, *Arabidopsis thaliana* (mustard plant), was completed in 2000. The *Arabidopsis* genome is relatively small, only 125 million base pairs containing about 25,500 genes, roughly 39% of the sequence. Among mobile elements, retroelements are numerically the most abundant class, but all of the major types of eukaryotic transposons are also present. A roughly 620-kb segment of mitochondrial DNA was found embedded in the centromere of chromosome 2, providing some of the most dramatic data that organelle sequences can become incorporated into the nuclear genome. As discussed in Chapter 12, this is probably a major pathway of lateral transfer into eukaryotic genomes.

Genome sizes even among organisms vary over a very wide range that seem to display at least roughly similar degrees of complexity. This surprising observation, the C paradox, can be attributed in some cases to proliferation of retrotransposons. The *Arabidopsis* genome (125 million bp) contains only 5.3% retroelements. Barley has a genome of 4,800 million bp and more than 70% retroelements. Rice and maize are intermediate in size and probably intermediate in retroelement content. Lily and fern have enormous genomes, 36,000 and 160,000 million bases, respectively. In these, the retroelement content is unknown, but upon analysis may provide a new record for retroelement density. Why parasites have proliferated to such levels in some genomes, including ours, remains unknown.

REFERENCES

1. Adams M.D., Celniker S.E., Holt R.A., Evans C.A., Gocayne J.D., Amanatides P.G., Scherer S.E., Li P.W., Hoskins R.A., Galle R.F., et al. 2000. The genome sequence of *Drosophila melanogaster*. *Science* **287**: 2185–2195.

2. Arabidopsis Genome Initiative. 2000. Analysis of the genome sequence of the flowering plant *Arabidopsis thaliana*. *Nature* **408**: 796–816.

3. Bennetzen J.L. and Kumar A. 1999. Plant retrotransposons. *Annu. Rev. Genet.* **33**: 479–532.

4. Boeke J.D. 1989. Transposable elements in *Saccharomyces cerevisiae*. In *Mobile DNA* (ed. D.E. Berg and M.M. Howe), pp. 335–374. American Society for Microbiology, Washington D.C.

5. Boissinot S., Chevret P., and Furano A. 2000. L1 (Line 1) retrotransposon

evolution and amplication in recent human history. *Mol. Biol. Evol.* **17:** 915–928.

6. The *C. elegans* Sequencing Consortium. 1998. Genome sequence of the nematode *C. elegans:* A platform for investigating biology. *Science* **282:** 2012–2018.

7. Cost G.J. and Boeke J.D. 1998. Targeting of human retrotransposon integration is directed by the specificity of the L1 endonuclease for regions of unusual DNA structure. *Biochemistry* **37:** 18081–18093.

8. Crick F.H.C. 1979. Split genes and RNA splicing. *Science* **204:** 264–271.

9. Dawkins R. 1976. *The selfish gene.* Oxford University Press, New York.

10. Dej K.J., Gerasimova T., Corces V.G., and Boeke J.D. 1998. A hotspot for the *Drosophila gypsy* retroelement in the *ovo* locus. *Nucleic Acids Res.* **26:** 4019–4025.

11. Doolittle W.F. and Sapienza C. 1980. Selfish genes, the phenotype paradigm and genome evolution. *Nature* **284:** 601–603.

12. Engels W.R. 1989. P elements in *Drosophila melanogaster.* In *Mobile DNA* (ed. D.E. Berg and M.M. Howe), pp. 437–484. American Society for Microbiology, Washington, D.C.

13. Fink G. 1987. Pseudogenes in yeast? *Cell* **49:** 5–6.

14. Fink G.R. and Boeke J.D. 1999. Yeast, an experimental organism for all times. *ASM News* **65:** 351–357.

15. Fisher G., James S.A., Roberts I.N., Oliver S.G., and Louis E.J. 2000. Chromosomal evolution in *Saccharomyces. Nature* **405:** 451–454.

16. Fitz-Gibbon S.T. and House C.H. 1999. Whole genome-based phylogenetic analysis of free-living microorganisms. *Nucleic Acids Res.* **27:** 4218–4222.

17. Flavell A.J. 1999. Long terminal repeat retrotransposons jump between species. *Proc. Natl. Acad. Sci.* **96:** 12211–12212.

18. Goffeau A., Barrell B.G., Bussey H., Davis R.W., Dujon B., Feldmann H., Galibert F., Hoheisel J.D., Jacq C., Johnston M., Louis E.J., Mewes H.W., Murakami Y., Philippsen P., Tettelin H., and Oliver S.G. 1996. Life with 6000 genes. *Science* **274:** 546–567.

19. Gregory T.R. 2001. Coincidence, coevolution, or causation? DNA content, cell size, and the C-value enigma. *Biol. Rev. Camb. Philos. Soc.* **76:** 65–101.

19a. Hogensch J.B., Ching K.A., Batalov S., Su A.I., Walker J.R., Zhou Y., Kay S.A., Schultz P.G., and Cooke M.P. A comparison of the Celera and Ensembl predicted gene sets reveal little overlap in novel genes. *Cell* **106:** 413–415.

19b. Jackson I.J. 2001. Mouse genomics: Making sense of the sequence. *Curr. Biol.* **11:** R311–R314.

20. Ji H., Moore D.P., Blomberg M.A., Braiterman L.T., Voytas D.F., Natsoulis G., and Boeke J.D. 1993. Hotspots for unselected Ty1 transposition events on yeast chromosome III are near tRNA genes and LTR sequences. *Cell* **73:** 1–20.

21. Kalendar R., Tanskanen J., Immonen S., Nevo E., and Schulman A.H. 2000. Genome evolution of wild barley (*Hordeum spontaneum*) by BARE-1 retrotransposon dynamics in response to sharp microclimatic divergence. *Proc. Natl. Acad. Sci.* **97:** 6603–6607.

22. Karpen G.H. and Alshire R.C. 1997. The case for epigenetic effects on centromere identity and function. *Trends Genet.* **13:** 317–320.

23. Kazazian H.H. 2000. L1 retrotransposons shape the mammalian genome. *Science* **289:** 1152–1153.

24. Kidwell M.G. and Kidwell J.F. 1975. Cytoplasm-chromosome interactions in *Drosophila melanogaster*. *Nature* **253:** 755–756.
25. Kidwell M.G. and Lisch D. 1997. Transposable elements as sources of variation in animals and plants. *Proc. Natl. Acad. Sci.* **94:** 7704–7711.
26. Kim J.M., Vanguri S., Boeke J.D., Gabriel A., and Voytas D.F. 1998. Transposable elements and genome organization: A comprehensive survey of retrotransposons revealed by the complete *Saccharomyes cerevisiae* genome sequence. *Genome Res.* **8:** 464–478.
27. Korenberg J.R. and Rykowski M.C. 1988. Human genome organization: Alu, Lines, and the molecular structure of metaphase chromosome bands. *Cell* **53:** 391–400.
28. Kornberg A. and Baker T. 1991. *DNA Replication,* 2nd edition. W.H. Freeman, New York.
29. Lander E., Linton L.M., Birren B., Nusbaum C., Zody M.C., Baldwin J., Devon K., Dewar K., Doyle M., FitzHugh W., et al. 2001. Initial sequencing and analysis of the human genome. *Nature* **409:** 860–921.
30. Levis R., Ganesan R., Houtchens K., Tolar L.A., and Sheen F.-M. 1993. Transposons in place of telomeric repeats at a *Drosophila* telomere. *Cell* **75:** 1083–1093.
31. Lin X., Kaul S., Rounsley S., Shea T.P., Benito M.I., Town C.D., Fujii C.Y., Mason T., Bowman C.L., Barnstead M., et al. 1999. Sequence and analysis of chromosome 2 of the plant *Arabidopsis thaliana*. *Nature* **402:** 761–768.
32. Malik H.S. and Eickbush T. 2000. NeSL-1, an ancient lineage of site-specific non-LTR retrotransposons from *Caenorhabditis elegans*. *Genetics* **154:** 193–203.
33. Moore J.K. and Haber J.E. 1996. Capture of retrotransposon DNA at the sites of chromosomal double-strand breaks (comments). *Nature* **383:** 644–646.
34. Orgel L.E. and Crick F.H.C. 1980. Selfish DNA: The ultimate parasite. *Nature* **284:** 604–607.
35. Orgel L.E., Crick F.H.C., and Sapienza C. 1980. Selfish DNA. *Nature* **288:** 645–646.
36. Pardue M.L., Danilevskaya O.N., Traverse K.L., and Lowenhaupt K. 1997. Evolutionary links between telomeres and transposable elements. *Genetica* **100:** 73–84.
37. Plasterk R., Izsvak Z., and Ivics Z. 1999. Resident aliens: The Tc1/mariner superfamily of transposable elements. *Trends Genet.* **15:** 326–332.
38. Prak E.T.L. and Kazazian Jr., H.H. 2000. Mobile elements in the human genome. *Nat. Rev.* **1:** 134–144.
39. Rezsohazy R., van Luenen H.G., Durbin R.M., and Plasterk R. 1997. Tc7, a Tc1-hitch hiking transposon in *Caenorhabditis elegans*. *Nuceic Acids Res.* **25:** 4048–4054.
40. Ricchetti M., Fairhead C., and Dujon B. 1999. Mitochondrial DNA repairs double-strand breaks in yeast chromosomes. *Nature* **402:** 96–100.
41. Salzberg S.L., White O., Peterson, J., and Eisen J.A. 2001. Microbial genes in the human genome: Lateral transfer or gene loss? *Science* **292:** 1903–1906.
42. SanMiguel P. and Bennetzen J.L. 1998. Evidence that a recent increase in maize genome size was caused by the massive amplication of intergene retrotranposons. *Ann. Bot.* **81:** 37–44.
43. SanMiguel P., Tikhonov A., Jin Y.K., Motchoulskaia N., Zakharov D.,

Melake-Berhan A., Springer P.S., Edwards K.J., Lee M., Avramova Z., and Bennetzen J.L. 1996. Nested retrotransposons in the intergenic regions of the maize genome (comments). *Science* **274:** 765–768.

44. Smit A.F. 1999. Interspersed repeats and other mementos of transposable elements in mammalian genomes. *Curr. Opin. Genet. Dev.* **9:** 657–663.

45. Smit A.F.A. and Riggs A.D. 1996. Tiggers and other DNA transposon fossils in the human genome. *Proc. Natl. Acad. Sci.* **93:** 1443–1448.

46. Sun, X., Wahlstrom J., and Karpen G. 1997. Molecular structure of a functional *Drosophila* centromere *Cell* **91:** 1007–1019.

47. Tarchini R., Biddle P., Wineland R., Tingev S., and Rafalski A. 2000. The complete sequence of 34kb of DNA around the rice Adh1-adh2 region reveals interrupted colinearity with maize chromosome 4. *Plant Cell* **12:** 381–391.

48. Venter J.C., Adams M.D., Myers E.W., Li P.W., Mural R.J., Sutton G.G., Smith H.O., Yandell M., Evans C.A., Holt R.A., et al. 2001. The sequence of the human genome. *Science* **291:** 1304–1351.

49. Vicient C.M., Suoniemi A., Anamthawat-Jónsson K., Tanskanen J., Beharav A., Nevo E., and Schulman A.H. 1999. Retrotransposon BARE-1 and its role in genome evolution in the genus *Hordenum. Plant Cell* **11:** 1769–1784.

50. Weil C.F. and Kunze R. 2000. Transposition of maize Ac/Ds transposable elements in the yeast *Saccharomyces cerevisiae. Nat. Genet.* **2:** 187–190.

51. Wendel J.F. and Wessler S.R. 2000. Retrotransposon-mediated genome evolution on a local ecological scale. *Proc. Natl. Acad. Sci.* **97:** 6250–6252.

52. Wolf Y., Kondrashov F.A., and Koonin E.V. 2000. No footprints of primordial introns in a eukaryotic genome. *Trends Genet.* **16:** 333–334.

A Transposon Progenitor of the Vertebrate Immune System

EVERY DAY OUR IMMUNE SYSTEM FIGHTS OFF AN onslaught of invaders attempting to make our bodies their homes. Patients lacking functional immune systems are quickly overwhelmed by infection unless isolated in sterile environments. One of our main defenses is the adaptive immune system, which responds with countermeasures specific to each invading pathogen. To protect us, the immune recognition molecules—antibodies and receptors on the surfaces of T cells—are produced that recognize distinctive structures on each invader and expand in response to their presence. Viruses, bacteria, fungi, and other pathogens present an enormous diversity of macromolecules that must be recognized by the antigen receptors. Immune recognition must also be highly selective, allowing molecules of invaders to be distinguished from our own tissues. Autoimmune disease is the result of a failure to control the immune response against self-molecules.

The diversity in binding specificities needed to recognize invading microbes is largely generated by rearrangements at the DNA level. One of the main mechanisms for generating diversity relies on combinatorial reshuffling of DNA segments, each of which encodes a part of these molecules. Remarkably, there is now good evidence that the system for rearranging these gene segments evolved from a captured DNA transposon.

Is the evolutionary appearance of the vertebrate immune system an example of lateral transfer? There is no way to reconstruct the evolutionary pathways with certainty. The recombination system that generates the antigen receptor is found only in the jawed vertebrates, suggesting that this system is a relatively recent innovation. Today there are no obvious relatives of the progenitor transposon in the vertebrate genome, consistent with recent capture by lateral transfer.

In this chapter, we begin with an overview of the function of the immune system, emphasizing the role of antibodies and T-cell receptors (TCRs). The mechanisms for generating diversity are then presented, followed by studies documenting the link to transposon function. The chapter concludes with a brief survey of the other side of the struggle, the use of DNA rearrangements by pathogens to evade the immune system.

OVERVIEW OF THE ADAPTIVE IMMUNE SYSTEM (22)

It has been recognized for centuries that people exposed to a pathogen developed resistance to later infections. Those who survived a case of smallpox, for example, did not develop the disease upon subsequent exposure. The response was specific, however, to each pathogen. Surviving a case of smallpox did not confer resistance to influenza or measles.

The branch of the vertebrate immune system responsible is often called the adaptive immune system, because it adapts the host to resist specific invaders. This is distinct from the innate immunity conferred by systems that destroy pathogens nonspecifically.

Two types of adaptive immunity can be distinguished: the humoral (blood-borne) and cell-based responses. Each relies on dedicated sets of immune cells—the humoral response involves B cells, which produce immunoglobulin (antibody) molecules, and the cell-based response involves T cells, which use the TCR. These cells of the immune system are organized into several organs, including the thymus, spleen, lymph nodes, bone marrow, and Peyer's patches. Together these comprise one of the largest organ systems in the body. Other immune system cells circulate in the blood and lymph. Antibodies are produced by the B cells, which develop in the bone marrow and then move to other sites, including blood and lymph nodes. The T cells are produced in the bone marrow and mature in the

Innate immunity is conferred by a collection of mechanisms, such as natural killer cells, which probably recognize altered forms of normal structures, and phagocytes, which engulf and destroy infecting microorganisms. Most vertebrates possess both adaptive and innate immunity, whereas invertebrates possess only the more primitive innate system. Aspects of the innate immune system are described in Chapter 13.

thymus. T cells then emigrate from the thymus to circulate in the blood and congregate in the immune organs.

The broad topic of immunology is not covered in detail below, only points necessary for the discussion of the DNA rearrangements that generate diversity. The discussion focuses on the mammalian immune system and not that of fish and birds, which differ in ways not central to our story.

Humoral Immunity (18, 22, 50)

Central to humoral immunity is the recognition by antibodies of an antigen, named for *anti*body *gen*erator. An antigen can be any of myriad proteins, polysaccharides, or other molecules found on an infecting microbe. Antigens are recognized with exquisite specificity. For example, related proteins with only single amino acid substitutions can be readily distinguished by antibodies.

The structure of a typical antibody is diagramed in Figure 11.1 and shown as a molecular structure in Figure 11.2. The molecule adopts an overall Y shape. The antigen-binding sites are located at the tips of the arms of the Y, so each antibody has two antigen-combining sites. Antibodies are composed of four polypeptide chains, two copies each of a larger "heavy" chain and a smaller "light" chain. Each light chain is composed of two protein domains, the variable (V_L) region and the constant (C_L) region. The heavy chain is composed of four to five domains, a variable (V_H) region and several constant (C_H) regions. Each domain has a related structure, composed of two layers of antiparallel β sheets—the "immunoglobulin fold." The detailed amino acid sequence in the variable domains differs from antibody to antibody, explaining the different specificities. Three protein loops extend from the tip of the variable domain of each chain to bind the antigen (Figs. 11.1, 11.2). The constant regions are so named because they do not vary in amino acid sequence from antibody to antibody. The character of the heavy-chain constant region determines the class of the antibody. There are five classes, IgA, IgD, IgG, IgE, and IgM. There are two types of light chains, named κ and λ.

Once an antibody binds to an antigen, the constant regions recruit other immune functions to eliminate the invader. For example, immunoglobulin G (IgG) molecules bound to an invading bacterium direct the complement system to lyse the foreign cell. IgE molecules, once they bind antigen, signal specialized mast cells to release histamine, which provokes the tissue responses associated with hay fever. This is thought to recruit effector cells to the site of infection and to activate their function.

Each antibody-producing cell makes only one specificity of antibody. Each cell makes a single type of light chain, either κ or λ, and one heavy-chain type. The system for controlling antibody production is thus tightly regulated, as each diploid cell contains two genes for heavy chains and multiple genes for light chains.

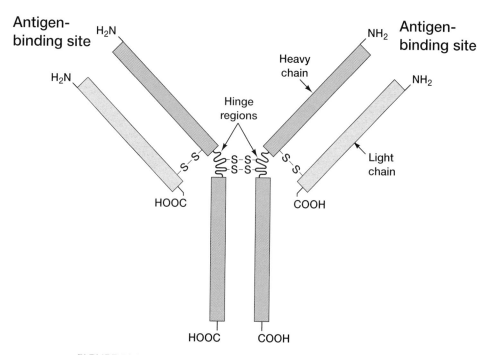

FIGURE 11.1. An antibody molecule. The heavy chain is shown as green, the light chain as gray. Disulfide bonds (–S–S–) connect the heavy chains to the light chains and to each other. NH$_2$ and COOH indicate the amino and carboxyl termini of the protein. The antigen-binding site is at the tips of the Y-shaped molecule. (Modified from Alberts et al. 1994 by permission of Routledge, Inc., part of The Taylor & Francis Group. [3])

Cell-based Immunity (22)

Lymphokines secreted by the T-helper cells include interleukins-2, -3, -4, -6, and -7, interferon-γ, and granulocyte-macrophage colony-stimulating factor.

Antigen-binding proteins are also present on the surface of T cells, where they contribute to several aspects of immune function. These T-cell receptors, like the antibodies, contain antigen-binding sites that vary in amino acid sequence from T cell to T cell. TCRs always act while bound to the cell surface. Antibodies can also bind to cell surfaces but exert much of their effector functions during circulation in the blood and lymph.

Two main types of T cells are the cytotoxic T cells and helper T cells. Cytotoxic T cells are important for eliminating normal cells that have become infected by a microbe, typically a virus. These cells recognize that cells are infected and initiate their destruction, thereby preventing the infection from spreading. Helper T cells, on the other hand, upon binding antigen send out signals that recruit other immune effector cells to the site of infection and activate their function. Helper T cells achieve this by synthesizing extracellular signaling molecules that bind appropriate receptors on immune effector cells and direct their activities.

The TCR does not bind free antigen, but only binds to antigen-derived peptides in the proper context. Immune system cells such as macrophages process foreign antigens by engulfing them, cleaving the

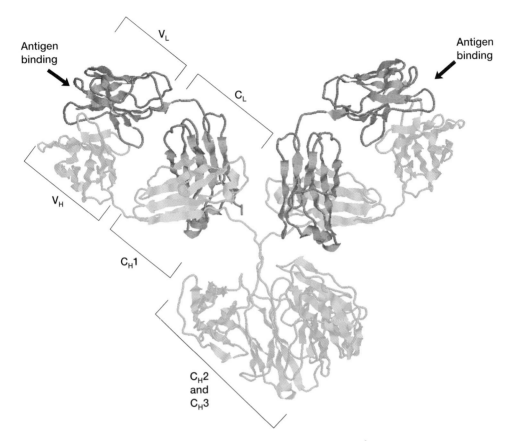

FIGURE 11.2. Structure of an antibody molecule. The light chains are shown as gray, the heavy chains as green. The antigen-binding site is as marked. Image rendered from structural coordinates determined by Harris et al. (1998). (18)

proteins into short peptides, and displaying these peptides on the cell surface. Normal cells also display peptides on the cell surface, allowing immune surveillance for virus infection.

Antigenic peptides are displayed on either of two specialized proteins, called major histocompatibility group (MHC) class I and MHC class II. The name derives from the initial discovery of these proteins as major determinants of whether grafts would be accepted or rejected during transplantation. MHC molecules bind to antigenic peptides inside cells, then the complex translocates to the cell surface. Antigen bound to the MHC molecules can then bind to TCRs on cytotoxic or helper T cells (see Fig. 4.24, left side). Several other proteins also participate in receptor binding and initiation of the signaling cascade that ensues. For example, the CD4 protein, prominent in Chapters 6 and 7 as the receptor for HIV, resides on helper T cells and participates in engaging MHC class II. In summary, the TCR, like antibodies, must have many different recognition specificities, although antigen binding takes place in the framework provided by MHC and other cell-surface proteins.

COVALENT REARRANGEMENT OF THE GENES FOR THE ANTIGEN RECEPTORS (8, 21, 27, 50)

The diverse binding specificities of the antibody and TCR variable regions are generated in part by covalent rearrangements of the genes encoding these proteins. The structure of the genomic loci encoding antigen receptors exists in an unrearranged form in all cells of the body except those destined to become B cells or T cells. The unrearranged or germ-line forms of these genes do not encode functional antigen receptors. In appropriate immune progenitor cells, the gene segments encoding the antigen-binding proteins are rearranged, so that full-length proteins can be produced. The products of the rearrangement reaction differ from cell to cell, generating the different antigen-binding specificities.

The genomic region encoding the κ light chain of the mouse is shown in Figure 11.3. The locus is shown at the top in its germ-line, or unrearranged, form. Starting from the left, an array of V regions, which encode the variable regions in the antibody structure, can be seen. There are about 300 V regions in the mouse κ locus. Proceeding rightward, we encounter the four J regions, named for "joining" segments. Right of the J regions is the single C region, which encodes the constant segment of the κ light chain.

Rearrangement proceeds with the joining of a V region to a J region. There are 300 V regions and 4 J regions, which yield 1,200 possible coding region sequences. The fused VJ region is then transcribed, producing a message that encodes the VJ and C regions. RNA splicing then removes the extra J regions and intron sequences between the VJ and C gene segments. Translation yields the κ light-chain protein.

Gene rearrangement at the heavy-chain locus is conceptually similar, but an additional gene segment is involved. The mouse heavy-chain locus is shown in Figure 11.4. The locus encodes between 100 and 1000 V regions. These V regions become joined to short segments called D regions, for "diversity," which are present at about 12 copies per locus. The D regions become joined to J regions, of which there are about four. Rightward of these are the C regions that determine the type of antibody produced. The heavy-chain locus is rearranged to fuse V, D, and J segments, thereby forming the variable portion of the coding region. Transcription, splicing, and translation yield heavy-chain proteins.

Each antibody is composed of a pair of heavy chains and a pair of light chains, yielding a very large number of combinatorial possibilities for antibody protein sequences. Rearrangement at the light-chain locus yields 1,200 possible coding sequences. Rearrangement at the heavy-chain locus yields 4,800–48,000 possible combinations. Because each antibody is composed of heavy and light chains, the total number of possibilities is the product (4,800 to 48,000 x 1,200) = 6 to 60 million

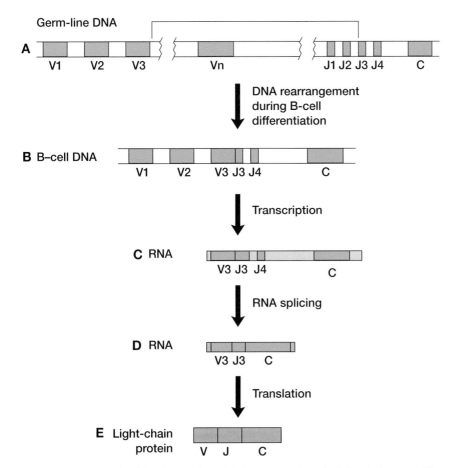

FIGURE 11.3. Role of DNA recombination in construction of a light-chain gene. (*A*) The light-chain locus in its germ-line arrangement is made up of multiple V (variable) and J (joining) segments and a single C (constant) region. In cells destined to produce antibodies (the B-cell lineage), V and J segments are recombined (*B*) to yield a functional light-chain coding region. Transcription of this message (*C*) yields an RNA containing the rearranged VJ segment, potentially unrearranged J regions, and the C region. RNA splicing (*D*) removes introns and unrearranged J regions, yielding the continuous light-chain coding region. Translation (*E*) yields the light-chain protein. (Modified from Alberts et al. 1994 by permission of Routledge, Inc., part of The Taylor & Francis Group. [3])

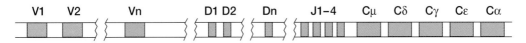

FIGURE 11.4. The heavy-chain locus. In addition to the V, J, and C gene segments found in the light-chain locus, the heavy-chain locus contains D (diversity) regions, which are assembled by recombination into VDJ gene segments. Five different C regions are found in different types of antibodies. A different type of DNA recombination (class switching) swaps C regions to yield antibodies with different immune functions. (Modified from Alberts et al. 1994 by permission of Routledge, Inc., part of The Taylor & Francis Group. [3])

possible combinations. Diversity is further expanded by additional mechanisms during the immune response, including base addition and deletion at the recombination junctions and somatic hypermutation (discussed below).

Gene rearrangements generate diversity in a similar fashion at the TCR loci. In this case, there are four loci, α, β, γ, and δ. The first two are found on cytotoxic and helper T cells, and the latter only on a small minority of cells present at epithelial surfaces and other locations. The mechanism of rearrangement involves the same signals and enzymes as in assembly of antibody gene segments. In the following sections, we use the antibody loci as examples with the understanding that the mechanistic points apply to the TCR loci as well.

DNA SIGNALS MEDIATING VDJ RECOMBINATION (19, 25)

We now look more closely at the machinery mediating VDJ recombination. The gene segments that become rearranged during construction of the antigen-binding genes are flanked by distinctive DNA sites that direct the gene assembly process, called recombination signal sequences (RSS). An example from the κ light-chain locus is shown in Figure 11.5.

Just abutting the upstream edge of the V_κ segment is a signal composed of a conserved 9-bp segment (nonamer), a spacer, and a conserved 7-bp segment (heptamer). The spacer is conserved in length at

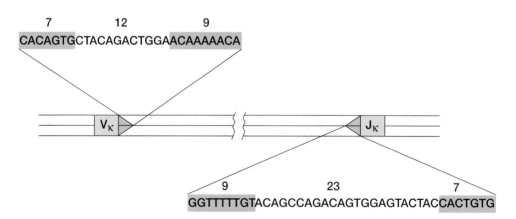

FIGURE 11.5. Recombination signal sequences (RSSs) recognized during VDJ recombination. The RSSs are shown as green triangles. Note that the signals are in inverted orientation, allowing deletion of the intervening DNA (*white*) to fuse the V and J gene segments (*green*). The RSS is composed of two subsequences 7 bp (heptamer) and 9 bp long (nonamer). Recombination takes place between one RSS with 12 bp between the heptamer and nonamer and a second with a spacing of 23 bp (the 12-23 rule).

12 bp, but not the DNA sequence. At the upstream edge of the J_κ partner is a related DNA sequence, composed of identical 9-bp and 7-bp sequences, but with a different spacer, in this case 23 bp. The orientation of the conserved heptamer and nonamer sequences is opposite in the two cases, so these sequences form an inverted repeat.

DNA breaking and joining always takes place at the same location relative to the conserved sequences, at the border between the heptamer and the coding segment. Rejoining of the separate coding region segments yields the completed VJ region. Recombination in the heavy-chain locus proceeds similarly, but with the addition of D regions as well. D regions are also flanked by heptamer and nonamer repeats. Two recombination events, V to D and D to J, assemble the VDJ sequence of the heavy chain.

RECOMBINATION ACTIVATING GENES (11, 24, 28, 31, 35, 37, 54)

The conserved heptamer/nonamer recognition sequences are the sites of action of the *r*ecombination *a*ctivating *g*ene (RAG) proteins, the enzymes responsible for the initial DNA breakage steps. The RAG genes were initially identified in a daring experiment in which mammalian DNA was introduced into cells incapable of VDJ recombination, and rare recombination events were assayed. Cells that had carried out VDJ recombination were isolated and the newly introduced genes—which potentiated recombination—were reisolated. This allowed David Schatz, Margorie Oettinjer, and David Baltimore to isolate a DNA segment that contained genes directing VDJ recombination.

Remarkably, this DNA segment contained not one but two genes, both of which were required for VDJ recombination. Had these two genes been at separate locations in the chromosome, the functional isolation strategy would never have worked. The two genes were named RAG1 and RAG2. Often the amino acid sequences of newly isolated genes resemble known proteins, which allows aspects of function to be inferred. For both of the RAGs, however, no strong sequence similarities were initially detected.

A key advance in understanding VDJ recombination and RAG gene function came with the establishment of a test tube system for RAG recombination. This allowed models for function to be examined using reactions reconstructed with purified components. The demonstration that model reactions containing only the RAG proteins could carry out the initial DNA breaking reactions of VDJ recombination established that the RAGs carried out a key enzymatic step. From the earlier genetic data alone, the RAGs could instead have encoded required activators of as yet unidentified recombination enzymes.

More recent studies have emphasized the similarities between RAG1 and transposase proteins. For example, RAG1 contains three acidic residues required for recombination as with the bacterial transposases and retroviral integrases.

THE VDJ RECOMBINATION
PATHWAY (10, 12, 14, 26, 28, 32, 35, 47, 48, 51–54)

The pathway of VDJ recombination is summarized in Figure 11.6. The summary is based on genetic studies in vivo that identified required components and biochemical studies in vitro that probed their function. The two types of studies yield a consistent and, in some respects, surprising picture.

The DNA between the signal and coding sequences is cleaved to yield free DNA ends, but in a two-step reaction. Initially RAG 1 and 2

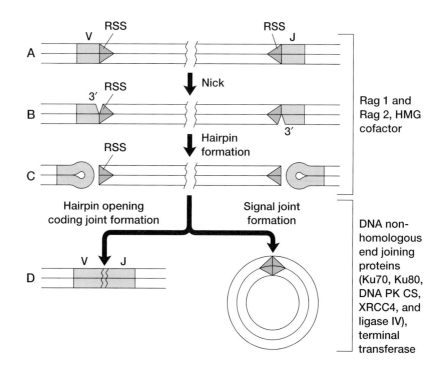

FIGURE 11.6. The pathway of VDJ recombination. (*A*) The germ-line arrangement of a V and J segment. VDJ recombination begins (*B*) with nicking at the RSS signals. The free 3′ end then attacks the other DNA strand, forming a hairpin in the coding ends (*C*). The nicking and hairpin formation reactions are carried out by RAG-1 and RAG-2 in association with HMG proteins. For coding joint formation, the hairpins are opened (either by the RAG proteins or cellular proteins), the DNA ends are processed by cellular repair enzymes and terminal transferase, and the ends are joined by the nonhomologous DNA end joining (NHEJ) proteins Ku70 and Ku80, DNA PK CS, ligase IV, and XRCC4. For signal joint formation, NHEJ proteins but not DNA PK are needed for the ligation reaction.

proteins cleave one DNA strand at the junction between the recombination signals and the coding segments. The resulting free 3′ DNA ends then attack the other DNA strand. A single-step reaction (transesterification) both breaks the second strand and links it to the first, yielding one free DNA end and one DNA hairpin.

The reaction takes place in an organized structure containing not just the RAG proteins, but also small DNA-binding proteins that act as architectural elements. Many of the reactions discussed in previous chapters involved small DNA-binding proteins that bind the reactant DNAs and help them adopt the active conformation, like IHF in phage λ integration (Chapter 4). In VDJ recombination, the cellular HMG 1 and 2 proteins assist the reaction, apparently by binding to the 23-mer spacer, allowing it to bend into the conformation required for cleavage.

The DNA double-strand breaks generated by the RAG/HMG proteins are then joined in new arrangements. The two coding ends are joined to assemble the antigen-binding gene, and the two signal ends are joined to form an extrachromosomal circle. For signal joint formation, the double-strand breaks are joined by the cellular double-strand break repair proteins, Ku, DNA-dependent protein kinase, DNA ligase 4, and its cofactor XRCC4. Ku protein binds tightly to DNA ends, apposing free DNA ends and allowing them to be joined by the action of ligase 4 and XRCC4. The DNA-dependent protein kinase is also important, serving as regulator or scaffold for the other end-joining proteins.

Joining of the coding ends requires first opening the hairpins and cleaning up the DNA ends prior to ligation. The enzymes responsible for hairpin opening are at present a matter of controversy. Candidates include the RAG proteins themselves and the cellular repair proteins Rad50, MRE11, and NBS1. After the hairpins are opened by one of these sets of enzymes, the messy DNA ends need to be made flush for ligation. Several enzymes contribute to this job. Nucleases remove free single-stranded DNA ends and polymerases polymerize along single-stranded regions to make them double-stranded. Another enzyme, terminal deoxynucleotide transferase, attaches deoxynucleotides to the free 3′ ends. The action of these enzymes together creates neat coding ends that can be joined by ligase 4 and its cofactors.

GENERATING THE FULL COMPLEMENT OF ANTIGEN-BINDING PROTEINS (12, 22)

The rearrangement of antibody gene segments yields around 10^7 possible antibody gene sequences, but actually there are far more antibody specificities present in a healthy mammal. Several further mechanisms also contribute to the generation of diversity. One important source is the small sequence changes at the junctions between V, D, and J segments. Addition of nucleotides by terminal transferase can result in the addition of amino acids to the encoded sequence. The process of hairpin formation, followed by hairpin opening and repair, can yield short inverted repeats at junctions called N regions. This too can lead to addition of amino acids in the encoded proteins. DNAs are also sometimes deleted at junctions, again altering the encoded amino acid sequence.

These reactions can also inactivate an antigen receptor gene. To preserve the coding region, the reading frame of the antigen-binding protein must be preserved. Insertions or deletions that cause the gene to be read in the wrong reading frame result in the synthesis of a protein unrelated to the correct antigen receptor. The DNA addition and deletion mechanisms have no special proclivity for inserting base pairs in multiples of three, leading to unproductive rearrangements in many cases.

Cells have developed a feedback system to maximize the chances of productive gene rearrangement. The production of a functional antigen receptor is sensed by the cell. Cells lacking functional receptors rearrange the second allele, so each cell has two chances to form a functional gene. Cells that fail twice are eliminated.

At an early stage of the development of antibody-producing cells (B-cells), some of the encoded antibodies remain associated with the cell surface. Engagement of antigen by antibody on the cell surface signals the cell to proliferate and begin producing soluble antibodies. During growth, the antibody-coding genes are selectively subjected to a high rate of mutation. Those cells synthesizing antibodies with higher affinity for antigen are disproportionately stimulated to multiply. The result is an increase in affinity of the antigen-binding repertoire over time, a process called affinity maturation.

The type of antibody heavy chains synthesized is also diversified by another type of site-specific recombination called "class switching." Class switch recombination swaps the DNA segments encoding heavy-chain constant regions while preserving the same antigen-combining site, broadening the repertoire of immune effectors directed against a particular invader. Class switch recombination is not carried out by the RAG proteins, but by another less well-understood recombination system.

PARALLELS BETWEEN VDJ RECOMBINATION AND TRANSPOSITION [1, 2, 20, 31, 49]

With this background in place, we can appreciate how the VDJ system might have evolved from a transposon (Fig. 11.7). The process is hypothesized to have begun with two insertions of a transposon into a gene encoding a cell-surface protein. The progenitor gene would have encoded a part of an immunoglobulin fold domain, as all of the vari-

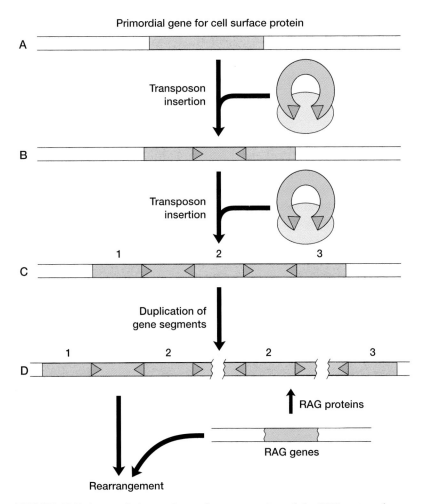

FIGURE 11.7. A speculative pathway for construction of the VDJ system from a transposon progenitor. (*A*) A primordial gene for a cell-surface protein (*gray*) becomes interrupted by insertions of a DNA transposon (*green*; sites of transposon action are shown by the dark green triangles) (*B*). Another insertion nearby disrupts the gene further (*C*). Many DNA transposons are known to favor local integration, lending plausibility to this idea. Duplication of DNA segments containing a part of the coding region and flanking sites of transposon action (*D*) could then build up the loci encoding the germ-line gene segments flanked by RSS signals (formerly transposon ends). Action of the RAG proteins, encoded from an element lacking sites of transposon action, could then mediate DNA rearrangements.

able domains of the adaptive immune system have this structure. Amplification of the DNA unit encoding the gene segment and the flanking transposon end sequences would create multiple gene segments each flanked by transposase-binding sites. Elsewhere in the genome, a transposon copy is hypothesized to exist that could supply transposase protein in *trans* to the array of gene segments. Transposase would then cleave at the transposon ends in the gene-segment array, yielding DNA double-stranded breaks. Joining of the broken double-strand ends could then join the gene segments, yielding the continuous coding region. The excised signal sequences would also be joined to form a circle, exactly as in VDJ recombination.

The involvement of DNA hairpins in VDJ recombination also parallels transposition. Excision of the prokaryotic transposons Tn5 and Tn*10* generates DNA hairpins at the ends of the element DNA (Chapter 3), whereas excision of eukaryotic *hAT* DNA transposons is inferred to leave hairpins in the target DNA (Chapter 9). The latter case exactly parallels the formation of DNA hairpins during VDJ recombination. If the excised signal sequence is taken as the transposon mimic, the flanking coding sequence would mimic the target site, and it is just this region that transiently forms the hairpin intermediate.

The linkage of the RAG1 and RAG2 genes is also a potential consequence of their transposon origin. There is no obvious reason that these genes need to be linked for function (although it was a crucial stroke of luck for the scientists who isolated the RAGs by their function!). However, if both genes were transported into the vertebrate genome on the same transposon, their linkage is much easier to understand.

These parallels set the stage for the stunning finding that RAG1 and RAG2 can actually catalyze DNA transposition in vitro. The groups of Martin Gellert and David Schatz found that purified RAG proteins could direct the insertion of excised signal sequences—transposon analogs—into a target DNA in vitro. The transposon mimic was found to be inserted neatly into the host, so that no sequences were lost at the integration junction. The products of this reaction also resembled transposon insertions in another respect. The points of attachment of the signal sequence ends in the target DNA displayed a defined spacing relative to each other. As a consequence, after repair of this intermediate, five base pairs of target sequence are duplicated at each target DNA junction, exactly as with transposons and retroviruses. This provides compelling evidence for the transposon origin of the adaptive immune system. The only piece missing is the demonstration of transposition in vivo, a goal now being pursued by several laboratories.

Which eukaryotic DNA transposon family gave rise to the VDJ system? At present there is too little evidence to pinpoint a particular candidate, but features of several are prominent in the VDJ system. Hairpin formation in the flanking target DNA most closely resembles the *hAT* family. Another parallel is the involvement of the Ku/DNA PK system in sealing double-stranded breaks, used in VDJ recombination and in resealing target DNA after excision of P elements. This suggests that a P-element-like transposon may have given rise to the VDJ system, although Ku/DNA PK could be implicated as important for other DNA transposons in further studies. P elements also mimic VDJ recombination in requiring different arrangements of sites at each end of the P-element DNA, possibly mimicking the 12/23 rule. Database searches for transposase proteins similar to the RAG proteins have so far yielded no compelling matches, leaving the exact progenitor uncertain.

Did the transposon progenitor of the immune system arrive in the vertebrate genome by lateral transfer? The adaptive immune system is a recent evolutionary development, present only in jawed fishes and higher taxa. No relatives of the RAG genes have been identified in the draft human genome sequence (Chapter 10) that might provide clues to their origin. Given the frequency of transfer of DNA transposons (Chapter 9), it seems reasonable to hypothesize that the progenitor transposon was captured by lateral transfer and persisted as the VDJ system, although the history cannot be reconstructed with certainty.

THE PATHOGEN SIDE: DNA REARRANGEMENTS TO EVADE THE IMMUNE SYSTEM

Given the effectiveness of DNA rearrangements in the vertebrate immune system, it is hardly surprising that many pathogens have fought back with DNA rearrangements of their own. Diversity in the antigen-binding repertoire is thus parried by diversity in the antigens bound. A few of the pathogen DNA rearrangements involve lateral DNA transfer, but many do not and so are off the main theme of this book. A few examples are presented below for perspective on the function of the immune system and for their general interest as mobile DNA systems.

Prokaryotic Phase Variation (13, 23, 40)

Many prokaryotic bacteria are able to switch between stable states of gene expression during growth, a process called "phase variation." Examples include the switching of surface proteins in *Salmonella typhimurium*, turning pilus expression on and off in pathogenic *Escherichia coli*, and swapping surface proteins in *Neisseria gonorrhoeae*. Many of these mechanisms involve covalent changes in DNA, although there is considerable variety in the details (see Table 11.1).

As early as 1922, it was recognized that *S. typhimurium* switched between two antibody-binding specificities or "serotypes." This was found to correspond to different flagellar antigens, encoded by genes named H1 and H2. Switching back and forth took place at a frequency of 10^{-3} to 10^{-5} per bacterial cell division. Evidently the flagellar proteins were recognized by the immune system, and *S. typhimurium* responded by switching between either of two types.

Modern studies have revealed that DNA inversion mediates the alternate expression of H1 and H2 (Fig. 11.8). The invertible segment contains a promoter, which in one orientation directs expression of the nearby H2 gene. A second gene, rH1, which encodes a repressor of the H1 gene, is also transcribed with H2. Also encoded in the invertible

TABLE 11.1. Some Examples of Phase Variations in Prokaryotes

Observation	Organism	Mechanism	References
Flagellar variation	*Salmonella typhimurium*	site-specific inversion of DNA segment	Glasgow et al. (1989) (13)
Fimbrial variation	*Escherichia coli*	site-specific inversion of DNA segment	Glasgow et al.(1989) (13)
Pilin variation	*Moraxella bovis*	site-specific inversion of DNA segment	Glasgow et al. (1989) (13)
Pilin variation	*Neisseria gonorrhoeae*	duplicative trans-position of genes (gene conversion) (1985, 1986) (45, 46)	Haas and Meyer (1986) (16); Segal et al. (1985, 1986) (38,39); Swanson et al.
Opacity protein variation	*N. gonorrhoeae*	duplication/deletion of nucleotides in leader peptide sequence	Stem et al. (1984, 1986) (42, 43)
Flagellar variation	*Campylobacter coli*	DNA rearrangement	Harris et al. (1987); Guerry et al. (1988) (15, 17)
Vi surface antigen variation	*Citrobacter freundii*	precise/insertion/ deletion of IS1	Brinton (1959) (7); Snell-ings et al. (1981) (41); Baron et al. (1982) (4)
Extracellular polysaccharide variation	*Pseudomonas atlantica*	precise/insertion/ deletion of DNA sequence	Bartlett et al. (1988) (5)
Serotype variation	*Borrelia hermsii*	duplicative trans-position of genes (site-specific recombinations?)	Meier et al. (1985) (30); Plasterk et al. (1985, 1986) (33, 35)
Agglutinogen variation	*Bordetella pertussis*	?	Robinson et al. (1986) (36)
Tail fiber (host range) variation	Bacteriophages Mu and P1	site-specific inversion of DNA segment	Glasgow et al. (1989) (13)

(Modified, with permission, from Glasgow et al. 1989. [13])

region is an invertase, named Hin for *H in*version. The action of Hin at sites at the edge of the invertible region (*hix* sites) catalyzes DNA inversion. Inversion of the promoter turns off expression of H2 and rH1, allowing the H1 gene to be expressed.

The mechanism of inversion has been worked out in detail in a beautiful series of studies by Reid Johnson, Nick Cozzarrelli, Mel Simon, and coworkers. Figure 11.9 illustrates the protein–DNA complexes inferred to mediate inversion. The DNA sites involved include not only the two *hix* sites, but also an enhancer sequence located within the invertible region. The enhancer binds the host-encoded Fis protein, named for *f*actor for *i*nversion *s*timulation. Fis introduces a sharp bend into DNA, thus serving as an architectural element for complex

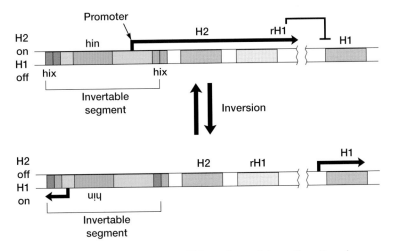

FIGURE 11.8. Hin-mediated inversion. Either of two *Salmonella* cell-surface proteins H1 and H2 (*dark gray*) are expressed alternatively as a result of DNA inversion. The invertible segment (*green*) encodes the Hin recombinase (*dark green*) and a promoter reading outward from the locus. In the orientation shown at the top, the internal promoter directs expression of the H2 gene and a repressor of H1 expression (rH1). the rH1 product represses the H1 gene located elsewhere in the chromosome. Inversion (*bottom*) turns off expression of H2 and rH1, allowing expression of the unlinked H1 gene. Another cycle of inversion returns the system to the H2-expressing top configuration. (Modified, with permission, from Glasgow et al. 1989. [13])

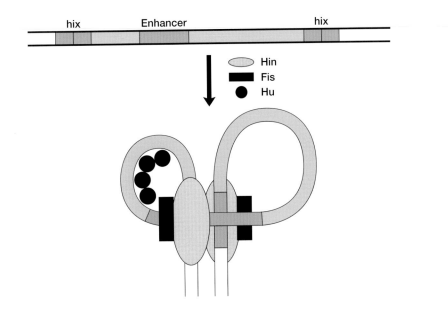

FIGURE 11.9. The invertase synaptic complex. The *hin-hix* DNA (*top*) adopts a specific wrapped configuration when complexed with Hin (*gray ovals*) and the *E. coli*-encoded DNA-binding proteins Fis (*black rectangles*) and Hu (*black circles*). The Hu protein assists DNA bending when the *hix* sites are close to the internal recombinational enhancer that binds Fis. The path of the DNA through the complex has been specified precisely by studies of changes in DNA topology accompanying the reaction in vitro. (Modified, with permission, from Glasgow et al. 1989. [13])

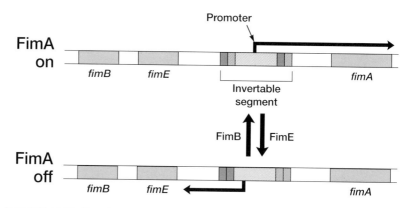

FIGURE 11.10. Regulation of expression of *E. coli fim* gene expression by DNA inversion. The invertible segment contains a promoter, which in one orientation (*top*) directs expression of *fim*A and in the other, expression of *fim*E. (Modified, with permission, from Glasgow et al. 1989. [13])

formation. The Fis/enhancer complex assists the binding of Hin to the *hix* sites with the proper geometry for subsequent recombination. Even the motions in the intermediates leading to the final products has been deduced from the topology of DNA products formed in vitro.

The Hin system is closely related to a number of other DNA inversion systems found in prokaryotes, including the Gin inversion system of phage Mu described in Chapter 4. Many other molecular switches for oscillating between stable states are known to employ closely related inversion systems. The high degree of similarity among these systems suggests that they have been transferred widely among prokaryotic cells and viruses, as with other operons and phage gene cassettes.

E. coli pathogenesis is also regulated by a DNA inversion system. Pathogenic *E. coli* infections often involve adherence of the bacterium to epithelial or blood cells of the host. A bacterial-encoded pilus, also called a "type 1 fimbriae," tethers the bacteria to the cell by binding mannose sugars. The bacterium can switch expression of the encoded pilus protein on and off by inverting a DNA segment (Fig. 11.10).

N. gonorrhoeae achieves phase variation by two different mechanisms, involving covalent DNA changes but not DNA inversion. The first involves the genes encoding the pilus, which promotes pathogenesis by allowing the bacterium to persist in the urethra by adhering to tissue. *N. gonorrhoeae* switches between pilus⁺ and pilus⁻ states. In addition, the genes expressed in the pilus state can encode diverse pilus proteins. These rearrangements may help evade the host immune response. Typically, only one pilus gene is expressed, although many copies reside in the *N. gonorrhoeae* chromosome. Many of the nonexpressed copies are incomplete, either deleted or truncated by stop codons. The expressed gene copy is periodically altered by gene conversion events. If an inactive sequence is copied into the expressed locus, for example a sequence containing a stop codon, then the pilus⁺ strain is converted to pilus⁻. If a functional coding region is

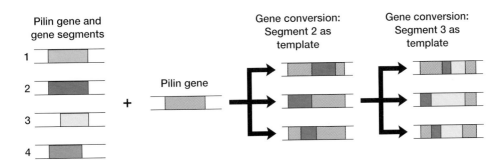

FIGURE 11.11. Pilin gene sequence variation mediated by gene conversion. The *N. gonorrhoeae* genome contains many inactive pilin gene copies (numbered 1–4) in addition to the active copy (*green*). In the example, gene conversion transfers information from gene 2 to the expressed pilin gene copy; the changes are heterogeneous within the *N. gonorrhoeae* population. Further conversion, using gene 3 information, substitutes additional new sequences in the expressed locus. (Modified, with permission, from Swanson and Koomey 1989. [44])

copied in, the gene remains pilus[+], but the encoded protein will reflect the new hybrid coding region. Analysis of the active pilus gene coding regions over time shows an accumulation of overlapping conversion tracts, resulting in a complex chimeric pattern (Fig. 11.11). Curiously, the conversion of pilin gene sequences may be linked to DNA transformation. *Neisseria* is naturally competent for transformation throughout its life cycle. A few cells in the population undergo autolysis, releasing DNA that can be taken up by neighboring cells. Studies by H. Seifert and coworkers have established that pilin gene conversion is linked to DNA transformation, at least under some conditions in the laboratory.

The second means of immune evasion involves switching of the P.II outer membrane protein. Individual bacteria in a population can express either no P.II variants, one, or more than one. In a typical bacterial population, 0.1–1% of the members will express a variant different from the majority. In vivo the minority variant would presumably be periodically selected when the bacteria expressing the majority form are cleared by a successful immune response. There are estimated to be 7–10 genes for P.II proteins that are turned on and off periodically. The mechanism for gene activation involves localized covalent changes in DNA at the start of the P.II coding regions. These regions contain pentanucleotide sequences (5′-CTTCT-3′) typically repeated 4–27 times (Fig. 11.12). The number of repeats in front of a given gene is variable, reflecting probable polymerase "slipping" during DNA synthesis. Changing the number of repeats results in changing the reading frame in the downstream gene coding region. Because only one of the three reading frames encodes the correct P.II gene, expansion or contraction of the repeated region results in activation or inactivation of P.II gene expression. It appears that the pentanucleotide repeat sequence is particularly prone to expansion or contraction during replication, ensuring frequent variation of the P.II surface protein.

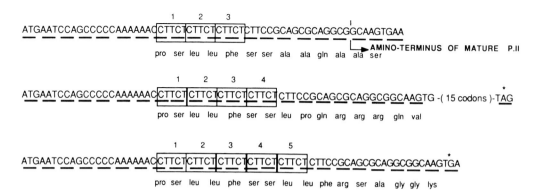

FIGURE 11.12. Pentanucleotide repeat variation results in activation or inactivation of copies of the gene for the P.II surface protein. Errors during replication result in changes in the number of pentanucleotide repeats in the 5′ coding region of the P.II gene. Because the 5′-CTTCT-3′ repeat is not a multiple of 3 nucleotides, altering the number of repeats alters the reading frame of the downstream gene. Only the top gene encodes the correct P.II protein. This process, active over many P.II genes in the population, results in continuous replacement of surface antigens. (Reprinted, with permission, from Swanson and Koomey 1989. [44])

Analysis of genome sequences suggests that the pathogens *Neisseria meningitidis* and *Campylobacter jejuni*, among others, also use this strategy for immune evasion.

Switching of Trypanosome Variant Surface Glycoproteins (6, 29)

Trypanosomes, unicellular protozoal parasites, inhabit the bloodstream of their eukaryotic hosts and so must also elude immune clearance. Trypanosomes do so by periodically switching their variant surface glycoprotein (VSG) coats.

Early studies of trypanosomes revealed that the numbers in the bloodstream of an infected person or animal fluctuated dramatically (Fig. 11.13). Initially the parasite grows to high levels but then plummets in abundance. The fall parallels the development of the host immune response, accounting for the clearance of the trypanosome. A week or two later, the trypanosomes achieve their former abundance, but again fall. Each new peak in abundance is accompanied by the appearance of a new form of the VSG coat, and each fall is paralleled by an immune response against that coat. On average, *Trypanosoma brucei* spontaneously switches VSG coat 10^{-6} to 10^{-7} times per cell division. Thus, given the numbers of trypanosomes present in a large animal, there will be a few trypanosomes with switched VSG coats present at any given time to reseed an infection after immune clearance. In one case, trypanosomes in an experimentally infected rabbit were shown to express more than 100 different VSG coats in succession.

About 10^7 copies of the VSG protein coat the surface of the trypanosome, masking it from the host immune system. The try-

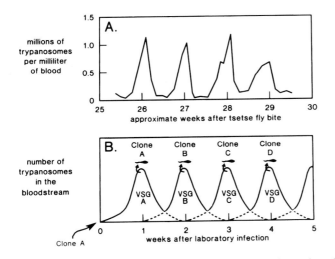

FIGURE 11.13. Abundance of trypanosomes in the blood of an infected person (*A*) or laboratory animal (*B*) as a function of time after infection. The numbers of trypanosomes increase dramatically initially, but then fall as the concentration of antibodies against the initial VSG protein increases. One or more trypanosomes bearing new VSG proteins are present in the population due to VSG switching, allowing trypanosomes bearing the new VSG to increase in numbers. Eventually the antibody response develops against the second VSG, causing the numbers of trypanosomes to fall, until a new variant again arises. This process can be repeated many times during the course of an infection. (Reprinted, with permission, from Donelson 1989. [9])

panosome genome is estimated to contain up to 1000 different genes for VSG proteins, but typically only one is expressed at any one time. The VSG variant expressed is switched periodically by a unique mechanism involving covalent rearrangements of DNA.

The expressed VSG gene copy is found near the telomere of a trypanosome chromosome. Mechanisms by which VSG and flanking genes are switched are diagramed in Figure 11.14. The simplest mechanism involves replacement of the expressed telomeric genes by genes from a silent site by gene conversion (Fig. 11.14A). The transfer of information is nonreciprocal, because the silent copy is transferred to the expression site, but the formerly expressed copy is lost. There are actually several telomeric expression sites, although usually only one is active (Fig. 11.14, B–D). This creates the possibility for several further modes of VSG gene activation. In telomere conversion, a silent telomere site may recombine with the active telomere site to turn on a previously silent telomere-associated site. This is a nonreciprocal event, because the formerly expressed sequence at the active site is lost. Activation can also take place via reciprocal telomere exchange, in which both copies of the gene are retained. Finally, the active telomeric site may become inactive and another site may be activated, a mechanism called in situ switching. All of these reactions serve to emphasize the importance of immune evasion by trypanosomes.

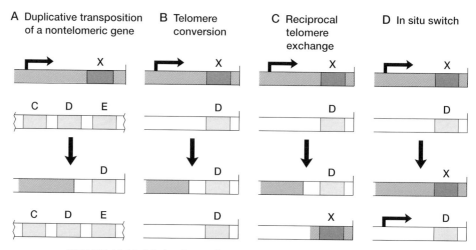

FIGURE 11.14. Mechanisms of VSG gene switching. The expressed gene copy is shown in green; other VSG genes elsewhere in the genome are shown as gray. (*A*) Duplicative transposition of an inactive VSG gene located at an internal chromosomal site into the expressed telomeric site allows a new gene to be expressed (gene D replaces gene X in the example). (*B*) Conversion of a VSG gene at an inactive telomere site (D) to the active site (X) results in expression of gene D. (*C*) Reciprocal exchange between telomeres, one baring an active gene (X) and the other an inactive gene (D), results in activation of D. (*D*) Two VSG genes resident at chromosomal termini can switch in activity, so that one becomes active and the other inactive. (Modified, with permission, from McCulloch et al. 1997 [© American Society for Microbiology]. [29])

Summary

Microbial pathogens would quickly overwhelm us if not for the intervention of our immune systems. Recognition of the myriad antigens (antibody-generating molecules) of invaders encountered in the life of an organism requires that antibodies and TCRs possess very large numbers of binding specificities. The numbers required are much too large for single genes to encode each molecule individually. Rather, sets of DNA segments are rearranged to form antigen-binding sites, greatly increasing the numbers by combinatorial diversity. Similar machinery is used to assemble gene segments at both the antibody and TCR loci. Remarkably, this machinery is very likely derived from a DNA transposon, potentially captured by lateral DNA transfer.

Antigen-binding proteins are assembled from V region (variable) and J region (joining) segments. For each, there are several to several hundred gene segments, but each mature antigen-binding protein typically has only one of each. In some of the loci, there is a third D region (diversity) that is incorporated between the V and J regions, contributing further diversity. Each gene segment is flanked by recombination sites, consisting of conserved heptamer and nonamer sequences. The spacers come in two lengths, 12 and 23 base pairs, and recombination only proceeds between a 12-mer and 23-mer signal.

Recombination is carried out by the RAG products RAG1 and RAG2. The RAGs bind to the conserved sites and cleave at the edges of the coding regions. Cleavage is a two-step process, in which one DNA strand is nicked, and then the free end attacks the other strand, forming a hairpin at one end while simultaneously breaking the DNA. Cleavage requires a cofactor, an architectural DNA-binding protein of the HMG 1 family, which helps shape the DNA into a conformation favorable for action of the RAG proteins. The free DNA ends are joined by the host cell nonhomologous DNA end-joining system.

The function of the VDJ system closely parallels the function of DNA transposons. The unrearranged loci resemble a cellular gene region containing inserted defective transposons. The involvement of DNA hairpins is common to VDJ recombination and transposition by a variety of transposons. The linkage of RAG1 and RAG2 in the chromosome is odd, but not if they entered the genome as components of a single transposon. Most strikingly, the RAG genes can carry out a reaction identical to transposition in vitro, in which a pair of excised signal ends are inserted into a DNA target exactly as with transposon DNA integration. Taken together, these data strongly argue that the progenitor of the vertebrate immune system was in fact a DNA transposon. The origin of the transposon is unknown. The adaptive immune system is only present in jawed fishes and their more advanced relatives, consistent with recent capture by lateral transfer. The VDJ recombination system provides a dramatic example of recruitment of mobile DNA machinery to generate evolutionary novelty.

The pathogens, the other side of the immune battle, respond with their own DNA rearrangements. Many bacteria engage in phase variation, the varying of surface proteins exposed to the immune system. *E. coli* and *S. typhimurium* use inversion systems to switch between stable states of gene expression. *N. gonorrhoeae* changes the expressed P.II antigen by a different mechanism, involving inaccurate replication of pentanucleotide repeats in the leader region of P.II genes, thereby switching the encoded reading frame. The *N. gonorrhoeae* pilus is also varied during infection by a nonreciprocal gene conversion mechanism, one that seems to be linked to capture of new sequences by transformation. The trypanosomes, protozoal parasites of mammals, encode on the order of 1000 genes for their surface protein VSG. Switching expression takes place by a complex group of mechanisms, including introduction of new sequences into a telomeric expression site by gene conversion.

REFERENCES

1. Agrawal A. and Schatz D. 1997. RAG1 and RAG2 form a stable post-cleavage synaptic complex with DNA containing signal ends in V(D)J recombination. *Cell* **89:** 43–53.
2. Agrawal A., Eastman Q.M., and Schatz D.G. 1998. Transposition mediated by RAG1 and RAG2 and its implications for the evolution of the immune system. *Nature* **394:** 744–751.
3. Alberts B., Bray D., Lewis J., Raff M., Roberts K., and Watson J.D. 1994. *Molecular biology of the cell*. Garland Publishing, New York.
4. Baron L.S., Kopecko D.J., McCowen S.M., Snelling N.J., Johnson E.M., et al. 1982. Genetic and molecular studies of the regulation of atypical citrate utilization and variable Vi antigen expression in enteric bacteria. In *Genetic engineering of microorganisms for chemicals* (ed. A. Hollander and R. DeMoss), pp. 175–194. Academic Press, New York.
5. Bartlett D.H., Wright M.E., and Silverman M. 1988. Variable expression of polysaccharide in the marine bacterium *Pseudomonas atlantica* is controlled by genome rearrangement. *Proc. Natl. Acad. Sci.* **85:** 3923–3927.
6. Borst P. and Fairlamb A.H. 1998. Surface receptors and transporters of Trypanosoma brucei. *Annu. Rev. Microbiol.* **52:** 745–778.
7. Brinton Jr., C.C. 1959. Nonflagellar appendages of bacteria. *Nature* **183:** 782–786.
8. Davis M.M., Calame K., Early P.W., Livant D.W., Joho R., Weissman I.L., and Hood L. 1980. An immunoglobulin heavy-chain gene is formed by at least two recombinational events. *Nature* **283:** 733–739.

9. Donelson J.E. 1989. DNA rearrangements and antigenic variation in African trypanosomes. In *Mobile DNA* (ed. D.E. Berg and M.M. Howe), pp. 763–781. American Society for Microbiology, Washington, D.C.

10. Eastman Q.M., Leu T.M., and Schatz D. 1996. Initiation of V(D)J recombination in vitro obeying the 12/23 rule. *Nature* **380:** 85–88.

11. Fugmann S., Villey I., Ptaszek L., and Schatz D. 2000. Identification of two catalytic residues in RAG1 that define a single active site within the RAG1/RAG2 protein complex. *Mol. Cell* **5:** 97–107.

12. Fugmann S., Lee A., Shockett P., Villey I., and Schatz D. 2000. The RAG proteins and V(D)J recombination: Complexes, ends, and transposition. *Annu. Rev. Immunol.* **18:** 495–527.

13. Glasgow A.C., Hughes K.T., and Simon M.I. 1989. Bacterial DNA inversion systems. In *Mobile DNA* (ed. D.E. Berg and M.M. Howe), pp. 637–659. American Society for Microbiology, Washington, D.C.

14. Grawunder U., Zimmer D., Fugmann S., Schwarz K., and Lieber M.R. 1998. DNA ligase IV is essential for V(D)J recombination and DNA double-strand break repair in human precursor lymphocytes. *Mol. Cell* **2:** 477–484.

15. Guerry P., Logan S.M., and Trust T.J. 1988. Genomic rearrangements associated with antigenic variation in *Campylobacter coli. J. Bacteriol.* **170:** 316–319.

16. Haas R. and Meyer T.F. 1986. The repertoire of silent pilus genes in *Neisseria gonorrhoeae:* Evidence for gene conversion. *Cell* **44:** 1007–1015.

17. Harris L.A., Logan S.M., Guerry P., and Trust T.J. 1987. Antigenic variation of *Campylobacter* flagella. *J. Bacteriol.* **169:** 5066–5071.

18. Harris L.J., Skaletsky E., and McPherson A. 1998. Crystallographic structure of an intact IgG1 monoclonal antibody. *J. Mol. Biol.* **275:** 861–872.

19. Hesse J.E., Lieber M.R., Gellert M., and Mizuuchi K. 1987. Extrachromosomal DNA substrates in pre-B cells undergo inversion or deletion at immunoglobulin V(D)J joining signals. *Cell* **49:** 775–783.

20. Hiom K., Melek M., and Gellert M. 1998. DNA transposition by the RAG1 and RAG2 proteins: A possible source of oncogenic translocations. *Cell* **94:** 463–470.

21. Hozumi N. and Tonegawa S. 1976. Evidence for somatic rearrangement of immunoglobulin genes coding for variable and constant regions. *Proc. Natl. Acad. Sci.* **73:** 3628–3632.

22. Janeway Jr., C.A., Travers P., Walport M., and Capra J.D. 1999. *Immunobiology: The immune system in health and disease*, 4th edition. Elsevier, London.

23. Johnson R.C., Bruist M.F., and Simon M.I. 1986. Host protein requirements for in vitro site-specific DNA inversion. *Cell* **46:** 531–539.

24. Landree M., Wibbenmeyer J., and Roth D. 1999. Mutational analysis of RAG1 and RAG2 identifies three catalytic amino acids in RAG1 critical for both cleavage steps of V(D)J recombination. *Genes Dev.* **13:** 3059–3069.

25. Lewis S., Gifford A., and Baltimore D. 1985. DNA elements are asymmetrically joined during the site-specific recombination of κ immunoglobulin genes. *Science* **228:** 677–685.

26. Lieber M.R. 1998. Pathological and physiolgical double-strand breaks. *Am. J. Pathol.* **153:** 1323–1332.

27. Max E.E., Seidman J.G., and Leder P. 1979. Sequences of five potential recombination sites encoded close to an immunoglobulin k constant region gene. *Proc. Natl. Acad. Sci.* **76:** 616–620.

28. McBlane J.F., van Gent D.C., Ramsden D.A., Romeo C., Cuomo C.A., Gellert M., and Oettinger M. 1995. Cleavage at a V(D)J recombination signal requires only RAG1 and RAG2 proteins and occurs in two steps. *Cell* **83:** 387–395.

29. McCulloch R., Rudenko G., and Borst P. 1997. Gene conversions mediating antigenic variation in *Trypanosoma brucei* can occur in variant surface glycoprotein expression sites lacking 70-base-pair repeat sequences. *Mol. Cell. Biol.* **17:** 833–843.

30. Meier J.T., Simon M.I., and Barbour A. 1985. Antigenic variation is associated with DNA rearrangements in a relapsing fever *Borrelia*. *Cell* **41:** 403–409.

31. Oettinger M., Schatz D., Gorka C., and Baltimore D. 1990. RAG-1 and RAG-2, adjacent genes that synergistically activate V(D)J recombination. *Science* **248:** 1517–1523.

32. Plasterk R. 1998. Ragtime jumping. *Nature* **394:** 718–719.

33. Plasterk R., Simon M.I., and Barbour A. 1985. Transposition of structural genes to an expression sequence on a linear plasmid causes antigenic variation in the bacterium *Borrelia hermsii*. *Nature* **318:** 257–263.

34. Plasterk R., Simon M.I., and Barbour A. 1986. Molecular basis for antigenic variation in a relapsing fever *Borrelia*. In *Antigenic variation in infectious diseases* (ed. T.H. Birkbeck and C.W. Penn), vol. 19, pp. 127–146. IRL Press, Oxford, United Kingdom.

35. Ramsden D.A., Paull T., and Gellert M. 1997. Cell-free V(D)J recombination. *Nature* **388:** 488–491.

36. Robinson A., Duggleby C.J., Gorringe A.R., and Livey I. 1986. Antigenic variation in *Bordetella pertussis*. In *Antigenic variation in infectious diseases* (ed. T.H. Birkbeck and C.W. Penn), vol. 19, pp. 147–163. IRL Press, Oxford, United Kingdom.

37. Schatz D., Oettinger M., and Baltimore D. 1989. The V(D)J recombination activating gene, RAG-1. *Cell* **59:** 1035–1048.

38. Segal E., Hagblom P., Seifert H.S., and So M. 1986. Antigenic variation of gonococcal pilus involves assembly of separated silent gene segments. *Proc. Natl. Acad. Sci.* **83:** 2177–2181.

39. Segal E., Billyard E., So M., Storzbach S., and Meyer T.F. 1985. Role of chromosomal rearrangement in *N. gonorrhoeae* pilus variation. *Cell* **40:** 293–300.

40. Seifert H., Ajioka R., Marchal C., Sparling P., and So M. 1988. DNA transformation leads to pilin antigenic variation in *Neisseria gonorrhoeae*. *Nature* **336:** 392–395.

41. Snellings N.J., Johnson E.M., Kopecko D.J., Collins H.H., and Baron L.S. 1981. Genetic regulation of variable Vi antigen expression in a strain of *Citrobacter freundii*. *J. Bacteriol.* **145:** 1010–1017.

42. Stern A., Brown M., Nickel P., and Meyer T.F. 1986. Opacity genes in *Neisseria gonorrhoeae:* Control of phase and antigenic variation. *Cell* **47:** 61–71.

43. Stern A., Nickel P., Meyer T.F., and So M. 1984. Opacity determinants of *Neisseria gonorrhoeae:* Gene expression and chromosomal linkage to the gonococcal pilus gene. *Cell* **37:** 447–456.

44. Swanson J. and Koomey J.M. 1989. Mechanisms for variation of pili and outer membrane protein II in *Neisseria gonorrhoeae*. In *Mobile DNA* (ed. D.E. Berg and M.M. Howe), pp. 743–761. American Society for Microbiology, Washington, D.C.

45. Swanson J., Bergstrom O., Barrera O., Robbins K., and Corwin D. 1985. Pilus gonococcal variants. Evidence for multiple forms of piliation control. *J. Exp. Med.* **162:** 729–744.

46. Swanson J., Bergstrom S., Robbins K., Barrera O., Corwin D., and Koomey J.M. 1986. Gene conversion involving the pilin structural gene correlates with pilus+ pilus– changes in *Neisseria gonorrhoeae. Cell* **47:** 267–276.

47. Taccioli G.E., Rathbun G., Oltz E., Stamato T., Jeggo P.A., and Alt F.W. 1993. Impairment of V(D)J recombination in double-strand break repair mutants. *Science* **260:** 207–210.

48. Taccioli G.E., Gottlieb T.M., Blunt T., Priestley A., Demengeot J., Mizuta R., Lehmann A.R., Alt F.W., Jackson S.P., and Jeggo P.A. 1994. Ku80: Product of the XRCC5 gene and its role in DNA repair and V(D)J recombination. *Science* **265:** 1442–1445.

49. Tonegawa S. 1983. Somatic generation of antibody diversity. *Nature* **302:** 575–581.

50. Tonegawa S., Maxam A.M., Tizard R., Bernard O., and Gilbert W. 1978. Sequence of a mouse germ-line gene for a variable region of an immunoglobulin light chain. *Proc. Natl. Acad. Sci.* **75:** 1485–1489.

51. van Gent D.C., Mizuuchi K., and Gellert M. 1996. Similarities between initiation of V(D)J recombination and retroviral integration. *Science* **271:** 1592–1594.

52. van Gent D.C., Ramsden D.A., and Gellert M. 1996. The RAG1 and RAG2 proteins establish the 12/23 rule in V(D)J recombination. *Cell* **85:** 107–113.

53. van Gent D., Hiom K., Paull T., and Gellert M. 1997. Stimulation of V(D)J cleavage by high mobility group proteins. *EMBO J.* **16:** 2665–2670.

54. van Gent D.C., McBlane J.F., Ramsden D.A., Sadofsky M., Hesse J., and Gellert M. 1995. Initiation of V(D)J recombination in a cell-free system. *Cell* **81:** 925–934.

DNA Transfer among the Domains of Life

P REVIOUS CHAPTERS PRESENTED EXAMPLES of horizontal transfer between distantly related organisms, such as transfer of conjugative transposons between gram-positive and gram-negative bacteria or movement of *mariner* elements among eukaryotes. This chapter investigates the most extreme cases of all, the transfer of DNA between the domains of life.

All life is divided into three domains, the Bacteria, Eucarya, and Archaea. The Bacteria encompass all prokaryotes (organisms without a defined nucleus) except the Archaea. Included in the Bacteria are gram-positive and gram-negative bacteria, mycoplasms, cyanobacteria, and a few others. Archaea (or Archaebacteria) are prokaryotes that differ from the Bacteria in their ribosomal RNA sequences, membrane lipids, prominent introns, and a variety of other features. The Eucarya consist of organisms with cells containing nuclei and a suite of other characteristic features. The Eukaryotes include the fungi, plants, and animals. As discussed in earlier chapters, the boundaries between taxonomic groups are increasingly blurred by the recognition of extensive lateral transfer. Divisions at the domain level remain relatively secure, although here too, lateral transfer complicates the picture.

This chapter begins with an account of DNA transfer into plants by the Agrobacteria, which program plants to grow structures the bacteria then inhabit. Studies of gene transfer between domains by conjugation, a related mechanism, are then covered. Later sections treat DNA transfer events that are less common, so that sequence accumulation is typically apparent only on evolutionary time scales. Genes are known to flow from the genomes of eukaryotic organelles, the mitochondria and chloroplasts, to the nuclear genome. This provides a means for the Bacteria-derived organelle sequences to become incorporated into the eukaryotic nucleus. The remarkable final picture is one of robust gene flow even among creatures of different domains.

DNA TRANSFER FROM *AGROBACTERIUM* TO PLANTS (8, 15, 28, 36, 39)

Crown Gall Formation

The results of DNA transfer from bacteria to plants can be seen on a walk through an orchard. Many fruit trees have swellings, typically where the trunk enters the soil, called crown galls (Fig. 12.1). The galls occur at sites where wounds to the tree became infected with a soil bacterium called *Agrobacterium tumefaciens*. A complex interplay between infecting *Agrobacterium* and the host plant results in the development of the gall. Central to this process is the transfer of DNA from the bacteria to the plant nuclear genome. A related bacterium, *Agrobacterium rhizogenes*, also transfers DNA into plants, which in this case genetically modifies root cells and causes hairy root disease.

Gall formation is mediated by transfer of a segment of DNA, called the T-DNA, from *Agrobacterium* to the plant. The name derives from "*Tumor*," reflecting the resemblance between plant galls and vertebrate tumors. The steps leading to T-DNA transfer require intricate signaling between the bacteria and the plant. Following transfer, the T-DNA becomes incorporated into the plant cell. Overproduction of the T-DNA-encoded gene products, which include plant growth hormones, together with plant cell genes, then directs the development of

FIGURE 12.1. Crown galls on a willow tree. (Courtesy W. Merrill–from the Department of Plant Pathology teaching collection, Pennsylvania State University; see Atlas 1984. [1])

the gall. T-DNA transfer results in phenotypic transformation of infected plant cells into crown gall tumor cells. As with cells from many mammalian tumors, T-DNA-transformed cells can grow and divide in culture with reduced dependence on growth factors (auxins and cytokinins in the case of plants).

Gall formation benefits *Agrobacterium* in at least two ways. The environment of the gall provides a safe haven for *Agrobacterium* cells, which otherwise would need to compete with myriad bacteria in the soil. The gall also synthesizes the nutrients octopine or nopoline under the direction of genes encoded in the T-DNA. *Agrobacterium* thus genetically engineers its host to provide high-end housing and meals as well!

T-DNA and Genetic Engineering

Today *Agrobacterium* provides a key tool for genetic engineering of plants. New genes can be incorporated into the T-DNA in *Agrobacterium* by recombinant DNA techniques, then the engineered *Agrobacterium* is exposed to the plant cells of interest. The engineered T-DNA becomes transferred to plant cells as with normal T-DNA transfer, resulting in the stable introduction of new genes in the plant cells. Under laboratory conditions, new plants can be grown from the genetically modified cells. Many crop plants have been modified by this method to express desirable traits such as increased disease resistance or improved nutritional value.

The Ti Plasmid

Much of the machinery for *Agrobacterium* DNA transfer is encoded on a plasmid, called the Ti plasmid for "*T*umor *i*nducing" (Fig. 12.2). The Ti plasmids are large, ranging in size from about 140 to 500 kilobases.

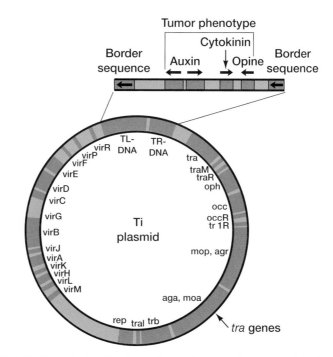

FIGURE 12.2. Ti plasmid, showing the coding regions for *vir* genes and the location of the T-DNA. (Modified, with permission, from Kaper and Hacker 1999. [16])

Encoded on the plasmid are the T-DNA itself and genes for T-DNA transfer functions, called *vir* genes (for *vir*ulence).

The T-DNA encodes several genes important for tumor development, including genes influencing the auxin- and cytokinin-mediated signaling pathways that regulate plant growth. These genes contain regulatory regions that permit them to be active after transfer to the plant cell, a remarkable adaptation for a bacterial DNA sequence. Ti plasmids also encode genes for synthesis of either of two opines, octopine or nopoline, that cause the plant cell to synthesize nutrients used by the *Agrobacterium*. Curiously, the octopine plasmids actually have multiple clustered T-DNA regions instead of the single one found in the nopoline plasmids.

The T-DNA region is flanked by 25 bp of directly repeated border sequences that serve as binding sites for the proteins that direct DNA transfer. A DNA site just outside the T-DNA region increases the efficiency of transfer. This site has been given the colorful name *overdrive* to reflect this stimulatory role.

The *vir* genes are clustered on the Ti plasmid near the T-DNA region. Many of the *vir* genes are collected in six operons, *virA–virE* and *virG*. The *virA* and *virG* products encode a sensor–activator pair responsible for detecting the presence of a wounded plant and activating expression of the *vir* genes in response. The *virD* and *virB* gene products are the machinery for DNA transfer itself. The *virE* and *virC* gene products contribute to virulence and influence the range of

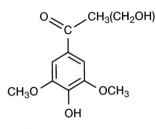

Acetosyringone
(hydroxy-acetosyringone)

FIGURE 12.3. The inducers of *vir* gene expression AS and hydroxy-AS. These or related compounds are secreted by wounded plants, signaling to *Agrobacteria* an opportunity to invade the wound and initiate crown gall formation. (Redrawn, with permission, from Zambryski 1989. [39])

species to which transfer can take place, but are not strictly required for the transfer process. A few other genes have also been found that contribute functions important for transfer.

Also carried on the Ti plasmid are genes for more conventional conjugative transfer between bacterial cells. These can direct transfer of the whole Ti plasmid between *Agrobacterium*, providing a means of mobilizing the Ti plasmid machinery.

Still another set of genes encoded on the Ti plasmid is responsible for catabolism of plant products induced by opines. These allow the bacterium to digest the compounds synthesized by the plants under the direction of the T-DNA.

Plant–Bacterial Recognition

Crown gall formation begins with the detection of a wounded plant. Plant cells are surrounded by a tough cell wall, composed in part of polymers of phenolic compounds called lignins. Wounding leads to a breakdown of some of these compounds, yielding acetosyringone (AS) and hydroxy-AS among other products (AS; Fig. 12.3). AS activates *vir* gene expression in *Agrobacterium* through the *virA* and *virG* protein products. The *virA* and *virG* products comprise a "two-component system," a type of genetic switch common in bacteria. Two-component systems are composed of a sensor protein (the *virA* product) that usually resides in the cell membrane. Upon receipt of an appropriate signal, the sensor protein, which also possesses kinase activity, phosphorylates the response protein (the *virG* product). Phosphorylation takes place on histidine residues, an unusual reaction characteristic of these two-component systems. The phosphorylated *virG* protein is activated to bind DNA and stimulate transcription of *vir* genes, which contain appropriate binding sites. The *virA* kinase probably binds AS and phosphorylates the *virG* protein, causing it to bind upstream of the *vir* genes and activate transcription of the *vir* gene products.

For transfer to take place, *Agrobacterium* must also bind to a target cell and transfer DNA through a cell-surface pore. *Agrobacteria* encode pili responsible for binding the two partner cells. The pilus and DNA pore are formed primarily by the *virB*-encoded proteins, which are abundant and located at the cell membrane. Other *vir* genes contribute to transfer, but their roles are less well understood. The *virF*- and *virH*-encoded proteins are required for transfer to some but not all plant species. The *virH* protein bears some resemblance to the P450 cytochrome enzymes, which are known to detoxify poisons in other organisms. The *virH* protein has been proposed on this basis to detoxify plant compounds otherwise poisonous to *A. tumefaciens*.

DNA Breaking and Joining Reactions Mediating T-DNA Transfer

The enzymatic reactions mediating T-DNA transfer have many parallels with bacterial conjugation (Fig. 12.4 and Plate 9). Transfer initiates with nicking at one border of the T-DNA, at the opposite end of the T-DNA from the *overdrive* element. After nicking, the T-DNA strand becomes unpaired, probably aided by DNA helicases and single-stranded DNA-binding proteins that stabilize single-stranded DNA. *Agrobacterium* encodes a special single-stranded DNA-binding protein, *virE*, which serves this role.

As with conjugation, the transfer mechanism involves concomitant replication of the T-DNA. DNA synthesis initiates at the free 3′ end generated by nicking, allowing replacement of the DNA strand that is transferred to the plant cell. The DNA synthesis reaction may also help displace the strand to be transferred from the complementary DNA, as a DNA polymerase traversing a DNA sequence can displace a base-paired sequence ahead of it. Transfer takes place in the 5′ to 3′ direction, the same polarity of DNA transfer as with conjugation.

The products of the *virD* region are central to T-DNA transfer. *virD* encodes four proteins, VirD1–VirD4, all of which are required. VirD2 binds to the 25-bp repeats at the border of the T-DNA, then VirD1 and VirD2 together cleave one DNA strand. The VirD2 protein remains associated with the cleaved DNA 5′ end after cleavage, much as with the "relaxosome" proteins that mediate conjugation. The bound VirD2 protein probably serves to direct the DNA to the appropriate sites on the bacterial cell surface, and then escorts the T-DNA into the plant cell. The *virD3* and *virD4* products are also known to be required, but their roles are less clear. The VirC proteins contribute by binding to the *overdrive* enhancer and increasing the efficiency of the cleavage reaction.

The displacement–resynthesis reaction terminates when the second T-DNA border is reached. A second nicking reaction then takes place, freeing the T-DNA strand. This termination step differs from bacterial conjugation, because in the latter transfer there is no special termination step, and transfer of much larger DNAs can take place. For

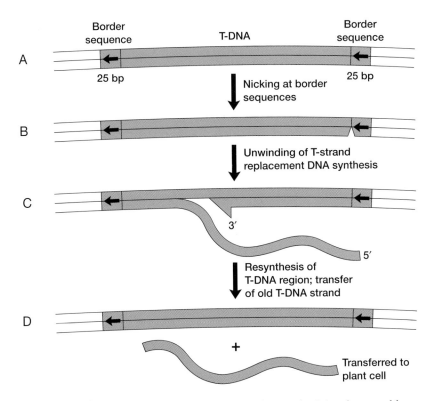

FIGURE 12.4. Schematic diagram of T-DNA transfer, emphasizing the resemblance to bacterial conjugation. (*A*) The T-DNA region resident in the Ti plasmid. (*B*) Transfer is initiated by nicking at the right border sequence. (*C*) DNA unwinding "peels off" the single strand that becomes transferred to the recipient cell, and replacement DNA synthesis restores the displaced strand. (*D*) The free single strand becomes transferred to the recipient cell. (Modified, with permission, from Zambryski 1989. [39])

Agrobacterium, too, there are probably infrequent cases in which termination does not take place correctly, leading to the transfer of longer sequences containing the T-DNA flanking region.

Incorporating T-DNA into the Plant Cell

Once a DNA molecule has exited the *Agrobacterium* cell and entered the plant, the single-stranded T-DNA must be incorporated into a chromosome of the plant host. This is one of the least understood aspects of T-DNA transfer. Following entry into the plant cell, the covalently bound virD2 product accompanies the T-DNA to the nucleus. The virD2 product appears to encode a nuclear localization signal, which serves as a "zip code" that routes the virD2/T–DNA complex to the nucleus. Throughout the process, the transferred strand is coated by the virE2 protein, which also forms a pore in the plant cell membrane through which the T-DNA passes on the way in.

This process has been named "illegitimate" recombination to reflect the lack of sequence homology between the incoming DNA and the site of integration.

The T-DNA sequences are generally integrated into the plant DNA as intact strands and not rearranged during transfer. The sequences at the right 25-bp border repeat tend to be intact, containing at most deletions of a few base pairs. Sequences at the left end often contain larger deletions, possibly reflecting different processing in the absence of the covalently bound VirD protein. There is no strong preference for specific target site sequences, ruling out site-specific recombination as the mechanism of integration. Different numbers of T-DNA segments are found in transformed plant cells, and the flanking plant DNA occasionally sustains rearrangements. These features mimic those seen upon incorporation of naked DNA into eukaryotic cells after transfection, suggesting that host cell systems recognize and integrate the T-DNA as with any newly introduced DNA.

Much remains to be understood about the nuclear localization of the T-DNA, joining of T-DNA strands to host chromosomes, synthesis of the second DNA strand, and repair of integration intermediates.

T-DNA Transfer and Type IV Secretion Systems (2a, 4a)

The T-DNA transfer system is a member of a large group of prokaryotic machines dedicated to transferring macromolecules into cells. The type IV export systems are virulence factors for a number of pathogens, such as *Brucella suis* (brucellosis), *Bordetella pertussis* (whooping cough), *Helicobacter pylori* (gastric ulcers), *Legionella pneumophila* (Legionnaire's disease), and *Rickettsia prowazekii* (epidemic typhus). For some of these, the mechanism of pathogenesis has been traced to transport of toxins into the host cells, similar to transfer of T-DNA into plants. In many cases, the detailed mechanism is unknown but is assumed to involve transport of some material. The type IV secretion systems are recognizable in DNA sequence as clustered groups of related genes similar to the *Agrobacteria vir* genes (Plate 9).

Emphasizing the similarities, Frank Cannon and coworkers demonstrated that substituting the *oriT* and relaxosome machinery of the conjugative plasmid pRSF1010 for that of the Ti plasmid allowed the T region to be transferred to plants. The reaction still required the Ti *vir* genes. This study argued strongly that the *virD* DNA nicking/transfer system is functionally identical to that of conjugative plasmids, whereas other *vir*-encoded proteins dictate the specificity for plants.

T-DNA in the Germ Line of Modern Plants? (11)

So far, we have reviewed examples of agrobacteria-mediated gene transfer into somatic tissues, but what about stable modification of the germ line? Agrobacteria-mediated gene transfer is used routinely in the laboratory to create heritable modifications in plants, but has its T-

DNA become integrated into the plant germ line in natural settings? Several studies suggest that it has, at least in a few cases.

Closely related sequences have been found in the *Agrobacterium rhizogenes* T-DNA and in plant nuclear genomes. *A. rhizogenes* contains an Ri plasmid ("root inducing") that is functionally similar to the *A. tumefaciens* Ti plasmid. Transfer of the Ri T-DNA results in formation of root teratomas at the infection site, which are associated with a syndrome that includes wrinkled leaves and small flowers. Genes that cause the hairy root disease—*rolA, rolB, rolC, rolD, orf1, 3*, and *orf14*—are encoded in a cluster in the Ri T-DNA.

Similar gene clusters have been found in the genomes of diverse plants, including petunia, carrot, and six *Nicotiana* (tobacco) species. The encoded proteins have sequence identity ranging from 55% to 89%, well above the range expected for accidental matches. The gene order seen in the T-DNA is also preserved in the plant gene clusters, leaving little doubt that the *rol* gene clusters were dispersed by horizontal gene transfer.

But in which direction were the genes transferred? *Agrobacterium* might have picked up a set of plant genes, or the plant species might have independently received new sequences from *Agrobacterium*. There is evidence to support both possibilities. Potentially in support of transfer from plants to *Agrobacterium* is the observation that the *rol* genes in T-DNA have promoters active in plants, and one gene, *rolA*, even has an intron, all characteristics of eukaryotic genes. In favor of transfer from *Agrobacterium* to plants is the apparent lack of *rol* genes in at least one tobacco species, seemingly simpler to explain as independent recent transfer into the germ lines of different plant species, and the known ability of T-DNA to transfer from bacteria to plants.

In principle, this question could be resolved by mapping the integration sites of the *rol* gene clusters in each plant species. Because T-DNA is integrated apparently randomly into the host chromosome, independent transfer events should result in integration at different sites. If the *rol* genes are ancient plant genes inherited by descent, they should reside in the same chromosomal region in many plant species. Whichever the direction, it seems clear that the *rol* gene cluster has been transferred between domains of life and incorporated as a heritable component.

Transfer of DNA by *Agrobacterium* across Very Large Phylogenetic Distances (3, 5, 17, 24, 27, 32)

The Agrobacteria display a very diverse host range, causing crown gall disease in at least 93 different families of plants. This observation, together with the studies described below, prompted investigators to explore the limits of T-DNA transfer. Remarkably, these studies have revealed that *Agrobacterium* can transfer genes to an extremely wide variety of cells, including cells from humans.

Experiments were designed in which T-DNA was modified to contain markers active upon transfer to new types of cells. *Agrobacterium* was then incubated with target cells, and transfer of the engineered T-DNA was assessed. To promote stable incorporation, in some cases the selectable marker was flanked by sequences present in the new host, promoting DNA incorporation by homologous recombination. Such a strategy documented T-DNA transfer to a gram-positive bacterium (*Streptomyces lividans*), three yeasts (*S. cerevisiae, Schizosaccharomyces pombe,* and *Kluyveromyces lactis*), and seven filamentous fungi, including a commercially important mushroom (*Aspergillus awamori, Aspergillus niger, Fusarium venenatum, Trichoderma reesei, Collectotrichum gloeosporioides, Neurospora crassa,* and *Agaricus bisporus*). These findings document transfer to two of the three domains of life (transfer to Archaea has not yet been reported) and to very different cell types within each domain.

Strikingly, a recent report documented that transfer was possible to human cells as well. Talya Kunik, Vitaly Citovsky, and coworkers engineered the T-DNA to contain a marker detectable in human cells, and then incubated the modified *Agrobacterium* with several types of human cell cultures. Transfer of the marker was detected, and characterization of the transferred region in one case showed that it had a structure consistent with *vir*-mediated T-DNA transfer. Further studies showed that the *vir* genes were required for transfer to human cells. These studies establish that *Agrobacterium* can transfer DNA to human cells, and they raise the question of whether it ever happens in natural settings. One possibility involves *Agrobacterium* "yellow group," which has been associated with infections in dialysis patients, Hodgkin's disease, and septic arthritis following transplantation. Further studies should reveal whether these diseases involve the transfer of bacterial DNA segments from the infecting *Agrobacterium* to human cells.

CONJUGATIVE TRANSFER BETWEEN DOMAINS OF LIFE (14)

Conventional conjugative transfer has also been shown to mediate transfer of DNA between domains. George Sprague and coworkers reported that *E. coli* could transfer a conjugative plasmid to the yeast *S. cerevisiae* with high efficiency. This reaction requires all of the machinery of conventional conjugation, including the *oriT* origin of transfer, the relaxosome site-specific DNA-nicking proteins, and the pilus and membrane pore for DNA transfer. Recovery of the transferred DNA was particularly efficient when the newly introduced DNA could be stabilized by homologous recombination with endogenous sequences. Conjugative transfer is so efficient that today it provides a useful technology for introducing new DNA into yeast cells.

TRANSFER OF DNA TO ANIMAL CELLS BY INTRACELLULAR BACTERIA (4, 13)

Patrice Courvalin and coworkers have reported another means by which bacteria can deliver new genes into animal cells. They found that under certain circumstances, bacteria can enter cultured mammalian cells and burst, thereby delivering DNA into the host cell. Under the right circumstances, the bacteria-derived DNA can persist in the mammalian cell.

A number of pathogenic bacteria are known to penetrate into mammalian cells and replicate in the intracellular environment. For *Shigella flexneri*, the genes responsible for cell entry are carried on a 200-kb plasmid, pWR100. Introduction of this plasmid into *E. coli* transferred the ability to invade mammalian cells. Courvalin and coworkers introduced the required genes into an *E. coli* strain that had been further engineered to contain a weakened cell wall. The resulting bacteria could enter mammalian cells, but they lysed due to the cell wall defect, spilling the bacterial DNA into the mammalian cell. For experiments to test gene transfer, the *E. coli* was further engineered to contain DNA sequences detectable in mammalian cells, such as genes allowing growth in the presence of inhibitory drugs. Plasmids were readily detected after exposure to the engineered *E. coli*, particularly if the transferred DNA was capable of replication in mammalian cells. Protocols requiring stable incorporation of a bacterial plasmid into the mammalian chromosome also documented transfer.

These observations provided compelling evidence that transfer can take place from bacteria to mammalian cells in the laboratory. In nature it may be more difficult, because in most cases the bacterial cell wall will remain intact after transfer. Furthermore, to modify the mammalian genome in a heritable fashion, the process would need to take place in germ-line cells. This may well be infrequent, but the efficiency of gene transfer by this method in the laboratory bolsters the idea that it may happen in natural settings, at least on evolutionary time scales.

GENE EXCHANGES BETWEEN BACTERIA AND ARCHAEA (6, 23, 37)

We now turn to examples of cross-domain gene transfer detected by analyzing primary DNA sequences. Cross-domain gene transmission is generally expected to be infrequent, and transferred sequences become fixed in the new host species only rarely. In general, the accumulation of new sequences from other domains is expected to be slow, but the data suggest the rates can be significant over geological time scales.

Several studies of DNA sequences of prokaryotic and eukaryotic organisms have led to proposals for gene transfer from one to the other. These studies, relying solely on DNA sequence comparisons, are more uncertain than many of the above cases because the DNA transfer event cannot be observed directly. Many issues complicate analysis based on sequence alone. When partial genome sequences are compared, there is always the danger that the gene of interest will have as-yet-unidentified relatives that change the interpretation. For example, two genes may seem to be closely related in a prokaryote and a eukaryote, but lacking in intermediate species, supporting lateral transfer. However, discovery of a previously unknown homologous gene in the intermediate species would suggest simple inheritance of a conserved function by descent. Further complicating analysis is the fact that in many cases the inferred transfers took place very long ago, resulting in considerable divergence between the genes in question even though they may share a common ancestor. This reduces the degree of similarity and raises the degree of uncertainty.

As discussed in Chapter 5, there are many bacterial or archaeal genes that have the strongest similarities to genes from another domain, suggestive of lateral transfer. Relatively clear-cut cases have been found where (1) sequence composition and the codon usage of genes matches that of the proposed domain of origin and not the present host; (2) the gene blocks are associated with sequences suggestive of mobility, such as transposon, phage or tRNA sequences; and (3) blocks of multiple genes are colinear in the two domains, suggesting transfer of multigene regions. By the above reasoning, *Aquifex* and *Thermatoga*, both extremeophile Bacteria, contain 16% and 24% archaeal genes. Going the other way, the Archaea *Archaeoglobus fulgidis* appears to have captured and incorporated a large set of bacterial genes for fatty acid metabolism.

Today there is abundant evidence for gene transfer between the Bacteria and Archaea domains, although the mechanisms are unclear. Are Archaea naturally competent for transformation? Are there phage that can infect both Bacteria and Archaea? Can the gene transfer agents found in Bacteria and Archaea (Chapter 4) cross domain boundaries? Further studies of mobile DNA mechanisms should answer some of these questions and specify potential routes of transmission.

GENE TRANSFER FROM BACTERIA TO EUKARYOTES VIA ENDOSYMBIONTS (7, 9, 12, 18–21, 31, 34, 35)

The newly completed eukaryotic genome sequences have provided dramatic examples of gene transfer from Bacteria to Eucarya. Many of these reflect a flow of genes to the nuclear genome from the cytoplasmic organelles, mitochondria (Fig. 12.5), and chloroplasts (Fig. 12.6), which are themselves believed to be derived from Bacteria. The analysis of these events relies in part on the endosymbiotic theory of the origin of eukaryotic cells, so we next briefly review this theory and then return to gene transfer.

The endosymbiotic theory holds that eukaryotic cells are composites built up from a primordial cell that accumulated additional bacterial cells as symbionts (used here as a synonym for mutualists). Eukaryotic cells, according to the theory, established an endosymbiotic relationship with a bacterium and then took over the bacterial oxidative phosphorylation system to generate energy. Many believe that this occurred when high levels of oxygen first appeared in the earth's atmosphere, probably about 1.5×10^9 years ago.

Origin of the Endosymbiotic Theory

The earliest proposal for a form of lateral transfer appears to have been from C. Mereschowsky, who posited circa 1910 that cyanobacteria were the prototype of chloroplasts in plants. Several other authors expanded on this in subsequent decades, adding the proposal that mitochondria, as well, were derived from bacteria. R. Stanier and Lynn Margulis reintroduced the theory in the 1970s in its present form, in which mitochondria and chloroplasts were postulated to have formerly been free-living bacteria. Initially, this theory was not popular with the scientific community, but a series of findings over the ensuing years provided strong support. Margulis is now collecting evidence to expand the idea further; for example, trying to establish a spirochete origin of cilia and eukaryotic flagella.

Strong support for the endosymbiotic theory comes from the observation that the mitochondria and chloroplasts are membrane-enclosed and contain their own genomes, presumably relics of the chromosomes of the original invader. The relationship is hypothesized to be beneficial for both bacteria and host, because the bacteria get a comfortable home and the host gets nutrients in return. Many contemporary pathogenic bacteria are known to invade eukaryotic cells, potentially mimicking the entry of the primordial symbiont, or perhaps bacteria were ingested by proto-eukaryotes as food but persisted internally.

The nature of the cells that fused to form the modern eukaryote is a topic of energetic debate. Many favor the view that chloroplasts arose from cyanobacteria and mitochondria from α-proteobacteria (see Chapter 5, Fig. 5.6). The origin of the primordial eukaryotic host cell is more controversial, but many believe that it was probably related to the Archaea. In support of this, many of the core "informational" enzymes involved in transcription and translation in eukaryotes resemble those of the Archaea, although there are many exceptions and the debate continues.

If we accept that the progenitors of the eukaryotic organelles were Bacteria, and that the progenitor of the host cell was an Archaea, then Bacteria-derived genes in the contemporary nucleus may have arisen as imports from organelle genomes. Gene transfer from Bacteria to Eucarya, according to this view, took place through the intermediate endosymbiont.

W. Ford Doolittle has proposed that a "ratchet" mechanism may actually favor nuclear accumulation of symbiont genes. There are many copies of organelles in cells, and therefore many copies of their genomes. Lysis of an organelle, followed by transfer of DNA to the nucleus, provides an opportunity for incorporation of the organelle

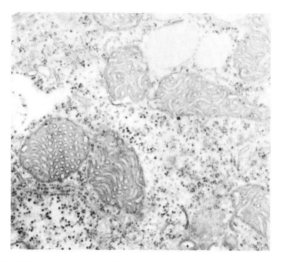

FIGURE 12.5. Electron micrograph showing mitochondrion. (Courtesy Bijan Ghosh, UMDNJ, Rutgers University; see Atlas 1984. [1])

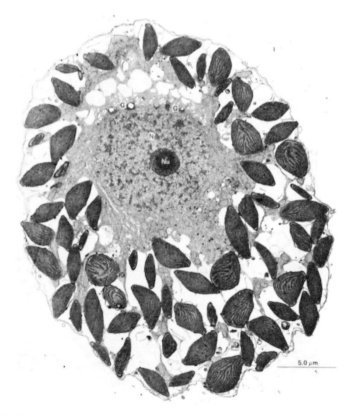

FIGURE 12.6. Electron micrograph of a *Vacularia virescens* cell showing multiple chloroplasts (the dark objects clustered around the cell periphery). (BPS–Stanley C. Holt, University of Washington; see Atlas 1984. [1])

gene copy. Probably transport and incorporation are rare, but repeated lysis and transport can supply a steady "rain" of organelle genes on the nucleus. Once an organelle gene is incorporated into the nucleus, it may duplicate an existing nuclear gene. If so, one or the other gene can then be lost without particular harm to the organism. The key to Doolittle's idea is that loss of the nuclear copy is irreversible, whereas loss of the organellar copy is inconsequential. Thus, a ratchet process may result in a steady replacement of nuclear genes with organelle genes.

Organelle Imports in the *Arabidopsis* Genome (34)

The hypothesis that genes can flow from organellar genomes to the nuclear genome received dramatic support from the sequence of the *Arabidopsis thaliana* chromosome 2. The centromeric region of the chromosome was found to contain an insert, >270 kilobases in length, that was 99% identical to the mitochondrial genome. The insert differed from the mitochondrial DNA only by an internal deletion of about 100 kb. This finding provides unambiguous evidence for the transfer of a large segment of an organellar genome to the nuclear genome. Furthermore, there are very few point mutations in the mitochondrial DNA segment, suggesting that the transfer must have been recent. Whether the mitochondrial sequence now contributes to the function of chromosome 2 is unknown.

The sequences of the *Arabidopsis* chromosomes also provide evidence for the transfer of chloroplast genes to the nuclear genome. Of the genes in chromosome 2, for example, 135 *Arabidopsis* genes were found to be most similar to those of Cyanobacterial relatives. If one accepts that Cyanobacteria were the progenitors of chloroplasts, these genes are likely derived from the chloroplast endosymbiont. Presumably many such genes would still be important for chloroplast function. Consistent with this, 71 of these genes (52.5%) were found to have recognizable signals for localization of the encoded protein to chloroplasts, greatly strengthening the proposal that these genes were originally captured from chloroplasts and their products now must return there to function. Plate 8 indicates some of the clearest organelle imports identified in the full *Arabidopsis* sequence.

Organelle Imports in the Vertebrate Genome (2, 9, 18, 26, 35)

Many vertebrate genes are proposed to have originated as imports from bacteria. An extensive literature accumulated on this subject prior to the report of the draft human genome sequence. For example, R. Doolittle and coworkers compared the sequences of 57 enzymes found in both bacteria and eukaryotes and found that, on average, the protein sequences were 37% identical, with 20% to 57% identity. Thus, to build a strong case for horizontal transfer, one wants to see sequence identity above the 57% match that represents the high end of the accidental distribution. A number of vertebrate genes were found to fit that criterion, including, for example, genes for glucose phosphate isomerase (*gpi*) and glyceraldehyde 3-phosphate dehydrogenase (*gapdh*). Some of the data are quite striking. For example, the *gapdh* of *E. coli* is 81% identical to that of trypanosomes, a value much higher than that expected by accident. Similarly, the *gpi* genes of *E. coli* and mammals are 70% identical.

The fact that transfer may often be from the bacteria (or endosymbiont) genome to the eukaryotic genome does not rule out that transfer may occasionally go the other way. As described above, there are some hints that plant genes may have been transferred to *Agrobacteria*. There is even an intriguing suggestion that genes encoding proteins of the immunoglobulin superfamily may have been transferred from eukaryotes to bacteria, because they are greatly expanded in the vertebrate lineage but present only infrequently in bacterial species.

Analysis of the draft human genome sequence by Eric Lander and coworkers led to a dramatic proposal for bacterial imports, which has since unraveled. Fully 223 human genes were identified in the sequence that had the strongest similarity not to genes in yeast, worms, flies, or *Arabidopsis*, but rather bacteria, suggesting that these genes were imported from bacteria into the vertebrate lineage. The data are so striking that stringent precautions had to be taken to exclude possible contamination of the sequenced human DNA with bacterial genes, but contamination could be ruled out experimentally. Surprisingly, the analysis of these bacterial imports did not suggest any single bacterium as the primary source of the new genes. The putative donors were diverse, with some genes acquired from gram-positive species and others from gram-negatives. These data suggested that, in vertebrates, the organelle import pathway described above is only one of many for gene acquisition from prokaryotes.

A reanalysis of the data has called this conclusion into question, however. Steven Salzberg, Jonathan Eisen, and coworkers have pointed out that addition of more eukaryotic sequences from unfinished genome sequences to the analysis reveals that many of the 223 human genes do in fact have close matches to other eukaryotic genes. Evidently gene loss in some of the eukaryotic lineages accounts for the patchy phylogenetic distribution, rather than lateral transfer. Reanalysis of the 223 initially proposed examples of lateral transfer against the larger collection of eukaryotic sequences leaves only about 40 genes with bacterial closest relatives. As more eukaryotic sequences are determined, it seems likely that this number will decline still further. Many of the reduced set of 40 genes will probably also be excluded as candidates for lateral transfer as more data accumulate.

At least occasional lateral transfer from the mitochondrial genome to the human nuclear genome remains likely, however. The earlier suggestions for mitochondrial imports still stand, and two of the remaining 40 genes in the set of human genes with closest relatives to bacterial genes do have closest matches to α-proteobacteria, consistent with lateral transfer from mitochondria. In addition, Arian Smit has observed that fragments of the mitochondrial genome are detectable in the human genome, although analysis of these sequences is just beginning. Stepping back, it now seems likely that the human genome will be found to resemble other eukaryotic genomes in containing bacterial sequences derived primarily from resident organelles.

Organelle Imports in Other Genomes (10, 22, 25, 38)

The idea that genes can be transferred from the mitochondrial genome to the nuclear genome has received support from functional studies in *S. cerevisiae* (also discussed in Chapter 10). This yeast repairs DNA double-strand breaks typically by recombination, and, in rare instances, new sequences can appear at the sites of repair. Evidently two recombination events connect broken double-stranded ends to a third sequence, which in the repaired product is found inserted between the two ends. Ty1 sequences, for example, occasionally are found at the sites of repaired double-stranded breaks. Strikingly, segments of mitochondrial sequences are also sometimes found at repair sites, indicating that mitochondrial DNA is accessible to the repair machinery and providing a specific mechanism for assimilation of new genes in the yeast nuclear genome.

An experiment designed to quantify DNA transfer from the yeast mitochondrial genome to the nuclear genome revealed a remarkably high rate. The marker gene *URA3* was introduced into the *S. cerevisiae* mitochondrial genome, where it was not functional. This allowed a selection for transfer to the nuclear genome, yielding a whopping rate of one in 10^5 per cell per generation. If all mitochondrial sequences are transferred at this frequency, the process is very efficient indeed.

A possible example of transfer from bacteria to yeast involves the origin of actin and tubulin genes, which are ubiquitous in eukaryotes but unknown in bacteria. Recent structural studies revealed that the bacterial *ftsA* and *ftsZ* proteins structurally resemble actin and tubulin, although there is little resemblance in the primary sequence. Gerry Fink and Jef Boeke noted that the genes for *ftsA* and *ftsZ* are linked in bacteria, and the genes for actin and tubulin are also linked in some yeast, supporting the idea that linkage has been maintained after lateral transfer. Whether this example involved transfer from organelle genomes is unknown.

Probable bacterial imports have also been identified in a number of other eukaryotic genomes, although the analysis is still in its infancy. In yeast and *C. elegans*, for example, candidate imports from bacteria have been proposed. Interestingly, some of the worm genes have acquired introns, suggesting that intron insertion followed transfer from bacteria.

At this writing, probable bacterial imports have not been fully cataloged in the sequenced eukaryotic genomes. Such a comparison will allow the timing of probable imports to be modeled, potentially allowing analysis of the contribution of bacterial imports to the evolutionary history of each lineage. It will be fascinating to see whether import of specific new genes into the nuclear genome can be correlated with particular evolutionary changes in the host species.

GENE TRANSFER BY EATING DNA? (7, 29, 30, 33)

The studies described above in which ingested intracellular bacteria transferred genes to eukaryotic cells indicate that DNA may be transferred by eating, at least at the cellular level. What about gene transfer into animals by ingesting DNA in food? Transfer by eating would help explain cases where prokaryotic-like genes are found in eukaryotes that do not resemble the genes of the expected progenitors of the organelles. A surprising series of experiments has suggested that it may be possible to transfer genes into mammalian tissues simply by eating DNA in food. Ranier Schubbert, Walter Doerfler, and coworkers fed M13 DNA to mice and later detected the DNA not only in the gut but in circulating blood cells. They then attempted to clone copies of M13 sequences integrated into the mouse genome and actually succeeded in finding several possible examples. The study was incomplete in an important respect—association of the new sequences with mouse genomic DNA was not demonstrated unambiguously—but further studies should clarify whether eating actually resulted in DNA transfer. The implication is that newly introduced genes could be transferred out of the gut at least partially intact and incorporated into cells of other tissues. So much DNA is consumed as a normal component of food that transfer by this mechanism, if confirmed, could be very frequent.

Summary

DNA can be transferred even between organisms of different domains, the greatest phylogenetic distances. In some cases, transfer is carried out by dedicated biological machinery, as with T-DNA transfer by *Agrobacterium*. Other cases of transfer rely on systems not specially evolved for interdomain transfer, such as natural transformation or intracellular invasion by bacteria.

For the case of *Agrobacterium*, the T-DNA transfer mechanism allows the bacterium to genetically engineer its host to provide a comfortable new niche. The T-DNA encodes genes affecting the auxin and cytokinin signaling systems, which direct the plant to grow the tumorous crown gall that the bacteria inhabit. Other genes encoded on T-DNA cause the plant to synthesize opines consumed by *Agrobacterium*.

Transfer is effected by *vir* gene products encoded together with the T-DNA on the Ti (tumor-inducing) plasmid. The *virA* and *virG* genes encode a signaling system that detects nearby wounded plants and induces the other *vir* genes in response. VirB and other proteins form the pilus that stabilizes the interacting cells and permits DNA transfer. Proteins encoded in the *virD* operon direct DNA nicking and escort the DNA into the plant cells, as with relaxosome-mediated transfer during conjugation. In the donor *Agrobacterium*, single-stranded DNA is displaced from the duplex plasmid concomitant with resynthesis of the T-DNA region. Transfer is initiated and terminated at defined border sequences, resulting in transfer of the T-DNA segment only and not the whole plasmid. Once in the plant cell, the T-DNA is integrated into the host chromosome by poorly defined plant illegitimate recombination enzymes.

Agrobacterium-mediated transfer or conventional conjugation can result in transfer of DNA to cells of different domains. *Agrobacterium* can introduce T-DNA not only into plants, but also into bacteria, fungi, and even human cells. Conjugation is actually an efficient way of introducing DNA into yeast for genetic manipulation. Each of these transfer reactions requires the appropriate plasmid origin of nicking and the *vir* or *tra* gene products.

Other less specific means also appear to mediate genetic transfer between domains. For example, penetration of bacteria into mammalian cells, followed by lysis of the bacteria, can also transfer genes from bacteria to mammals.

Larger-scale gene transfers between domains have taken place over evolutionary time. The primary sequences of Bacteria and Archaea suggest robust exchanges between the two on evolutionary time scales. Sequence comparisons suggest that eukaryotic cells have imported many genes from Bacteria, some via bacteria-derived organelle genomes. Although once controversial, this mode of transfer has been documented unequivocally by the finding in *Arabidopsis* chromosome 2 of a >270-kb region 99% identical to mitochondrial DNA. In yeast, transfer from the mitochondria to the nucleus can be demonstrated directly. Transfer from the mitochondrial to human genomes is also suggested by comparison of DNA sequences. Today there is no doubt that the DNA sequences of many contemporary organisms contain a rich record of imports from other domains.

REFERENCES

1. Atlas R.M. 1984. *Microbiology: Fundamentals and applications*. Macmillan, New York.
2. Bateman A., Eddy S.R., and Chothia C. 1996. Members of the immunoglobulin superfamily in bacteria. *Protein Sci.* **5:** 1939–1941.
2a. Buchanan-Wollaston V., Passiatore J.E., and Cannon F. 1987. The mob and oriT mobilization functions of a bacterial plasmid promote its transfer to plants. *Nature* **328:** 172–175.
3. Bundock P., Dulk-Ras A., Beijersbergen A., and Hooykaas P.J.J. 1995. Trans-kingdom T-DNA transfer from *Agrobacterium tumefaciens* to *Saccharomyces cerevisiae*. *EMBO J.* **14:** 3206–3214.
4. Courvalin P., Goussard S., and Grillot-Courvalin C. 1998. Gene transfer from bacteria to mammalian cells. In *Horizontal gene transfer* (ed. M. Syvanen and C.I. Kado), pp. 107–116. Chapman & Hall, London.
4a. Covacci A., Telford J.L., Giudice G.D., Parsonnet J., and Rappouli R. **19??.** *Helicobacter pylori:* Virulence and genetic georgraphy. *Science* **284:** 1328–1333.
5. de Groot M.J.A., Bundock P., Hooykaas P.J.J., and Beijersbergen A.G.M. 1998. *Agrobacterium tumefaciens*-mediated transformation of filamentous fungi. *Nat. Biotechnol.* **16:** 839–842.
6. Deckert G., Warren P.V., Gaasterland T., Young W.G., Lenox A.L., Graham D.E., Overbeek R., Snead M.A., Keller M., Aujay M., Huber R., Feldman

R.A., Short J.M., Olsen G.J., and Swanson R.V. 1998. The complete genome of the hyperthermophilic bacterium *Aquifex aeolicus. Nature* **392:** 353–358.

7. Doolittle W.F. 1998. You are what you eat: A gene transfer ratchet could account for bacterial genes in eukaryotic nuclear genes. *Trends Genet.* **14:** 307–311.

8. Dumas F., Duckely M., Pelczar P., Van Gelder P., and Hohn B. 2001. An *Agrobacterium* VirE2 channel for transferred-DNA transport into plant cells. *Proc. Natl. Acad. Sci.* **98:** 485–490.

9. Feng D.-F., Cho G., and Doolittle R.F. 1997. Determining divergence times with a protein clock: Update and reevaluation. *Proc. Natl. Acad. Sci.* **94:** 13028–13033.

10. Fink G.R. and Boeke J.D. 1999. Yeast, an experimental organism for all times. *ASM News* **65:** 351–357.

11. Frundt C., Meyer A.D., Ichikawa T., and Meins F. 1998. Evidence for the ancient transfer of Ri-plasmid T-DNA genes between bacteria and plants. In *Horizontal gene transfer* (ed. M. Syvanen and C.I. Kado), pp. 94–103. Chapman & Hall, London.

12. Gray M.W., Burger G., and Lang B.F. 1999. Mitochondrial evolution. *Science* **283:** 1476–1481.

13. Grillot-Courvalin C., Goussard S., Huetz F., Ojcius D., and Courvalin P. 1998. Functional gene transfer from intracellular bacteria to mammalian cells. *Nat. Biotechnol.* **16:** 862–866.

14. Heinemann J.A. and Sprague G.F. 1989. Bacterial conjugative plasmids mobilize DNA transfer between bacteria and yeast. *Nature* **340:** 205–209.

15. Kado C.I. 1998. Evolution of the selfish Ti plasmid of *Agrobacterium tumefaciens* promoting horizontal gene transfer. In *Horizontal gene transfer* (ed. M. Syvanen and C.I. Kado), pp. 63–72. Chapman & Hall, London.

16. Kaper J.B. and Hacker J. 1999. *Pathogenicity islands and other mobile virulence elements.* American Society for Microbiology, Washington, D.C.

17. Kunik T., Tzfira T., Kapulnik Y., Gafni Y., Dingwall C., and Citovsky V. 2001. Genetic transformation of HeLa cells by *Agrobacterium. Proc. Natl. Acad. Sci.* **98:** 1871–1876.

18. Lander E., Linton L.M., Birren B., Nusbaum C., Zody M.C. Baldwin J., Devon K., Dewar K., Doyle M., FitzHugh W., et al. 2001. Initial sequencing and analysis of the human genome. *Nature* **409:** 860–921.

19. Margulis L. 1971. Symbiosis and evolution. *Sci. Am.* **225:** 48–57.

20. Martin W. and Muller M. 1998. The hydrogen hypothesis for the origin of the first eukaryote. *Nature* **392:** 37–41.

21. Mereschowsky C. 1905. Uber Natur und Ursprung der Chromatophoren in Pflanzenteilen Bio. *Zentbl.* **25:** 593–635.

22. Moore J.K. and Haber J.E. 1996. Capture of retrotransposon DNA at the sites of chromosomal double-strand breaks (comments). *Nature* **383:** 644–646.

23. Ochman H., Lawrence J.G., and Groisman E.A. 2000. Lateral gene transfer and the nature of bacterial innovation. *Nature* **405:** 299–304.

24. Piers K.L., Heath J.D., Liang X., Stephens K.M., and Nester E.W. 1996. *Agrobacterium tumefaciens*-mediated transformation of yeast. *Proc. Natl. Acad. Sci.* **93:** 1613–1618.

25. Ricchetti M., Fairhead C., and Dujon B. 1999. Mitochondrial DNA repairs double-strand breaks in yeast chromosomes. *Nature* **402:** 96–100.

26. Salzberg S.L., White O., Peterson J., and Eisen J.A. 2001. Microbial genes

in the human genome: Lateral transfer or gene loss? *Science* **292:** 1903–1906.

27. Sauter C. 1995. Is Hodgkin's disease a human counterpart of bacterially induced crown-gall tumours? *Lancet* **346:** 1433.

28. Scheiffele P., Pansegrau W., and Lanka E. 1995. Initiation of *Agrobacterium tumefaciens* T-DNA processing. *J. Biol. Chem.* **270:** 1269–1276.

29. Schubbert R., Hohlweg U., Renz D., and Doerfler W. 1998. On the fate of orally ingested foreign DNA in mice: Chromosomal association and placental transmission to the fetus. *Mol. Gen. Genet.* **259:** 569–576.

30. Schubbert R., Renz D., Schmitz B., and Doerfler W. 1997. Foreign (M13) DNA ingested by mice reaches peripheral leukocytes, spleen, and liver via the intestinal wall mucosa and can be covalently linked to mouse DNA. *Proc. Natl. Acad. Sci.* **94:** 961–966.

31. Stanier R. 1970. Organization and control in prokaryotic and eukaryotic cells. *Symp. Soc. Gen. Microbiol.* **20:** 1–38.

32. Swann R.A., Foulkes S.J., Holmes B., Young J.B., Mitchell R.G., and Reeders S.T. 1985. "*Agrobacterium* yellow group" and *Pseudomonas paucimobilis* causing peritonitis in patients receiving ambulatory peritoneal dialysis. *J. Clin. Pathol.* **38:** 1293–1299.

33. Syvanen M. 1999. In search of horizontal gene transfer. *Nat. Biotechnol.* **17:** 833.

34. The Arabidopsis Genome Initiative 2000. Analysis of the genome sequence of the flowering plant *Arabidopsis thaliana. Nature* **408:** 796–815.

35. Venter J.C. 2001. The sequence of the human genome. *Science* **291:** 1304–1351.

36. Ward D.V. and Zambryski P.C. 2001. The six functions of *Agrobacterium* VirE2. *Proc. Natl. Acad. Sci.* **98:** 385–386.

37. Nelson K.E., Clayton R.A., Gill S.R., Gwinn M.L., Dodson R.J., Haft D.H., Kickey E.K., Peterson J.D., Nelson W.C., Ketchum K.A., McDonald L., Utterback T.R., Malek J.A., Linher K.D., Garrett M.M., Stewart A.M., Cotton M.D., Pratt M.S., Phillips C.A., Richardson D., Heidelberg J., Sutton G.G., Fleischmann R.D., Eisen J.A., White O., Salzberg S.L., Smith H.O., Venter J.C., and Fraser C.M. 1999. Evidence for lateral gene transfer between Archaea and bacteria from genome sequence of *Thermotoga maritima. Nature* **399:** 323–329.

38. Wolf Y., Kondrashov F.A., and Koonin E.V. 2000. No footprints of primordial introns in a eukaryotic genome. *Trends Genet.* **16:** 333–334.

39. Zambryski P. 1989. *Agrobacterium*-plant cell DNA transfer. In *Mobile DNA* (ed. D.E. Berg and M.M. Howe), pp. 309–333. American Society for Microbiology, Washington, D.C.

Controlling Mobile Element Activity

CELLS MAINTAIN A TENSE RELATIONSHIP WITH mobile DNA. Newly introduced sequences can sometimes be beneficial, conferring new traits on cells that allow them to persist in otherwise inaccessible niches. A conjugative plasmid bearing genes for detoxifying pollutants, for example, can allow the host bacterium to persist in a polluted pond.

Mobile DNAs can also harm host cells, as in the case of infection by lytic viruses. Accordingly, cells express diverse systems to control mobile element activity, and the elements themselves encode considerable regulator apparatus as well.

In this chapter, we review the systems by which DNA transfer is regulated. Regulation generally differs between prokaryotic and eukaryotic cells, so the two are treated separately. Several mechanisms for controlling gene transfer have been documented but are not yet well understood. For these, the puzzles are presented, together with hypotheses that are still being tested. The systems that control mobile element activity within a cell overlap with those that control transfer between cells, so the two are considered together. The chapter concludes with a review of some of the mechanisms that mobile elements themselves encode for regulating their activity and so optimizing their relationship with host cells.

PROKARYOTIC STRATEGIES FOR CONTROLLING DNA TRANSFER

Restriction/Modification Systems (9, 33, 34)

Restriction-modification systems have not been found in eukaryotes, although site-specific nucleases are known (e.g., the maturases of mobile introns discussed in Chapter 8).

Bacterial restriction enzymes are potent machines for destroying newly introduced DNA. Restriction enzymes bind to specific sequences on DNA and cleave it at or near the binding site (Fig. 13.1). A new DNA entering a cell, such as the chromosome of a lytic phage, will thus be fragmented by a restriction enzyme and prevented from initiating an infection.

Why don't restriction enzymes also fragment the genome of the host bacteria? Restriction enzymes, it turns out, are always paired with DNA methylase enzymes that render the recognition sites in the bacterial DNA insensitive to digestion. The methylase binds to the same recognition sequence as does the restriction enzyme, but instead of cleaving the DNA, it attaches methyl groups to the DNA bases, thereby preventing the restriction enzyme from cleaving at its recognition site.

As often happens in host–parasite interactions, the host cell restriction systems are sometimes countered by phage-encoded inhibitors. Phage T4 modifies its DNA by substituting cytosine with glucosylated deoxymethylcytosine, *Bacillus subtilis* phages replace thymine with hydroxymethyluracil and uracil, and cyanophage S-2L replaces adenine with 2-amino adenine. These modifications protect the phage from host-encoded restriction systems, and also allow phage to degrade host DNA specifically with phage-encoded nucleases that do not act on the modified phage DNA. Another strategy is adopted by phage T7, which encodes a protein (the product of gene 0.3) that inhibits the *Eco*B restriction endonuclease, apparently by binding to the subunit responsible for site recognition.

Restriction Endonucleases

Restriction enzymes can be grouped into three families. The type I enzymes are composed of three subunits, which independently contribute endonuclease, methylase, and DNA-binding activities. The complex binds to the recognition site if it is unmethylated, then threads the adjacent DNA through the enzyme, ultimately cleaving at some distance from the site of binding. If the site is methylated, the enzyme releases without cleavage of the DNA. If the site is methylated on one strand but not the other, as is the case immediately following DNA replication, the enzyme methylates the unmodified site. This restores full protection to the DNA, and the enzyme subsequently releases from its binding site. These enzymes consume large amounts of chemical energy by ATP hydrolysis—as many as 10,000 ATP molecules are hydrolyzed per cleavage event—apparently in connection with the translocation of the enzyme along DNA. The type I enzymes are found in enteric bacteria; examples include *Eco*K, *Eco*B, and *Hinf*III.

The type II enzymes bind to DNA as homodimers, and cleavage typically takes place within the recognition site. The recognition sequences are rotationally symmetric, analogous to palindromes, which are phrases that read the same forward as backward (such as "madam I'm Adam"—the world's first words!) (Fig. 13.1). Such a DNA palindrome typically binds a twofold symmetric protein dimer, with one subunit bound to each half of the repeated sequence. Only Mg^{++} is required for the reaction; there is no requirement for ATP or any other high-energy cofactor. The type II enzymes are highly active—*Eco*RI, for example, cleaves four DNA molecules per enzyme molecule per minute in test tube studies. Some type II restriction enzyme recognition sites and their sites of cleavage are shown in Table 13.1. The cognate methylases are encoded on separate proteins.

The type II enzymes are exceptionally useful in genetic engineering, because they can precisely break DNA at predetermined locations, allowing DNA molecules to be rejoined in new arrangements by the action of DNA ligase. An unprecedented search has been conducted among diverse microorganisms for new type II restriction enzymes, yielding more than 400.

Type III enzymes, such as *Eco*PI, are somewhat less well studied, mainly because they are less useful in genetic engineering. The type III enzymes are composed of two subunits. As with the type I enzymes, this complex carries out both DNA cleavage and methylation. The type III enzymes bind ATP but do not cleave it.

A   B

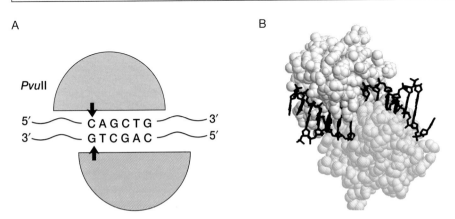

FIGURE 13.1. The restriction enzyme *Pvu*II. (*A*) Diagram of a dimer of *Pvu*II bound to its recognition site 5′-CAGCTG-3′. (*B*) X-ray structure of a *Pvu*II dimer bound to DNA. The arrows mark the points of DNA cleavage. The structure was rendered by F. D. B. from coordinates determined by Cheng et al. (1994) (11); Horton and Cheng (2000). (25).

TABLE 13.1. A Few Type II Restriction Endonucleases and Their Sites of Action

Enzyme	Host	Recognition sequence
*Alu*I	*Arthrobacter luteus*	5´-AGCT-3´ 3´-TCGA-5´
*Bam*HI	*Bacillus amyloliquefaciens*	5´-GGATCC-3´ 3´-GGATCC-5´
*Eco*RI	*E. coli*	5´-GAATTC-3´ 3´-CTTAAG-5´
*Fok*I	*Flavobacterium okeanokoites*	5´-GGATG(N)$_9$-3´ 3´-CCTAC(N)$_{13}$-5´
*Not*I	*Nocardia otitidiscaviarum*	5´-GCGGCCGC-3´ 3´-CGCCGGCG-5´
*Pvu*II	*Proteus vulgaris*	5´-CAGCTG-3´ 3´-GTCGAC-5´

The ● indicates the position of cleavage on each DNA strand.

Restriction enzymes may also provide additional functions to the cell. For example, it has been proposed that restriction enzymes may produce DNA breaks that are important in initiating recombination.

The Mismatch Repair System as a Barrier to Homologous Recombination (15, 20, 35, 39, 55, 66)

Many DNA sequences newly introduced into a cell must become incorporated into the genome by homologous recombination if they are to persist. Exceptions are those elements that can replicate as episomes (Chapter 3) or become integrated by site-specific recombination. Intriguingly, homologous recombination of imperfectly matched sequences is strongly inhibited by a host-cell DNA repair system, thereby blocking many forms of lateral transfer.

When two single DNA strands anneal, bases that do not match between the two form unpaired regions called "mismatches." Mismatches can also be generated by errors during DNA replication in which an incorrect base is incorporated during synthesis. Cells encode multienzyme DNA repair systems for identifying and removing base mismatches (Fig. 13.2).

In bacteria, the mismatch repair system must detect which DNA strand is newly synthesized, so that repair can restore the original

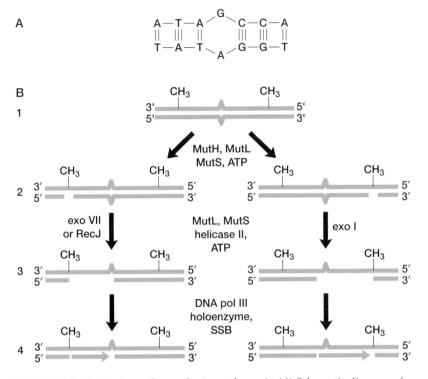

FIGURE 13.2. DNA mismatches and mismatch repair. (*A*) Schematic diagram of a DNA mismatch. (*B*) Pathway of methyl-directed mismatch repair in bacteria. (*1*) Diagram of a DNA strand containing a mismatch ("spikes" in the center of the green DNA). The DNA is shown with only one strand modified by methylation, as is the case immediately after DNA replication. (*2*) Mismatch repair begins with nicking of the unmethylated strand opposite the site of modification by the action of MutH, MutL, MutS, and ATP. Two different pathways (*right* and *left*) mediate subsequent repair depending on whether the nick is 5′ or 3′ of the mismatch. (*3*) In the former case (*left*), the nick is extended to the mismatch by exoVII and RecJ, and the gap is repaired by DNA pol III and SSB (*4*). In the latter case (*right*), the nick is extended by exo I, and subsequent repair proceeds similarly with DNA pol III and SSB. (*B*, Redrawn, with permission, from Grilley et al. (1993). [20])

sequence to both strands. The old and new strands are distinguished by the presence of methyl groups attached to the A residues of 5′-GATC-3′ sequences. Following replication, there is a lag before the newly synthesized DNA becomes methylated, thereby marking the undermethylated strand as newly synthesized. Mismatches are corrected so that the newly synthesized unmethylated strand is modified to match the methylated parent strand.

If a newly introduced DNA strand is to alter the phenotype of the host cell, the new DNA strand must be at least a little different from that already present. Thus, hybrids between donor and host sequences formed as intermediates during generalized transduction are potential substrates for mismatch repair. Miroslav Radman and colleagues have found that the mismatch repair system in fact inhibits homologous recombination of slightly different DNA strands, thereby blocking incorporation of new DNA.

This interference can be seen dramatically in matings between related bacterial species. *Escherichia coli* and *Salmonella typhimurium* are 16% diverged in primary DNA sequence. Normally, transfer of DNA from one to the other by conjugation does not result in incorporation of the newly transferred DNA. However, if key genes for methyl-directed mismatch repair are inactivated, conjugative transfer takes place efficiently, including DNA incorporation into the recipient cell genome by homologous recombination.

Studies of pathogenic *E. coli* and *S. typhimurium* strains reveal a high frequency of mutations that inactivate methyl-directed mismatch repair. Consequently, these bacteria are more likely to incorporate transduced DNA than are wild-type strains. It has been proposed that this may be important for the pathogenic phenotype, because new pathogenic determinants could be thus assimilated more readily. Increased mutation rates could also promote development of resistance to some antibodies. This proposal is somewhat controversial, because it can be challenging to categorize bacterial isolates as pathogenic or nonpathogenic, but the idea is appealing.

A recent study of a large collection of *E. coli* isolates by Erick Denamur and colleagues has suggested that the efficiency of mismatch repair may fluctuate during bacterial evolution, thereby modulating the efficiency of lateral transfer. Sequencing of key genes encoding mismatch repair proteins (Mut H, U, L, and S) revealed that each has a composite character. They appeared to be built up of repeated lateral transfer events, resulting in patchy resemblance to repair genes from diverse species. Other chromosomal genes did not display this mosaic character. These observations support a model in which the efficiency of mismatch repair is in a state of flux. In some environmental situations, it may be beneficial to incorporate new DNA more efficiently, although at the price of an increased mutation rate, favoring disruption of the mismatch repair genes. In other settings, it may be a better strategy to keep the genome stable, thereby selecting for efficient mismatch repair. Changes in the mismatch repair genes by lateral transfer are hypothesized to account for much of the sequence variation.

EUKARYOTIC STRATEGIES FOR CONTROLLING DNA TRANSFER

Eukaryotes display many mechanisms for down-modulating the efficiency of DNA transfer systems. Some mechanisms are known in only a few organisms. For example, repeat induced point mutation (RIPing) has been found in molds only, and the adaptive immune system is unique to vertebrates. Some mechanisms are used quite widely, such as the RNAi/cosuppression system, which is found in both plants and animals. This section reviews first relatively widely used mechanisms, then the more idiosyncratic control strategies.

Mismatches Antagonizing Homologous Recombination (47, 71)

A control mechanism common to both prokaryotes and eukaryotes is the regulation of homologous recombination by mismatch repair. As with prokaryotes, incorporation of new sequences into eukaryotic chromosomes is strongly affected by the presence of mismatches. In mice, for example, it is possible to direct recombination of newly introduced DNAs into homologous sequences experimentally. If the newly introduced gene contains even a few bases that differ from the endogenous copy, the frequency of incorporation by homologous recombination is greatly reduced. Similar observations have been made in *Drosophila*. In animals defective in mismatch repair, this effect is strongly reduced.

RNAi and Cosuppression (3a, 14, 16, 26, 29a, 38, 52, 63, 65, 70)

Upon entering a new host cell, a selfish DNA such as a transposon or virus often seeks to synthesize new copies of itself. As a result, the cell will contain multiple copies of a DNA sequence that was not previously present in the genome. Quite recently, it has been discovered that many cells have evolved systems for recognizing and genetically inactivating repeated DNA sequences, presumably a defense against parasitic DNAs. In this section, we review a collection of related mechanisms for turning off expression of repeated sequences, named RNA interference (RNAi), cosuppression, and quelling. The multiplicity of names reflects their independent discovery by several laboratories, and the fact that different parts of the system are more prominent in different experimental settings.

One of the first indications of cosuppression came from experiments attempting to engineer petunias to produce more intensely purple flowers. Researchers introduced extra copies of genes directing pigment synthesis into petunia plants and analyzed the effects. To their surprise, the flowers did not become more purple, and some actually became partially or completely white, despite the presence of the normal genes in addition to the extra ones. The implication was that the added gene copies somehow turned off the endogenous genes as well as failing to function themselves.

Related findings were also seen in experiments designed to reduce the sensitivity of tobacco plants to virus infection. David Baulcombe and colleagues overexpressed a gene for the replicase enzyme of potato X virus in the hopes of obstructing viral replication. They found, oddly, that cells which expressed the replicase poorly were less prone to infection, the opposite of their expectation. With time, this was recognized as another manifestation of cosuppression. Similar results were seen for filamentous fungi, where the phenomenon was named quelling. In this case, introduction of a gene for carotinoid synthesis was seen to down-modulate the endogenous copy.

Another line of experiments converged with the above studies to provide a window on the mechanism. Many studies have reported that expression inside cells of an RNA complementary to a cellular mRNA, so-called "antisense" RNA, can reduce expression of the targeted sequence. However, in a number of cases, tests with control RNAs yielded unexpected results. RNAs with sequences identical to mRNAs (instead of the complementary antisense) sometimes resulted in gene inactivation. Andrew Fire and colleagues further discovered that injecting double-stranded RNA into the worm *Caenorhabditis elegans* resulted in efficient inactivation of the targeted gene. Double-stranded RNAs were also found to inactivate target genes in fly embryos and later in mammalian embryos and cells. These effects of double-stranded RNAs were named "RNA interference," or simply RNAi.

Remarkably, RNAi is not restricted to the site at which it initiates. Injection of RNA into the gut of a worm results in RNAi throughout the animal. Even more remarkably, the effect is often heritable. Breeding of RNAi-treated worms or flies yielded offspring that also displayed reduced expression of the targeted gene. Similar results were seen in plants.

An important finding, connecting RNAi and cosuppression, was that the silenced gene was in fact transcribed. Thus, the RNAi/cosuppression effect did not inhibit transcription initiation, initially a possible explanation. Evidently the RNA, once synthesized, is somehow inactivated, and recent experiments indicate that RNA degradation mediates the effect (Fig. 13.3). Direct analysis of cells undergoing RNAi reveal that the double-stranded RNA is cleaved into 20- to 22-bp fragments. This observation provides a potential explanation for the origin of RNAi specificity, because the involvement of RNA fragments allows base-pairing to mediate selective recognition.

Genetic screens for defects in RNAi and cosuppression have yielded candidate protein cofactors, allowing a specific mechanism to be proposed. Genes encoding RNA-dependent RNA polymerase, two ribonucleases, and helicase proteins have been identified as components of the RNAi pathway. This supports a model in which the RNA-dependent RNA polymerase copies the RNAi fragments, explaining the amplification of the RNAi effect. The nuclease may be important for degrading the double-stranded RNA into the short pieces required for recognition, or for subsequent destruction of the targeted RNA. G. Hannon and coworkers have found that cells induced for RNAi produce a RNase enzyme associated with the short RNAs, supporting this model. The helicase may be important for unpairing the double-stranded RNA to allow recognition of complementary sequence in the targeted mRNA.

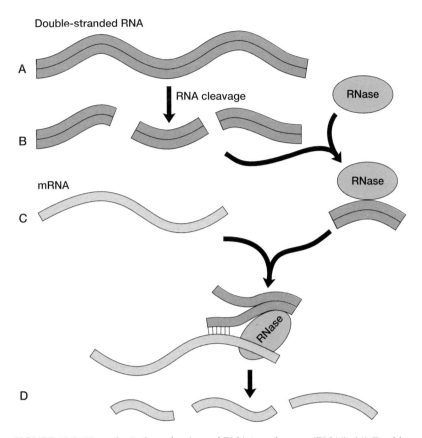

Double-stranded RNA

A

RNA cleavage

RNase

B

RNase

mRNA

C

RNase

D

FIGURE 13.3. Hypothetical mechanism of RNA interference (RNAi). (*A*) Double-stranded RNA (*dark green*) is recognized by an RNase enzyme and cleaved into short segments. Some of the segments bind the RNase enzyme (*B*), and so can use homology between the bound RNA and a cellular mRNA of the same sequence (*C*) to bind and cleave the cellular mRNA (*D*). (Modified, with permission, from Marx 2000 [© American Association for the Advancement of Science]. [38])

Tests with mutants defective in RNAi document its role in suppressing replication of genomic parasites. Worms mutant in the RNAi system display greatly increased transposon hopping, and mutant plants become more sensitive to some viruses.

Studies of RNAi are still in the early stages, and many intriguing questions remain unanswered. To what degree do RNAi and cosuppression use the same pathways? Already there are indications of at least partially separate pathways for each. Does RNAi feed back on the DNA, resulting in further gene silencing by other mechanisms? Again there is some evidence for this possibility. In plants, methylation of DNA often correlates with gene inactivation, and genes undergoing RNAi have been reported to become methylated selectively. Thus, RNAi may be a component of a still larger control system. Last, as yet uncertain is the intriguing question of whether RNAi is important in normal gene control during development.

Repeat-induced Point Mutations (RIP) (31, 69)

The filamentous fungi control proliferation of mobile elements by a remarkably direct mechanism—simply mutating them to death. Fungi such as *Neurospora* respond to the presence of repeated sequences such as mobile elements by inducing mutations in each of the repeat copies. The mechanism has been named RIP for *repeat-induced point* mutation to reflect the requirement for DNA repeats.

The RIP mechanism involves the replacement of C residues with T in repeated sequences. Such mutations are known to take place in two steps, methylation of C, followed by deamination of methylcytosine to deoxyuridine. This lesion is then repaired by cellular machinery, which replaces the U residue with T. Consistent with the involvement of this pathway in RIP, repeated sequences undergoing RIP are known to become methylated.

RIP has been shown directly to inactivate mobile elements. Certain rare *Neurospora* strains contain an active non-long terminal repeat (LTR) retrotransposon named *Tad*. Laboratory tests revealed that the *Tad* element was quickly inactivated during growth in conventional laboratory strains. Sequencing revealed that the inactivated *Tad* genomes contained the C to T mutations expected from RIP.

Examination of the chromosomes of *Neurospora* and related filamentous fungi revealed that all those studied contained recognizable *Tad*-related sequences, but all had sustained multiple C to T mutations. Figure 13.4 illustrates the mutations present in some representative *Tad* sequences. These findings suggest that *Tad* elements have been multiplying in filamentous fungi, but suffering death by mutation at a high rate.

DNA Methylation in Eukaryotes (5, 17, 42, 50, 57, 64, 68)

DNA in many vertebrate and plant cells is modified by attaching methyl groups to deoxycytosine residues at CpG dinucleotides (Fig. 13.5). In human cells, about 50–70% of C residues in CpGs are modified. Heavily methylated genes tend to be expressed at low levels, suggesting that methylation down-modulates gene expression. Transcriptional inhibition by this mechanism has been found to restrict the proliferation of transposons and endogenous viruses. CpG DNA methylation plays a different role than methylation in prokaryotes, where GATC methylation helps distinguish between parent and daughter DNA strands during replication, and methylation of restriction sites protects the host genome from cleavage. Invertebrates such as flies, worms, and yeast carry out little or no DNA methylation.

Vertebrate and many plant cells encode methyltransferase enzymes that attach the methyl groups at CpG dinucleotides. During replication, one DNA strand is synthesized using the other as template,

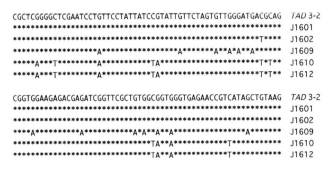

```
CGCTCGGGGCTCGAATCCTGTTCCTATTATCCGTATTGTTCTAGTGTTGGGATGACGCAG   TAD 3-2
************************************************************   J1601
*********************************************************T****   J1602
********************A*****************A********A**A**A******   J1609
*****A***T*********A***********TA*********************T*T**   J1610
*****A***T*********A***********TA*********************T*T**   J1612

CGGTGGAAGAGACGAGATCGGTTCGCTGTGGCGGTGGGTGAGAACCGTCATAGCTGTAAG   TAD 3-2
************************************************************   J1601
************************************************************   J1602
****A***********A***********A*A**A**A***************A******   J1609
*******************************TA**A***********T**********   J1610
*******************************TA**A***********T**********   J1612
```

FIGURE 13.4. Repeat-induced point mutations (RIP) in *Neurospora*. The top line shows a segment of sequence of the TAD retrotransposon of *Neurospora*. The lines below show the sequence of the same region from TAD elements exposed to RIP. The asterisks indicate bases that are the same as the initial element. (Reprinted, with permission, from Kinsey et al. 1994. [(31)])

resulting in a duplex in which one strand is methylated and the other is not (hemimethylated). The CpG methylase recognizes such sequences and attaches a methyl group to the unmethylated strand, thereby propagating the methylation state during cell division.

The correlation between methylation and reduction of gene expression has been recognized for some time, leading to the question of exactly which genes are regulated by methylation. Recently, several studies have documented correlations between undermethylation and activation of endogenous retroviruses and transposons.

In one curious case, a hybrid wallaby resulting from a cross between a swamp wallaby (*Wallabia bicolor,* Fig. 13.6) and a tammar wallaby (*Macropus eugenii*) was found to possess chromosomes that differed from either parent. The altered chromosomes contained expanded centromeres, regions that in many metazoan species contain large numbers of mobile element sequences (Chapter 10). The hybrid genome was also found to have greatly reduced CpG methylation compared to either parent. Highly repeated undermethylated sequences were cloned from the hybrid genome and found to be abundant at the expanded centromeres, probably accounting for their observed increase in size. Sequencing of the newly proliferated centromere-associated DNAs revealed that they were made of a retrovirus, named KERV-1, for *k*angaroo *e*ndogenous *r*etro*v*irus 1. Evidently the reduction in methylation in the interspecific hybrid allowed the KERV-1 retrovirus to proliferate. Why methylation was reduced by hybridization is unclear. The phenomenon of centromere expansion has been reported in a number of macropodid (kangaroo) hybrids, suggesting that the phenomenon may not be restricted to the animal studied.

Another example of activating a retroelement by reducing CpG methylation comes from studies of mice mutant in the CpG methyltransferase. T. Bestor and colleagues deleted the gene for murine DNA

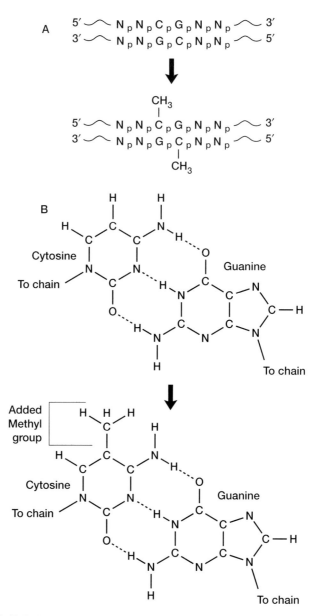

FIGURE 13.5. CpG methylation. (*A*) A segment of DNA sequence showing the points of attachment of methyl groups (CH$_3$) resulting from CpG methylation. (*B*) A C-G base pair (*top*) and the modification due to methylation (*bottom*).

methyltransferase-1 (*Dnmt-1*) from the mouse germ line and studied the effects on proliferation of the endogenous intracisternal A-type particles (IAP). As discussed in Chapter 6, the IAP elements are retrovirus-like sequences that may multiply by intracellular retrotransposition or pseudotype formation. Deletion of both copies of the *Dnmt-1* gene resulted in a 50- to 100-fold increase in IAP transcription. Possible

FIGURE 13.6. A swamp wallaby. (Reprinted, with permission, from Nowak 1999 [courtesy Zoological Society of San Diego]. [49])

effects on retrotransposition were not studied, but these data indicate that, at least for the IAP case, methylation controls the transcription of a genomic parasite.

Similar effects have been seen for the control of DNA transposons in plants. Studies going back to Barbara McClintock document that some elements, including *Ac* and *Spm*, can be reversibly inactivated. Modern studies of inactive and reactivated alleles reveal that these elements differ by the extent of DNA methylation, with the inactive form enriched for methyl-C residues. Mutations in the *Arabidopsis* gene *ddm1* (named for *d*ecrease in *D*NA *m*ethylation) were found to result in the formation of new mutations, which could be attributed to mobilizations of the *h*AT transposon *CAC1*. Another transposon, named Roberton's *mutator*, was also found to be activated by mutation of *ddm1*.

Other roles for methylation have been proposed in X-chromosome inactivation, a form of heritable gene inactivation called "imprinting," and gene control in general. Stepping back, it seems that a major role of CpG methylation is in controlling transposons and endogenous retroviruses, but that methylation is also important in regulation more widely. For all the genes subject to methylation, it is unclear why some sequences are selected for methylation, while others are not.

The Adaptive Immune System (18, 28)

Vertebrate cells have evolved diverse strategies to block the replication of invaders that import new nucleic acid sequences, including the adaptive immune system, the innate immune system, and the interferon system. The following sections summarize those systems, emphasizing their role in blocking lateral gene transfer, while recognizing that this is a fragment of the much larger subject of immunology. Many of these systems are important for controlling retroviral infection, one of the most prominent means of lateral gene transfer in vertebrates (Table 13.2). Several of these mechanisms were touched on in previous chapters and are treated here only briefly. Detailed descriptions can be found in Chapters 6, 7, and 11, and in the references in Table 13.2.

The most prominent host cell defense against viruses is the adaptive immune system. Following viral infection, antibodies or T-cell receptors (TCRs) of the correct specificity bind viral invader antigens and initiate an immune response. For the case of retroviruses, the cellular immune response appears to be the most effective, involving elimination of virus-producing cells by cytotoxic T lymphocytes, rather than antibody-mediated elimination of the virion particles themselves.

Once a virus infects a cell, it must begin producing viral proteins to replicate, and this offers an opportunity for detection by the immune system. Recognition of virus infection by CTL begins with the degradation of viral peptides inside the infected cell. Peptides are then loaded onto histocompatibility proteins (HLA in humans), which are routed to the cell surface. Probably peptides from all proteins synthe-

TABLE 13.2. Some Defenses against Retroviruses and Retrovirus-like Elements

System	Mechanism/comments	Review/Reference
Humoral immunity	antibody response often ineffective against lentiviruses	Poignard et al. (1999) (53)
Cell-mediated immunity	cytotoxic T-cells, other mechanism; active against lentiviruses	Walker and Goulder (2000) (67)
Interferon system	shuts down translation in viral producer cells; inhibits HIV cDNA production, late gene expression	Isaacs and Lindenmann (1957) (3); Baca-Regen et al. (1994) (27)
Fv1	antagonizes early infection by MoMLV strains; locus encodes endogenous provirus *gag*	Best et al. (1996) (4); Boeke and Stoye (1997) (7)
Fv4	blocks initial infection; locus encodes retroviral *env* that sequesters receptors	Limjoco et al. (1993) (36)
Vif-antagonized system	active in viral producer cell, blocks accumulation of cDNA in target cells	Madani and Kabat (1998) (37); Simon et al. (1998) (59)
Proteasome system	degrades HIV proteins, leads to cDNA destruction	Schwartz et al. (1998) (58); S.L. Butler, E. Johnson, and F.D. Bushman (in prep.) (8)
NER DNA repair pathway	reduces accumulation of Ty1 cDNA in *S. cerevisiae*	Curcio and Garfinkel (1999) (12)

sized in a cell are loaded into HLA and displayed on the cell surface, but "self" proteins are recognized as such. Only foreign proteins initiate an immune response. T-cell receptors on the surface of CTLs recognize the viral peptide–HLA complex on the surface of the infected cell. Several further molecules engage with this complex, initiating a cascade of signaling and effector functions. One of the most prominent effectors results in direct lysis of the infected cell, thereby preventing further virus production.

The importance of the cellular immune response has been demonstrated dramatically for the case of HIV. Antibodies capable of neutralizing infectious virions are common in infected patients, but do little to slow viral replication. However, using SIV-infected macaques as a model system, the cellular immune response has been demonstrated to keep the infection in check. CTLs were selectively depleted from SIV-infected monkeys and a large boost in the titer of virus in blood was observed. Restoring the original number of CTLs returned the viral load to the starting level. The immune response against HIV and other viruses is a topic of intense interest and experimentation. For more detailed treatments see the references in Table 12.2.

The Innate Immune System (22, 23, 28, 30)

The adaptive immune system requires some 4–7 days to mount the initial response to infection by a typical pathogen. During this period, cells of the innate immune system resist the infection and generate signals that activate the adaptive immune system. Invertebrates lack the adaptive immune system but have active innate immunity.

Cells of the innate immune system, such as macrophages and neutrophils, have surface receptors that bind molecules commonly found on invading pathogens. Engagement of pathogen molecules causes the phagocytic cells to engulf and destroy the invader. These cells also secrete signaling molecules—cytokines—in response to receptor engagement that cause inflammation in the local tissue, which further recruits immune cells and stimulates the development of an adaptive immune response.

Many of the receptors important for innate immunity are members of the Toll-like receptor family, named for the fruit fly protein that was the first member of this family identified. These receptors bind molecules unique to invading pathogens, such as lipopeptide and peptidoglycan (Toll-like receptor 2), lipopolysaccharide (Toll-like receptor 4), or bacterial flagellin proteins (Toll-like receptor 5).

Lateral transfer is clearly opposed by the ability of the innate immune system to clear bacteria and other infectious agents that might provide sources of new DNA. Another function of the innate immune response probably also inhibits lateral DNA transfer. DNA containing unmethylated CpG dinucleotides is typically foreign, because vertebrates methylate the majority of CpG sequences. Toll-like receptor 9 binds DNA containing unmethylated CpG sequences and activates an immune response, which clears the DNA and the invading pathogen. Many retroviral genomes have lower frequencies of CpG dinucleotides than expected from the base composition, probably as a defense against this response. Thus, the innate immune response against foreign DNA likely further reduces the opportunities for lateral gene transfer in vertebrates.

The Interferon System (3, 27, 28)

The interferon system is another prominent vertebrate defense against invaders. Interferon proteins are members of the cytokine family, a loosely defined group of intercellular signaling molecules. Cytokines are typically small proteins (MW less than or equal to 30 kD), usually active as dimers or trimers. Expression is typically induced by a signal such as viral infection, which causes cells to secrete cytokines that accumulate near the site of infection. Cytokines act by binding to high-affinity receptors on cell surfaces, thereby initiating a signaling cascade that ultimately affects nuclear gene expression. Among the genes turned on as a result are those encoding further cytokines, thereby amplifying and broadening the response.

The interferons, although one of many cytokines, are of special interest for their prominent role in blocking viral replication. Interferons were first discovered in experiments in which chick embryo cells infected with influenza virus were shown to have reduced sensitivity to subsequent infection with influenza and also other viruses. The activity was traced to factors secreted from the infected cells, which were later purified and named the interferons.

Three predominant types of interferons have been recognized: α, β, and γ. The first two can be produced from most cell types, whereas interferon γ is produced mainly from cells of the immune system. Induction of interferons α and β typically results from viral infection. Double-stranded RNA, presumably derived from viral replication intermediates, provides the inducing signal. Induction of interferon γ proceeds differently, arising as a result of immune detection of pathogens.

The interferon system blocks replication of viruses by several effector pathways. Two of the best understood are shown in Figure 13.7, both of which result in translational shutdown of the virus-infected cell. In each case, interferon induces the synthesis of enzymes that are activated by binding double-stranded RNA. In one branch (Fig. 13.7, left side), interferon treatment induces the synthesis of 2-5(A) synthetase, an unusual enzyme that connects ATP molecules into polymers with 2′ to 5′ linkages. The 2-5(A) polymer binds and activates the RNase L protein, which then degrades cellular mRNAs.

In the other pathway (Fig. 13.7, right side), interferon activates synthesis of the double-stranded RNA-dependent protein kinase (PKR). PKR binds double-stranded RNA, which results in autophosphoryla-

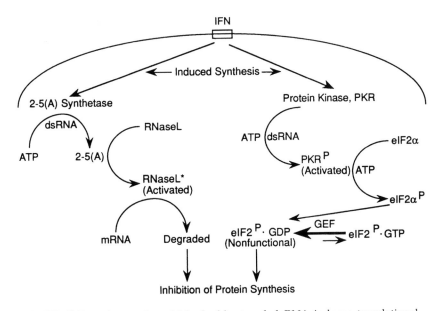

FIGURE 13.7. Pathways by which double-stranded RNA induces translational shutdown of vertebrate cells and its modulation by interferon. Interferon treatment of cells induces synthesis of two proteins that turn off translation, the 2-5(A) synthetase (*left*) and the double-strand RNA activated protein kinase (PKR) (*right*). 2-5(A) synthetase forms a polymerase of ATP units connected by 2´-5´ linkages. This polymer binds to RNase L, thereby activating the enzyme to cleave cellular mRNAs, resulting in turn-off of translation. PKR, upon activation by binding double-stranded RNA, phosphorylates eIF2α, thereby blocking its GDP–GTP exchange activity and blocking translation initiation. (Reprinted, with permission, from Fields and Knipe 1996. [18])

tion that activates the enzyme. Probably two PKR molecules bind to a single double-stranded RNA molecule and phosphorylate each other. Activated PKR phosphorylates eIF-2 α, a factor important for translation initiation. Normally the eIF-2 α protein hydrolyzes GTP to GDP to promote translation initiation. Phosphorylation freezes eIF-2 α in the GDP-bound form, thereby preventing turnover and blocking further translation. Thus, for both pathways in Figure 13.7, interferon treatment creates a sensitized state in which the presence of double-stranded RNA causes a translational shutdown, thereby blocking further viral production.

Studies of the effects of interferon on retroviral replication have revealed blocks at two steps in the viral life cycle. Several early studies showed that injection of anti-interferon antibodies could exacerbate retrovirus-induced diseases in mice, implicating interferon in suppressing viral pathogenesis. For HIV, pretreatment of cultured cells with interferon α caused a profound reduction in virus titer. This was due to a reduced accumulation of reverse transcription products and also a block late in the life cycle at particle assembly or release. It is at present unclear whether these effects can be accounted for by the known mechanisms of translational shutdown or instead represent the action of additional effector systems.

Further Mechanisms Active against Retroviruses and Retrotransposons (4, 7, 8, 12, 37, 59)

Intensive studies of retroviruses have identified several further cellular antiretroviral systems, although much of the work is still at an early stage. Several of these pathways were discussed in the chapters on retroviruses and retroelements, but are described here again briefly to complete our survey of the host armamentarium inhibiting lateral DNA transfer (Table 13.2).

Historically, some of the most intensively studied host cell restrictions on retroviruses are the Fv1 and Fv4 systems. The presence of the appropriate Fv1 allele in murine cells blocks replication of some murine leukemia virus (MLV) strains. The determinant on the viral side was mapped to the viral capsid coding region. Cloning of the gene for Fv1 revealed that it is, in fact, encoded by a defective endogenous retrovirus. How the presence of this sequence leads to the observed blockade to early MLV replication is unclear, but is under intensive investigation.

The Fv4 system also blocks early steps of infection, and the gene is also encoded by a defective retroviral fragment. In this case, the Fv4 gene encodes a retroviral envelope protein that sequesters the cognate receptor. Thus, the presence of the Fv4 allele obstructs infection of MLV viruses using the same receptor for entry.

Another candidate antiretroviral system has been identified through studies of the HIV-1 viral infectivity factor (Vif) protein. Detailed studies reveal that Vif blocks the action of an inhibitory activity present in virus producer cells. The nature of that inhibitor and the mechanism of Vif action are among the central mysteries in contemporary HIV research.

Several further studies indicate that the cellular proteosome system may antagonize the early steps of retroviral replication. The proteosome machinery is responsible for degrading damaged or unwanted cellular proteins. Addition of proteosome inhibitors to CD4-positive cells together with HIV leads to increased infection. The effect can be mapped to an increased accumulation of viral cDNA integration complexes. The implication is that degradation of viral replication intermediates by the proteosome reduces the amount of cDNA that accumulates.

Studies of the yeast Ty retrotransposons reveal that retroelement cDNA is degraded at a high rate by the cell, thereby blocking retrotransposition. Ty1 elements normally are transcribed to a high level in wild-type cells, accounting for up to 0.8% of the total RNA, but only one cDNA molecule is made for every 14,000 RNAs. David Garfinkel and coworkers identified mutations of *S. cerevisiae* that supported an increased rate of retrotransposition by Ty1 and found that the mutant strains showed increased accumulation of Ty1 cDNA. The mutants altered genes encoding products important in DNA repair, including components of the TFIIH complex involved in nucleotide excision repair and control of transcription. The implication is that the DNA repair systems affected normally suppress transposition by obstructing reverse transcription or destroying the Ty1 cDNA once formed. Whether related systems inhibit retroviral replication is an important question for future studies.

MOBILE ELEMENT-ENCODED MECHANISMS FOR CONTROLLING DNA TRANSFER

Mobile elements exercise fastidious control over their own replication. Excessive DNA breaking and joining activity can be lethal to a cell, so careful control of the element-encoded enzymes involved is crucial to evolutionary persistence.

In some cases, a mobile element may maximize its Darwinian success by replicating as rapidly as possible, but this need not always be the most effective strategy. A virus that produced 100 progeny over 1 hour, but killed its host in the process, would leave fewer offspring than one that produced only 50 copies per hour, but did so for 3 hours. Moreover, tight regulation allows a mobile element to sense and respond to environmental factors, permitting it to optimize its replication strategy for the prevailing circumstances.

Many of the element-encoded regulatory systems also act in *trans* to inhibit the movement of related mobile elements as well. This provides the dual benefit of optimizing the relationship with the host while suppressing the proliferation of competing elements.

In this section, we review some of the most prominent element-encoded regulatory mechanisms. Examples of regulatory systems were presented in many of the earlier chapters. Here a few widely used strategies are described to illustrate the diversity of mechanisms known.

Transcriptional Control (6, 29, 48, 54, 62)

Regulation at the level of transcription provides one of the most straightforward means for mobile elements to adjust their levels of activity. Increasing transcription typically results in production of enzymes for replication and mobilization, resulting in increased synthesis of new genomes and transmission to new locations. Most mobile elements contain *cis*-acting DNA sequences that bind the host RNA polymerase and host-encoded transcriptional control proteins. The affinity of these sequences has presumably been optimized by natural selection to optimize the transcription rate, balancing the requirement of producing new elements against the requirement to maintain the production capacity of the host.

A striking example is provided by phage λ, which uses λ repressor to down-regulate transcription in a lysogen (described in Chapter 4).

Repressor binds to DNA sites in the phage right and left operators, thereby turning off operons encoding proteins for lytic growth. The presence of repressor in a lysogenic cell also means that any superinfecting λ genome is unable to initiate either the lytic or lysogenic pathways because repressor binds and turns off transcription. Repressor also serves as a sensor, detecting DNA damage in the host cell and, in response, initiating phage induction and the return to lytic growth. Thus, the action of repressor provides λ with the ability to (1) adopt the quiescent lysogenic state, (2) exclude competitors, and (3) induce when the host cell is damaged. For a detailed description of regulation of λ growth and the role of repressor, see *A Genetic Switch* by Mark Ptashne.

Phage Mu also uses a repressor protein to control induction, but with a twist. Recall from Chapter 4 that Mu forms new genome copies by the unconventional means of replicative transposition. The Mu repressor controls transcription of the transposition genes MuA and MuB, so that induction allows synthesis of transposase and subsequent transposition.

For conjugation, rates of transfer are tightly controlled, often by two-component sensor kinase-transcription factor systems. An example is the *virAG* system described in Chapter 11, which senses AS from wounded plants and consequently induced *vir* gene expression. As a result of tight transcriptional control, the T-DNA transfer apparatus is synthesized only when the conditions are right for gene transfer. In many other cases, the genes for conjugative transfer systems are activated by two-component systems, although for some of these the stimulus for transfer is unknown.

The tetracycline-inducible conjugative transposons of the gram-negative bacterium *Bacteroides* (named Tc^r Em^r DOT and Tc^r Emr 7853) provide an intriguing example of mobility controlled by antibiotic concentration. The presence of low amounts of tetracycline in the growth media can stimulate conjugative transfer 1000- to 10,000-fold. The control mechanisms have not been worked out in detail, although it is clear that complex regulatory machinery is present. The property of inducible transfer may explain why the encoded resistance gene *tetQ* is now found nearly universally in human colonic *Bacteroides* strains–whenever a patient is treated with tetracycline, the resistance determinant is mobilized.

The rate of transcription initiation generally determines the rate of retrovirus production from animal cells harboring a provirus. For the case of HIV, infection of quiescent cells such as resting T cells leads to little or no virus production. However, upon stimulation of cells by mitogens, HIV can be activated and virus production increased. One of the pathways involves the host transcription factor NF-κB, which upon activation binds to the U3 transcriptional control region and stimulates HIV transcription.

For transposons, diverse studies document that transposons with increased transcription rates typically have increased transposition rates. Transposition of Ty elements in yeast, for example, is controlled in part by down-modulation of element transcription. Although Ty elements are normally transcribed at a relatively high rate in yeast cells, further boosting transcription by artificially fusing the Ty element to the strong Gal promoter can greatly increase transposition (under inducing conditions in the presence of the galactose).

Many more examples document the general conclusion that rates of transcription initiation often control the activity of mobile elements. This provides a general control step at which natural selection can tune the opposing needs to produce the most daughter elements while preserving the synthetic capacity of the host.

Regulation by Ordered Assembly (1, 10, 19, 43-46, 61)

Most carefully studied mobile DNA systems have been found to assemble precise nucleoprotein complexes that carry out DNA transfer. The construction of these protein–DNA machines involves an ordered series of steps, requiring that each assembly reaction be correctly completed before the next step is executed. This sequential dependence on correct assembly prevents defective complexes from becoming active, suppressing inappropriate DNA cleavage or joining reactions.

Many of the assembly steps are sensitive to regulatory inputs, coupling correct assembly to the condition of the host cell. For example, in the well-studied case of phage Mu, assembly of active transposition complexes requires the host HU protein and DNA supercoiling, both sensors of the state of the host cell. For transposition by Mu and Tn7, assembly of the active transposition complex requires ATP hydrolysis, linking DNA mobilization to host cell energy levels. Recombination by the resolvase enzyme and HIV integrase is also dependent on correct assembly. Most mobile DNA systems that have been analyzed in detail have been found to employ ordered assembly pathways that sense host cell inputs and suppress side reactions.

Phage-encoded Systems for Blocking Growth of Other Phages (9, 24, 51)

Systems that protect bacterial cells from infection by phages are themselves often encoded by temperate phages, reflecting the struggle among phages for bacterial hosts. One of the most prominent are the lambdoid phage repressors, which suppress growth of incoming phage of the same immunity specificity.

Many phages also encode "heteroimmune exclusion" functions. For example, *rexAB* of λ, the only λ gene expressed in a lysogen besides repressor, encodes a function that blocks replication of phage T4. Curiously, the λ-encoded *rex* system causes the T4-infected cell to die

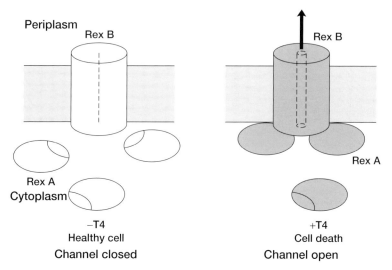

FIGURE 13.8. The *rexAB* heteroimmune exclusion system of phage λ. λ prophages encode two proteins in addition to repressor, RexA and RexB. RexB forms a channel in the bacterial membrane. RexA is required to open the channel. Infection by certain variants of phage T4 alters the activity of RexA, causing RexA to bind RexB, perhaps in multiple copies, permitting the RexB channel to open. Opening the channel leads to death of the infected cell, preventing T4 from multiplying and killing sister lysogens. Wild-type T4 encodes a system that counteracts Rex function, called rII, which allows T4 to replicate on λ lysogens. The name "Rex" derives from "rII exclusion" after its ability to block replication of T4 phages mutant in rII. (Modified, with permission, from Parma et al. 1992. [51])

before producing T4 phage (Fig. 13.8). This seems paradoxical but probably serves to protect the presumably related prophages in nearby cells. Many other phage-encoded heteroimmune exclusion functions are known, relatively few of which have been studied mechanistically.

Regulation of Transformation (2, 13)

Competence for transformation, the ability of bacteria to take up and assimilate DNA from the medium, is inducible in many bacterial species. An example familiar to many biologists is the induction of competence in *E. coli,* which is useful in the routine manipulation of plasmids in genetic engineering. *E. coli*, however, is not known to become competent for transformation during normal growth, but only under laboratory conditions.

Streptococcus species, in contrast, naturally become competent during their life cycles. Growing cultures secrete a factor, called competence factor (CF), that induces the competent state. *Streptococcus* does not take up DNA sequence specifically. The mechanism involves nicking and degradation of one strand, so that single-stranded DNA is taken in. The mechanism by which CF induces competence is not known in detail.

> **The Battles Recorded in Viral Genomes**
>
> The enemies of a virus are recorded in its genes. Bacteriophages are often present in nature in excess over bacteria, leading to a competition among phage for hosts. Consequently, many phage genomes encode functions that exclude other phages from an infected bacterial cell. Many phage also encode inhibitors of the cellular restriction enzymes. For animal viruses, the main enemy is not other viruses but the host immune system, so animal viruses encode diverse inhibitors of the immune response. Inhibitors of the interferon system are also widely seen. Plant viruses seem to have yet another enemy. Several plant viruses have been found to encode inhibitors of the RNAi/cosuppression system, indicating that RNAi is a major mechanism for suppressing viral replication in plants.

Hemophilus influenzae can be induced to become competent by switching from rich medium to minimal medium that fails to support cell division. Induction of competence correlates with multiple changes in cell physiology, including a change in membrane lipids and the appearance of new membrane structures. Cells also synthesize a set of membrane proteins that bind the 5′-AAGTGCGGTCA-3′ sequence that specifies uptake. In this case, the DNA is internalized in double-stranded form. *Neisseria gonorrhoeae* probably uses related machinery, although *Neisseria* are always in the competent state. As discussed in Chapter 11, antigenic switching in *Neisseria* may be linked to transformation and uptake of pilin genes from lysed neighboring cells.

For eukaryotes, transformation can be induced under laboratory conditions, for example, by subjecting cells to electric shocks in the presence of added DNA. Transformation is not known to take place naturally in eukaryotes, although future data could change this view.

Dominant Negative Transposase Proteins (21, 40, 41, 56)

For transposons, a widely used mechanism of regulation involves poisoning transposition with inactive transposase subunits. Biological functions carried out by multiprotein complexes can often be poisoned by incorporation of inactive components. Such mutations are genetically dominant, since the phenotype of a strain containing both mutant and wild-type alleles is mutant. Intriguingly, dominant negative transposase mutants also seem to have evolved to control transposition in nature.

Drosophila P elements, for example, encode two different forms of transposases, an 87-kD transposase and a 66-kD inhibitor. The two forms are produced from the same transposase gene but differ by alternate mRNA splicing. The transposase gene contains three introns. Removal of all three introns yields the message for the 87-kD transposase, whereas retention of the third intron produces the message for the 66-kD inhibitor. The splicing factors responsible for removing the third intron are found only in germ-line cells, thereby restricting transposition to the germ line. The presence of the 66-kD form may also be at least in part responsible for suppressing the excessive transposition seen in P-element hybrid dysgenesis. Thus, the dominant negative transposase form may be advantageous to the element by preventing undue transposition that would otherwise debilitate the host.

Regulation by dominant negative transposase forms has also been proposed for the bacterial transposon Tn5. Tn5 is a composite transposon, comprising IS50L on the left side and IS50R on the right. IS50L does not encode any proteins, but IS50R encodes two, a transposase protein and a *trans*-acting inhibitor. As with P elements, the inhibitor apparently acts as a poison subunit that inhibits transposition either by binding wild-type transposase monomers or by binding transposase sites in the element DNA. Production of the inhibitor protein is linked to methylation of the promoter, allowing transposition to be coupled to replication of the *E. coli* chromosome.

Transposition of *mariner* elements provides another example. Fly strains were studied that contained *mariners* with both wild-type and mutant transposase genes. Tests showed that the defective transposase proteins could obstruct the function of the wild type. The accumulation of such mutants may have contributed to the inactivation and ultimate death of *mariner* families seen in many organisms over evolutionary time.

The observation that transposons encode inhibitors of their own transposition emphasizes the importance of adjusting transposition rates to optimize the welfare of both the element and the host.

Translational Control of Transposase Synthesis (32, 60)

The bacterial transposon Tn10 displays a unique mechanism for control of transposition based on translational control of transposase synthesis. Tn10 is a composite transposon, composed of two IS elements, IS10L and IS10R, which flank genes conferring resistance to tetracycline. IS10R encodes transposase, IS10L is inactive. The transposase gene of IS10R is expressed from a promoter named pIN (see Fig. 3.10B). A second promoter, pOUT, lies within the transcription unit and reads in the opposite direction toward pIN. As a result of the relative positions of the promoters, RNA is synthesized from both DNA strands over a 35-bp region, allowing formation of double-stranded RNA in the overlapping bases. This overlap region includes the translation initiation codon for the transposase gene, and extensive studies establish that RNA–RNA pairing suppresses initiation of transposase synthesis. The pOUT promoter is actually stronger than that of pIN, and the pOUT message is more stable. The net result is multicopy inhibition: The greater the number of transposons in a cell, the less efficient the translation of transposase from each transposon copy. Thus, Tn10 itself encodes a mechanism to restrain its own transposition and so suppress accumulation of large numbers of element copies in a single cell. Tn10 also regulates transposition by several other means, including preferential *cis* action of transposase protein and methylation-dependent inhibition of transposition (Chapter 3), all of which together tightly limit the accumulation of element copies.

Summary

Newly introduced DNA can contribute to the welfare of a cell, as with transmission of useful new genes, or kill a cell, as with infection by a lytic virus. Consequently, cells devote considerable energy to controlling the fate of incoming DNA.

In prokaryotes, cells encode restriction enzymes that bind to specific DNA sequences and cleave the DNA at or near the recognition site. Cells avoid degrading their own DNA by pairing each restriction enzyme with a DNA methylase that modifies the DNA recognition site and prevents cleavage. The DNA of an incoming phage is not expected to be protected by methylation, so phage DNA will be cleaved by host restriction enzymes, and the cell will be protected from infection.

In those cases where a newly transferred DNA does not enter the cell equipped with machinery for integration or extrachromosomal replication, persistence requires integration into the host genome by homologous recombination. The DNA repair system responsible for correcting mismatches antagonizes the action of the homologous recombination system. In prokaryotes this is dramatically illustrated by the observation that *E. coli* and *Salmonella* can mate only if the mismatch repair system is inactivated. Similar results are seen in eukaryotes.

Eukaryotes have evolved a fascinating set of regulatory mechanisms for suppressing the activity of viruses and mobile elements. The cosuppression/RNAi pathway apparently involves a cellular RNase enzyme that binds a short double-stranded RNA. The RNase/dsRNA complex then uses sequence complementarity to recognize a specific mRNA target and destroy it nucleolytically. This response is even heritable, possibly because RNA-dependent RNA polymerase enzymes copy the targeting RNA.

The filamentous fungus *Neurospora* treats most repeated DNAs as unwelcome invaders, because most of its own genes are present as single copies and respond with repeat-induced point mutations (RIP). *Neurospora* methylates repeated sequences, which when followed by deamination of methyl-C, yield C-to-T substitutions. Thus, repeated sequences of mobile elements are mutated to destruction.

In vertebrates and many plants, the activity of mobile elements is controlled by DNA methylation. Attaching methyl groups to C residues at CpG dinucleotides in gene control regions results in a reduction of the transcription rate. Endogenous retroviruses in vertebrates and DNA transposons in plants are often selectively methylated and prevented from replicating.

In animals, the immune system provides a potent obstacle to DNA transfer. For eradication of virus infection, the cellular immune response is most prominent. Virus-infected cells are identified by the appearance of virus-derived peptides bound to histocompatibility proteins, which are then recognized by the T-cell receptor on cytotoxic T cells, triggering killing of the infected cells. Antibodies also bind viruses and program their destruction.

The innate immune system responds to infections by destroying pathogens and inducing the adaptive immune response. Cells of the innate immune system express surface receptors that recognize molecules characteristic of invaders, such as peptidoglycan or lipopolysaccharide. One receptor of the innate immune system binds and signals in response to unmethylated CpG sequences in DNA, which are also markers for invaders, providing another block to gene transfer to vertebrate cells.

Another antiviral response is that induced by interferon. Interferon treatment induces synthesis of 2′–5′ A synthetase and the double-stranded RNA-dependent protein kinase. In the presence of double-stranded RNA, presumably derived from an infecting virus, these effectors shut down translation in the infected cell, blocking further virus production and containing the infection.

Mobile elements encode diverse systems for regulating their own activity, allowing the replication rate to be optimized and the burden on the host to be minimized. One simple means of control is at the level of transcription initiation. Production of enzymes for mobilization is under transcriptional control for most elements, and experimental alteration of the transcription rates in many cases alters the rate of element movement. Transcriptional control also provides a relay by which cellular signals can be sensed and acted on by a resident mobile element.

Several types of element-encoded regulatory systems block replication of other mobile elements on the same cell. The lambdoid phage repressors provide a classic example, not only turning off replication of the prophage lytic genes, but also blocking replication of superinfecting phages of the same immunity type. Many phages also encode heteroimmune exclusion functions, which block replication of other types of phages, such as the λ *rex* system.

Several transposons, including Tn*5* and P elements, encode dominant negative transposase proteins that regulate transposition. These inhibitors reduce transposition directed by wild-type transposase, allowing transposition to be controlled by regulation of inhibitor production. Transposition in the case of Tn*10* is regulated at the level of translation of the transposase message by RNA–RNA interactions.

Transformation is also regulated in some prokaryotes. In *Neisseria*, switching of genes for surface antigens has been reported to be linked to transformation, aiding in immune evasion.

In summary, both the host cell and mobile DNA elements themselves encode diverse functions that modulate the rate of DNA transfer and thus their prospects for evolutionary persistence.

REFERENCES

1. Adzuma K. and Mizuuchi K. 1989. Interaction of proteins located at a distance along DNA: Mechanism of target immunity in the Mu DNA strand-transfer reaction. *Cell* **57:** 41–47.

2. Avery O.T., MacLeod C.M., and McCarty M. 1944. Studies on the chemical nature of the substance inducing transformation of pneumococcal types. Induction of transformation by a desoxyribonucleic acid fraction isolated from pneumococcus type III. *J. Exp. Med.* **79:** 137–159.

3. Baca-Regen L., Heinzinger N., Stevenson M., and Gendelman H.E. 1994. Alpha interferon-induced antiretroviral activities: Restriction of viral nucleic acid synthesis and progeny virion production in human immunodeficiency virus type 1-infected monocytes. *J. Virol.* **68:** 7559–7565.

3a. Bernstein E., Caudy A.A., Hammond S.M., and Hannon G. 2001. Role for a bidentate ribonuclease in the initiation step of RNA interference. *Nature* **409:** 295–296.

4. Best S., Le Tissier P., Towers G., and Stoye J.P. 1996. Positional cloning of the mouse retrovirus restriction gene *Fv*1. *Nature* **382:** 826–829.

5. Bestor T.H. 2000. The DNA methyltransferases of mammals. *Hum. Mol. Genet.* **9:** 2395–2402.

6. Boeke J.D. 1989. Transposable elements in *Saccharomyces cerevisiae*. In *Mobile DNA* (ed. D.E. Berg and M.M. Howe), pp. 335–374. American Society for Microbiology, Washington, D.C.

7. Boeke J.D. and Stoye J.P. 1997. Retrotransposons, endogenous retroviruses, and the evolution of retroelements In *Retroviruses* (ed. J.M. Coffin et al.), pp. 343–435. Cold Spring Harbor Laboratory Press, Cold Spring Harbor, New York.

8. Butler S.L., Johnson E., and Bushman F.D. 2001. HIV cDNA metabolism studied by fluorescence-monitored PCR: Notable stability of two-LTR circles. (in prep.)

9. Calendar R. 1988. *The bacteriophages*. Plenum Press, New York.

10. Chaconas G. 1999. Studies on a "jumping gene machine": Higher-order nucleoprotein complexes in Mu DNA transposition. *Biochem. Cell Biol.* **77:** 487–491.

11. Cheng X., Balendiran K., Schildkraut I., and Anderson J.E. 1994. Structure of PvuII endonuclease with cognate DNA. *EMBO J.* **13:** 3927–3935.

12. Curcio M.J. and Garfinkel D.J. 1999. New lines of host defense: Inhibition of Ty1 retrotransposition by Fus3p and NER/TFIIH. *Trends Genet.* **15:** 43–45.

13. Davison J. 1999. Genetic exchange between bacteria in the environment. *Plasmid* **42:** 73–91.

14. de Lang P., van Blokland R., Kooter J.M., and Mol J.N. 1995. Suppression of flavonoid flower pigmentation genes in Petunia hybrida by the introduction of antisense and sense genes. *Curr. Top. Microbiol. Immunol.* **197:** 57–75.

15. Denamur E., Lecointre G., Darlu P., Tenaillon O., Acquaviva C., Sayada C., Sunjevaric I., Rothstein R., Elion J., Taddei F., Radman M., and Matic I. 2000. Evolutionary implications of the frequent horizontal transfer of mismatch repair genes. *Cell* **103:** 711–721.

16. Elbashir S.M., Harborth J., Lendeckel W., Yalcin A., Weber K., and Tuschl T. 2001. Duplexes of 21-nucleotide RNAs mediate RNA interference in cultured mammalian cells. *Nature* **411:** 494–498.

17. Fedoroff N. 1989. Maize transposable elements. In *Mobile DNA* (ed. D.E. Berg and M.M. Howe), pp. 375–411. American Society for Microbiology, Washington, D.C.

18. Fields B.N., Knipe D.M., and Howley P.M. 1996. *Fields Virology*, 3rd edition. Lippincott-Raven, Phildelphia, Pennsylvania.

19. Gao K., Butler S.L., and Bushman F.D. 2001. Human immunodeficiency virus type 1 integrase: Arrangement of protein domains in active cDNA complexes. *EMBO J.* **20:** 3565–3576.

20. Grilley M., Griffith J., and Modrich P. 1993. Bidirectional excision in methyl-directed mismatch repair. *J. Biol. Chem.* **268:** 11830–11837.

21. Hartl D.L., Lozovskaya E.R., Nurminsky D.I., and Lohe A.R. 1997. What restricts the activity of mariner-like transposable elements? *Trends Genet.* **13:** 197–201.

22. Hayashi F., Smith K.D., Ozinsky A., Hawn T.R., Yi E.C., Goodlett D.R., Eng J.K., Akira S., Underhill D.M., and Aderem A. 2001. The innate immune response to bacterial flagellin is mediated by Toll-like receptor 5. *Nature* **410:** 1099–1103.

23. Hemmi H., Takeuchi O., Kawai T., Kaisho T., Sato S., Sanjo H., Matsumoto M., Hoshino K., Wagner H., Takeda K., and Akira S. 2000. A toll-like receptor recognizes bacterial DNA. *Nature* **408:** 659–660.

24. Hendrix R.W., Roberts J.W., Stahl F.W., and Weisberg R.A. 1983. *Lambda II*. Cold Spring Harbor Laboratory, Cold Spring Harbor, New York.

25. Horton J.R. and Cheng X. 2000. PvuII endonuclease contains two calcium ions in active sites. *J. Mol. Biol.* **300:** 1049–1056.

26. Hsieh H. and Fire A. 2000. Recognition and silencing of repeated DNA. *Annu. Rev. Genet.* **34:** 187–204.

27. Isaacs A. and Lindenmann J. 1957. Virus interference 1. The interferon. *Proc. R. Soc. Lond. B Biol. Sci.* **147:** 258–267.

28. Janeway C.A., Travers P., Walport M., and Capra J.D. 1999. *Immunobiology:*

The immune system in health and disease, 4th edition. Elsevier, London.

29. Kaiser A.D. and Jacob F. 1957. Recombination between related temperate bacteriophages and the genetic control of immunity and prophage localization. *Virology* **4:** 509–521.

29a. Ketting R.F., Haverkamp T.H., van Luenen H.G., and Plasterk R.H. 1999. Mut-7 of *C. elegans* required for transposon silencing and RNA interference is a homolog of Werner Syndrome helicase and RNase D. *Cell* **99:** 133–141.

30. Kimbrell D.A. and Beutler B. 2001. The evolution and genetics of innate immunity. *Nat. Rev. Genet.* **2:** 256–267.

31. Kinsey J.A., Garrett-Engele P.W., Cambareri E.B., and Selker E.U. 1994. The neurospora transposon Tad is sensitive to repeat-induced point mutation (RIP). *Genetics* **138:** 657–664.

32. Kleckner N. 1989. Transposon Tn*10*. In *Mobile DNA* (ed. D.E. Berg and M.M. Howe), pp. 227–268. American Society for Microbiology, Washington, D.C.

33. Kobayashi I., Nobusato A., Kobayashi-Takahasi N., and Uchiyama I. 1999. Shaping the genome–restriction modification systems as mobile genetic elements. *Curr. Opin. Genet. Dev.* **9:** 649–656.

34. Kornberg A. and Baker T. 1991. *DNA replication.* W.H. Freeman, New York.

35. LeClerc J.D., Li B., Payne W.L., and Cebula T.A. 1996. High mutation frequencies among *Escherichia coli* and *Salmonella* pathogens. *Science* **274:** 1208–1211.

36. Limjoco T.I., Dickie P., Ikeda H., and Silver J. 1993. Transgenic Fv-4 mice resistant to Friend virus. *J. Virol.* **67:** 4163–4168.

37. Madani N. and Kabat D. 1998. An endogenous inhibitor of human immunodeficiency virus in human lymphocytes is overcome by the viral Vif protein. *J. Virol.* **72:** 10251–10255.

38. Marx J. 2000. Interfering with gene expression. *Science* **288:** 1370–1372.

39. Matic I., Rayssiguier C., and Radman M. 1995. Interspecies gene exchange in bacteria: The role of SOS and mismatch repair systems in evolution of species. *Cell* **80:** 507–515.

40. McCommas S. and Syvanen M. 1988. Temporal control of transposition in Tn5. *J. Bacteriol.* **170:** 889–894.

41. Misra S. and Rio D.C. 1990. Cytotype control of *Drosophila* P element transposition: The 66 kd protein is a repressor of transposase activity. *Cell* **62:** 269–284.

42. Miura A., Yonebayashi S., Watanabe K., Toyama T., Shimada H., and Kakutani T. 2001. Mobilization of transposons by a mutation abolishing full DNA methylation in *Arabidopsis. Nature* **411:** 212–214.

43. Mizuuchi M., Baker T.A., and Mizuuchi K. 1992. Assembly of the active form of the transposase-Mu DNA complex: A critical control point in Mu transposition. *Cell* **70:** 303–311.

44. Mizuuchi M., Baker T.A., and Mizuuchi K. 1995. Assembly of phage Mu transpososomes: Cooperative transitions assisted by protein and DNA scaffolds. *Cell* **83:** 375–385.

45. Murley L.L. and Grindley N.D.F. 1998. Architecture of the γ-δ resolvase synaptosome: Oriented heterodimers identify interactions essential for synapsis and recombination. *Cell* **95:** 553–562.

46. Nash H. 1996. The HU and IHF proteins. In *Regulation of gene expression in* Escherichia coli (ed. E.C.C. Lin and A.S. Lynch), pp. 149–179. R. G.

Landes, Austin, Texas.

47. Nassif N. and Engels W.R. 1993. DNA homology requirements for mitotic gap repair in *Drosophila*. *Proc. Natl. Acad. Sci.* **90:** 1262–1266.

48. Nikolich M.P., Hong G., Shoemaker N.B., and Salyers A.A. 1994. Evidence for natural horizontal transfer of tetQ between bacteria that normally colonize humans and bacteria that normally colonize livestock. *Appl. Environ. Microbiol.* **60:** 3255–3260.

49. Nowak R.M. 1999. *Walker's mammals of the world*, 6th edition. Johns Hopkins University Press, Baltimore, Maryland.

50. O'Neill R.J., O'Neill M.J., and Graves J.A. 1998. Undermethlylation associated with retroelement activation and chromosome remodelling in an interspecific mammalian hybrid. *Nature* **393:** 68–72.

51. Parma D.H., Snyder M., Sobolevski S., Nawroz M., Brody E., and Gold L. 1992. The Rex system of bacteriophage lambda: Tolerance and altruistic cell death. *Genes Dev.* **6:** 497–510.

52. Plasterk R.H. and Ketting R.F. 2000. The silence of the genes. *Curr. Opin. Genet. Dev.* **10:** 562–567.

53. Poignard P., Sabbe R., Picchio G.R., Wang M., Gulizia R.J., Katinger H., Parren P.W., Mosier D.E., and Burton D.R. 1999. Neutralizing antibodies have limited effects on the control of established HIV-1 infection in vivo. *Immunity* **10:** 431–438.

54. Ptashne M. 1992. *A genetic switch*, 2nd edition. Cell Press and Blackwell Scientific, Cambridge, Massachusetts.

55. Rayssiguier C., Thaler D.S., and Radman M. 1989. The barrier to recombination between *Escherichia coli* and *Salmonella typhimurium* is disrupted in mismatch-repair mutants. *Nature* **342:** 396–401.

56. Reznikoff W.S. 1993. The Tn5 transposon. *Annu. Rev. Microbiol.* **47:** 945–963.

57. Riggs A.D., Xiong Z., Wang L., and LeBon J.M. 1998. Methylation dynamics, epigenetic fidelity and X chromosome structure. *Novaritis Found. Symp.* **214:** 214–225.

58. Schwartz O., Maréchal V., Friguet B., Arenzana-Seisdedos F., and Heard J.-M. 1998. Antiviral activity of the proteasome on incoming human immunodeficiency virus type 1. *J. Virol.* **72:** 3845–3850.

59. Simon J.H.M., Gaddis N.C., Fouchier R.A.M., and Malim M.H. 1998. Evidence for a newly discovered cellular anti-HIV-1 phenotype. *Nat. Med.* **4:** 1397–1400.

60. Simons R.W. and Kleckner N. 1983. Translational control of IS10 transposition. *Cell* **34:** 683–691.

61. Stellwagen A.E. and Craig N.L. 1998. Mobile DNA elements: Controlling transposition with ATP-dependent molecular switches. *Trends Biochem. Sci.* **23:** 487–490.

62. Stevens A.M., Shoemaker N.B., Li L.Y., and Salyers A.A. 1993. Tetracycline regulation of genes on *Bacteroides* conjugative transposons. *J. Bacteriol.* **175:** 6134–6141.

63. Svoboda P., Stein P., Hayashi H., and Schultz R.M. 2000. Selective reduction of dormant maternal mRNA in mouse oocytes by RNA interference. *Development* **127:** 4147–4156.

64. Thomas C.L., Jones L., Baulcombe D.C., and Maule A.J. 2001. Size constraints for targeting post-transcriptional gene silencing and for RNA-directed methylation of *Nicotiana benthamiana* using a potato virus X vector. *Plant J.* **25:** 417–425.

65. Tuschl T., Zamore P.D., Lehmann R., Bartel D.P., and Sharp P.A. 1999. Targeted mRNA degradation by double-stranded RNA in vitro. *Genes Dev.* **13:** 3191–3197.

66. Vulic M., Dionisio F., Taddei F., and Radman M. 1997. Molecular keys to speciation: DNA polymorphism and the control of genetic exchange in enterobacteria. *Proc. Natl. Acad. Sci.* **94:** 9763–9767.

67. Walker B.D. and Goulder P.J. 2000. AIDS. Escape from the immune system. *Nature* **407:** 386–390.

68. Walsh C.P., Chaillet J.R., and Bestor T.H. 1998. Transcription of IAP endogenous retroviruses is constrained by cytosine methylation. *Nat. Genet.* **20:** 116–117.

69. Watters M.K., Randall T.A., Margolin B.S., Selker E.U., and Stadler D.R. 1999. Action of repeat-induced point mutation on both strands of a duplex and on tandem duplications of various sizes in Neurospora. *Genetics* **153:** 705–714.

70. Zamore P.D., Tuschl T., Sharp P.A., and Bartel D.P. 2000. RNAi: Double-stranded RNA directs the ATP-dependent cleavage of mRNA at 21 to 23 nucleotide intervals. *Cell* **101:** 25–33.

71. Zijlstra M., Li E., Sajjadi F., Subramani S., and Jaenisch R. 1989. Germ-line transmission of a disrupted β 2-microglobulin gene produced by homologous recombination in embryonic stem cells. *Nature* **342:** 435–438.

Lateral DNA Transfer: Themes and Evolutionary Implications

THIS FINAL CHAPTER SUMMARIZES SOME of the main themes of the book, emphasizing the generalizations emerging from a broad overview, and then treats theories connecting DNA dynamics to larger issues in biology. The origin of introns, the origin of sex, and the early evolution of life have all been proposed to be associated with lateral DNA transfer and are covered in these sections. The book concludes with a few speculations on the future.

LATERAL DNA TRANSFER: THEMES AND GENERALIZATIONS

Lateral Transfer in the Different Domains of Life: A Question of Phenotype (5, 8, 17, 19, 29, 33, 38, 44, 51)

The frequency of lateral transfer and many of the mechanisms of lateral transfer differ among the domains of life. As a result, the genetic consequences differ as well. This section contrasts lateral DNA transfer in the Bacteria and Eucarya. The Archaea probably resemble the Bacteria, although there are some unique features such as the presence of split-*int* cassettes (Chapter 5).

One generalization is quite prominent in comparing prokaryotes to eukaryotes. Only in prokaryotes are gene sets responsible for whole attributes—genetic phenotypes—transferred in a single step. In eukaryotes, new sequences are transferred frequently but do not typically confer heritable phenotypes immediately upon incorporation.

The different rates of lateral transfer likely account for much of this difference. Lawrence and Ochman have argued that *Escherichia coli* captures at least 64.2 kilobases of new DNA by lateral transfer per million years, or about 1.4% of the genome. Of this, at least 16 kilobases of DNA per million years is estimated to become fixed and so persist during subsequent evolution. Studies of additional prokaryotes indicate that frequencies range from very little transfer to even more than in *E. coli*. Likely mechanisms can be specified—transduction, conjugation, and transformation—and each pathway is known to operate in nature. The overall picture is one of continuous high rates of lateral transfer, although of course in natural settings the rates probably vary widely.

The situation is much different with multicellular eukaryotes. There are a few systems that can transfer DNA between organisms efficiently, such as T-DNA transfer or retroviral infection, but these systems introduce new sequences into the germ line only infrequently. Consequently, heritable changes are rare, requiring evolutionary time scales for substantial accumulation of new DNA. Only eukaryotes have organelles, and so only this group can capture sequences by transfer from the organelle genome to the nuclear genome. This situation has the consequence of scattering bacterial-derived genes in the eukaryotic gene complement, but again, incorporation of new sequences into the metazoan germ line is slow.

As a consequence of the different mechanisms of lateral transfer, blocks of genes encoding complete evolutionary traits are often transferred among prokaryotes, but only rarely, if ever, among eukaryotes. For bacteria, as was discussed in Chapters 3–5, blocks of transferred genes allow evasion of antibiotic treatments, invasion of new hosts, or growth in the presence of pollutants. In eukaryotes the lower frequency and different character of lateral transfer leaves acquisition of genes encoding whole traits unlikely, and no examples are reported in the preceding chapters. Instead, eukaryotes seem to acquire junk DNA from genomic parasites and occasional new genes from bacteria, but few, if any, of these benefit the host eukaryote immediately, and none provides multiple genes for a new phenotype.

There are a few proposals for such massive gene transfer in eukaryotes, but the data are not persuasive. A few researchers have proposed that whole banks of genes may have been transferred between eukaryotes, resulting in the similar (homoplastic) structures, for example in marine larvae or angiosperm sexual structures. At present, it seems unlikely that these cases involve lateral transfer. Phylogenetic analysis of genes encoding components of the putative homoplastic structures should allow the issue to be settled experimentally.

Lateral transfer of phenotypes is one of the most dramatic differences in genetic exchange between domains. This difference has likely dictated in large measure the nature of evolutionary innovation in each domain and accelerated their divergence over time.

Selfish DNA and Evolutionary Innovation (1, 10–12, 14, 26, 32, 34, 40, 41, 47)

Prokaryotic genomes are quite streamlined, with roughly 85–95% of the genome encoding genes. In metazoans, genes are much more sparsely distributed, with only 1.5% of the human genome made up of protein coding regions. What is the origin of the rest of the eukaryotic DNA? Does it have a function?

As discussed in Chapters 6–9, much of the "junk" DNA in eukaryotes is derived from mobile elements, selfish DNAs that seek to achieve Darwinian success by expanding their numbers without contributing to the welfare of the host. In humans, at least 45% of the DNA arose from mobile elements or their action. LINE elements comprise roughly 20% of the genome, and *Alu* elements, likely formed by LINE element enzymes, another 11%. Miscellaneous additional endogenous proviruses, DNA transposons, and others make up the rest. In addition, the pseudogenes, derived from the action of retrotransposon machinery on cellular mRNA templates, add to the junk. The origin of the rest of the junk DNA is obscure, but it is likely to be largely ancient mobile DNA fragments that are just no longer recognizable due to genetic drift.

None of these observations proves that the junk DNA has no function. Evolution is inherently opportunistic, and it comes as no surprise that the exuberant sequence variation present in junk DNA is occasionally exploited by the host cell. In fruit flies, the telomeres are formed by LINE-like mobile elements. In many eukaryotes, the centromeres also contain abundant mobile element sequences. Here the junk DNA may act as an architectural element, a neutral spacer that is specified as a centromere by epigenetic mechanisms. Studies of the birth of genes have identified many cases of recruitment of probable junk DNA sequences to form new genes. At least 47 human genes appear to have evolved by incorporation of mobile element fragments, including the *RAG* genes that founded the adaptive immune system. Even *Alu* elements may have a function, possibly helping to regulate the cellular stress response by inhibiting the PKR kinase.

Selfish DNA can also contribute to the welfare of the host by seeding the host chromosomes with patches of DNA sequence homology that serve as sites for homologous recombination. These systems recognize the identical DNA sequences, break strands at each, and rejoin the DNAs in new arrangements. Homologous recombination is found in all organisms studied, perhaps because it is essential for restarting stalled replication forks and repairing damaged DNA. Homologous recombination reactions involving two sites on a single chromosome can generate DNA inversions or excision of the intervening DNA. Homologous recombination between sites on different chromosomes can lead to translocations (on linear chromosomes) or cointegrate formation (on circular chromosomes). Innumerable laboratory experiments document recombination at homologies supplied by transposons or integrating viruses. Naturally occurring examples include the two complete *Helicobacter pylori* sequences, which differ by inversions with mobile elements at the DNA borders. Similarly in eukaryotes, studies of Ty1 in different *Saccharomyces* species suggest that chromosomal rearrangements have been potentiated by Ty1 homology at the points of exchange. The draft human genome sequence revealed myriad segmental duplications of sequences; although the mechanism of their formation has not been established, it seems likely that homologous recombination at repeated transposon sequences has contributed.

Evolution is inherently opportunistic—consequently, variation due to mobile element activity can generate innovation, some of which may be positively selected. These points were anticipated by Leslie Orgel and Francis Crick in the initial articulation of the theory of Selfish DNA. Today we can flesh out this idea with many specific examples, including insect telomeres, new human genes, and the evolution of the human immune system.

Genome Structure and Lateral Transfer

Gene Clustering (13, 33, 38, 39)

Frequent lateral DNA transfer appears to promote clustering of genes that work together, at least in some cases and perhaps quite widely. J.G. Lawrence and J.R. Roth articulated this idea for prokaryotes as "The Selfish Operon Hypothesis." If genes are frequently subject to lateral DNA transfer, their likelihood of persistence in the new host may be related to how well they function there. A gene that arrives in a new host without required partners will not function. Such sequences will presumably be lost by genetic drift like other functionless DNA. However, if a newly introduced sequence is functional in the new host, it may be positively selected if the new activity is useful.

Many genes function as groups, such as the enzymes acting sequentially in a biosynthetic pathway. Transfer of single genes would usually not allow function, so single genes would not be positively selected. However, if all the genes in the pathway were transferred, there would be the opportunity for selection in the new host. Iterative cycles of gene reshuffling in the chromosome, lateral transfer, and selection could thus give rise to clustering of genes with related functions. Clustering of genes in operons is seen in many bacteria, probably due at least in part to this mechanism, although other forces may also contribute.

Gene clustering in eukaryotes may also be due to lateral transfer in at least a few cases. One notable example is the recombination activating genes (*RAGs*) 1 and 2. Both of these genes are needed for DNA rearrangements in the vertebrate adaptive immune system. The two genes are tightly linked, probably because they originated in an ancestral transposon that gave rise to the immune system. The origin of the progenitor transposon is obscure, but capture by lateral transfer is a reasonable guess. Other forms of eukaryotic gene clustering may be due to lateral transfer as well. For example, for genes captured by the nuclear genome from organelle chromosomes, clustering in the organelle genome may be preserved after transfer to the nuclear genome. Ongoing studies should clarify whether lateral transfer is a major force for clustering of eukaryotic genes.

Genome Size (4, 16, 25, 50)

Most bacteria have extremely compact genomes, unlike their eukaryotic relatives. Bacterial genomes are typically 85–95% coding, whereas eukaryotes range from 70% coding (the yeast *Saccharomyces cerevisiae*) to 1.5% coding (human). Bacteria seem to be under intense pressure to minimize the sizes of their genomes, possibly to allow rapid replication. Some bacteria are extremely streamlined, such as *Mycoplasma genitalium*, which has a mere 470 genes in 580,070 bases of DNA (Plate 1), and no detectable mobile elements. Yeasts may be under similar pressures to minimize genome size.

Humans are much less compact, and a few other eukaryotes are even worse. The human genome comprises some 3.4 billion base pairs. Gorillas, fish, and corn plant genomes are similar in size. Lily plants, in contrast, have genomes some ten times larger, fully 36 billion base pairs, and some *Amoebae* species are nearly tenfold larger still. This long-standing mystery has been named the C-value paradox.

A comparison of rice, maize, and barley, with genomes of 0.4, 3.2, and 4.8 billion base pairs, respectively, indicates that retrotransposons are responsible for much of the difference in size, accounting for 14%, 50–80%, and >70% of the genomes, respectively. Thus, the C-value paradox seems to find at least a partial explanation as a proliferation of mobile elements.

Of course, one then wonders what made the retroelements increase in number so dramatically. The answer is unknown, and speculation ranges widely. Models can be divided between those that invoke a function (sequence-independent) for some or all of the "junk" DNA, and those that attribute the expansion to pure selfishness on the part of genomic parasites. Some models in the functional class invoke the well-established correlation between genome size, nuclear size, and cell size. For example, it has been suggested that nuclei require a certain amount of DNA to "fill them up," so that small nuclei require small genomes and large nuclei require larger genomes. According to this idea, external forces specified nuclear size or cell size, and the genomes expanded as needed to fill the available space. Evolution of a larger organ or tissue might proceed by an increase in cell size, requiring an increase in DNA content. Another class of functional models suggests that the junk DNA may act as a buffer—for example, as a sink for destructive activated oxygen species. According to this idea, junk DNA would serve to soak up damaging chemicals, thereby protecting the coding DNA. Other models hold that genome expansion is evolutionarily neutral or only slightly negative. Some of these explanations look to explain the C-value paradox by forces affecting the activity of mobile elements, particularly retrotransposons. For example, many elements are induced to transpose by stress, so perhaps stressful conditions will induce an increase in genome size. This idea finds some support in studies of wild barley plants, in which the BARE transposons expanded in plants subject to drought. More research on the genome structure and biology of organisms with the most expanded genomes may shed new light on this intriguing puzzle.

Is there a link to lateral DNA transfer? One speculation holds that mobile element proliferation following introduction by lateral transfer, as with hybrid dysgenesis in fruit flies, could cause an increase in genome size. At present there is no evidence for or against this view. Future studies should help clarify the intriguing question of whether lateral transfer and genome expansion are linked.

Regulation of DNA Transfer (6, 7, 20, 27, 28, 36, 42, 43, 46, 48)

Rates of DNA mobility and DNA transfer are tightly regulated by both the host and the mobile element. From the host's point of view, many incoming sequences are damaging or fatal, as with lytic viruses, so a great deal of machinery has evolved to suppress DNA transfer. On the other hand, some types of DNA transfer are beneficial, such as acquisition of new surface antigen genes by transformation, and special machinery has evolved for this job, too.

From the point of view of a DNA parasite, it might initially seem that maximizing the rate of reproduction would lead to greater Darwinian success, but this need not always be the case. A relatively benign parasite replicating at a low rate might produce more offspring over time in the host genome than a parasite replicating at a high rate but quickly killing its host. Optimizing the production of offspring requires balancing the rate of reproduction with the welfare of the host. Tight regulation also allows cellular signals to be sensed by a mobile element and exploited to optimize their replication strategy. Accordingly, the systems that mobilize DNA tightly regulate their own activity.

The host-encoded regulatory systems involved generally differ between prokaryotes and eukaryotes. Prokaryotes defend themselves primarily with restriction enzymes, which cleave unmodified foreign DNA, and methylases, which modify and protect their own genomes. Eukaryotes have evolved a diverse array of systems, including RNAi/cosuppression, RIPing, methylation, the interferon system, and the adaptive and innate immune systems. The host response to genomic parasites is an area of intense research, and it is likely that further work will expand this list.

One host defense seems to be similar between the prokaryotes and eukaryotes. The mismatch repair system obstructs homologous recombination of related but nonidentical systems. *E. coli* will not normally mate with *S. typhimurium*, and the block is at the level of DNA incorporation in the recipient cell. However, cells of the two species can mate and incorporate new sequences if the mismatch repair system is eliminated by mutation. Similarly in eukaryotes, homologous recombination is inefficient if the incoming DNA has a few mismatches with the recipient site in the chromosome but much more efficient if they are identical. This aspect of DNA metabolism is apparently evolutionarily ancient, indicating its importance. Factors that modulate the activity of the mismatch repair system, in turn, have been proposed to modulate the frequency of lateral transfer as well.

On the element side, the rate of transcription initiation provides a control point for sensing cellular inputs and directing life-style decisions. Many mobile DNA systems display different transcriptional activity depending on the state of the host, allowing optimization of their replication strategy. The phage λ right operator provides a classic example, controlling whether a prophage chooses lysogeny or induction and lytic growth in response to cellular signals. The *Agrobacteria vir* genes responsible for T-DNA transfer are induced at the transcriptional level by AS, a plant compound emitted from wounded tissue at potential sites of invasion. Myriad other prokaryotic and eukaryotic genomic parasites are known to control replication at the level of transcription.

A variety of other mechanisms have been evolved by mobile elements to control gene transfer. P elements of flies and Tn5 of *E. coli* encode transposase derivatives that inhibit instead of stimulate transposition. Tn*10* encodes an elaborate antisense RNA system that controls transposase synthesis. Archaea "split-int" systems divide the integrase gene into two pieces upon recombination, thereby controlling excision. The diverse means of controlling DNA mobility highlight the opportunistic nature of biological regulation.

A sophisticated mode of regulation involves control of assembly of the nucleoprotein complexes involved in DNA transfer. Phage λ integration, for example, requires the assembly of multiple λ integrase monomers with the host protein IHF in a specifically wrapped complex, the intasome. Several points in the assembly process provide an opportunity for regulatory inputs, such as binding of IHF and the requirement for DNA supercoiling. The formation of complexes for each step in the reaction pathway requires the successful completion of the preceding steps, preventing the formation of aberrant and potentially damaging side products. Similar control at the level of assembly has been documented for complexes involved in phage Mu transposition, retroviral integration, and RAG-mediated recombination, and probably represents a general means of regulation.

Simple Reactions with Complicated Consequences (9, 18, 35, 45, 53)

Although the DNA rearrangements accompanying gene transfer can be complex, much of the process can be understood by a few simple principles. Iterative action of these mechanisms can build up complicated final products, such as the large R plasmids transporting antibiotic-resistance genes.

The action of a few types of enzymes explains the structures of many of the rearranged DNA products. The λ integrase system serves as the classic example of site-specific recombination, a reaction involving reciprocal exchange between two DNA sites of specific and often related sequence. Integration of the circular phage chromosome into the *E. coli* chromosome proceeds by such a reaction, resulting in formation of a simple circular DNA containing both genomes. Repeated iteration of this reaction can yield structures that are not so simple, such as multiple lysogens integrated at *attB*. Often, side products of simple reactions yield new DNA arrangements. For example, aberrant excision of a λ lysogen can yield specialized transducing phage capable of transporting nearby *E. coli* sequences to a new cell.

Some DNA transposons yield Shapiro structures as intermediates, in which only one strand at each end of the transposon becomes attached to new sequences and the other strand is still attached to the old flanking DNA (see Fig. 3.12). The multiple possible fates of this structure explain many of the once-baffling products of transposition. Replication through the element can yield a cointegrate, with transposon sequences at the borders joining two formerly separate replicons. Cleavage to remove the old flanking DNA yields a simple insert. Retroviral integration resembles the simple insert pathway. A theme uniting this group of reactions is the structure of the transposase and (retroviral) integrase enzymes, which are members of the DDE superfamily.

The machinery responsible for conjugation also forms a large group together with the T-DNA transfer system of *Agrobacteria* and the type IV secretion systems of bacterial pathogens (Plate 9). In all cases, a projection from the bacterial cell binds the target cell and brings the two into close proximity. DNA transfer then takes place, involving extrusion of a single DNA strand concomitant with replication. For the pathogens, the type IV secretion systems are also sometimes used to introduce protein toxins into the target cells.

A few other widely used mechanisms include (1) target-primed reverse transcription used by LINE elements and mobile introns, (2) recombinational repair used to incorporate new sequences in chromosomes, and (3) nonhomologous DNA end joining involved in VDJ recombination, P-element transposition, and retroviral replication. Understanding the function of these systems and their most common side products explains many of the specific DNA structures found in studies of mobile DNA and lateral transfer. Iterative application of these mechanisms explains much of the structure of modern genomes.

A "Diversity Arms Race" between the Adaptive Immune Response and Infectious Agents (20)

Several examples of DNA exchange considered in the previous chapters illustrate the "diversity arms race" between pathogens and their hosts. The immune system generates diversity in antigen-binding molecules by VDJ recombination and other mechanisms. The resultant antibodies or T-cell receptors produced from the covalently rearranged genes recognize antigens derived from invaders and program their destruction. Vast numbers of antigen-binding specificities are required to identify the myriad possible molecular configurations present on the surfaces of pathogenic invaders.

The pathogens, of course, fight back. Diverse systems have evolved to change the surface coats of viruses or bacteria, allowing them to evade the immune response as it develops. *Salmonella*, for example, switches between two surface proteins, a phenomenon called phase variation, by inverting a segment of the genome containing a transcription initiation sequence. Genes for surface antigens are switched on and off as a result. Trypanosomes switch among many variant surface glycoprotein genes, expressing only one or a few at a time. Populations of trypanosomes expand in an infected individual; their numbers fall with the rise of the immune response, then rise again, this time with a new surface protein. The cycle can be repeated many times, involving sequential expression of many surface glycoproteins.

Many aspects of the diversity arms race are connected to lateral DNA transfer, because it provides such an effective means for generating biological novelty. The adaptive immune system is based on a captured DNA transposon, which probably entered the vertebrate genome by lateral transfer. For *Neisseria gonorrhoeae*, switching expression to new surface antigen genes appears to be coupled to uptake of DNA by transformation. In retroviruses and other retroelements, cycles of replication and infection of new hosts are accompanied by very high rates of mutation due to the low fidelity of reverse transcription. For HIV, the high rate of mutation is important in evading the host immune response and thereby allowing the high levels of replication that lead to disease.

Another diversity arms race is currently under way between pathogens and biomedical science. Academic and corporate laboratories are working to create new antimicrobial chemotherapies, but pathogens are developing resistance at such a high rate that new drugs are only effective for a few years. Enormous efforts to develop inhibitors of HIV replication have conferred some benefit on patients, but drug-insensitive viruses are emerging at a high rate in response to selection by therapy. At present, clinicians are reporting that up to 20% of patients harbor drug-insensitive HIV variants in some patient groups. Genetic diversity, generated and expanded by lateral transfer, is at the heart of several of our most severe medical crises.

LATERAL TRANSFER AND EVOLUTION: INTRONS, SEX, AND THE EARLY EVOLUTION OF LIFE

The next sections review some of the more interesting speculations on the effects of DNA mobilization, focusing on processes affecting evolution quite broadly.

Did RNA Splicing Arise as a Defense against Mobile DNA? (10, 31)

One of the more curious proposals for the effect of mobile DNA on genome structure concerns the origin of RNA splicing. The discovery of genes in pieces in the late 1970s stunned molecular biologists. Why should genes be broken up into introns and exons? It is probably fair to say that the answer to this question is no clearer today than it was right after the initial discovery. However, one of the more durable speculations holds that splicing evolved to minimize the burden of transposable elements.

Integration of a transposable element into the chromosome of a cell has the potential to disrupt a cellular gene. For example, it has been estimated that 5% of novel insertions of retroviruses into the mouse genome result in detectable mutations. The burden of new mutations is greatly reduced, however, if some of the insertions can be removed after transcription by RNA splicing. By this means, gene activity could be preserved despite the insertion mutation, to the benefit of both the mobile element and the host organism.

The degree to which this view explains the origin of RNA splicing is uncertain. Other potential explanations are also possible. For example, one idea holds that introns are mostly ancient remnants of the initial assembly of genes. Many exons encode independently folded protein domains, perhaps reflective of an earlier stage of gene evolution. Examples have also been found in insects of relatively recent birth of new genes by exon shuffling. According to this view, RNA splicing would have arisen concomitant with exon shuffling as a part of the process that created modern genes.

The example of the mobile introns strengthens the idea that RNA splicing reduces the burden of mobile elements. The mobile introns are both transposons and introns, as demonstrated by striking laboratory experiments in yeast, protozoa, and bacteria. The insertions of these elements are phenotypically silent, because after transcription the mobile intron RNAs are removed from RNA transcripts precisely. This example establishes that splicing can mitigate the burden of mobile element insertions, although the evolutionary origin of the splicing system remains uncertain.

Sex (2, 24, 49)

Evolution has preserved sexual reproduction in almost all eukaryotic lineages, despite the finding that asexual reproduction is sustainable for many taxa. Rare examples of species lacking sex are known, but most of these are believed to be evolutionary "dead ends" on the brink of extinction. Many models have been proposed to explain the benefits of sex, with no clear consensus yet reached. One idea, for example, is that sex serves to increase genetic diversity, allowing organisms to adapt more effectively to a changing environment. An intriguing alternative is that sex exists at least in part to control genomic parasites.

Sexually reproducing organisms typically contain two copies of each chromosome, although organisms with different (usually even) numbers of chromosomes are known. Simple sexual reproduction involves first meiosis, in which the chromosome number is reduced to form a gamete with one copy of each chromosome per cell ($1n$). Fertilization follows, in which a $1n$ egg and a $1n$ sperm fuse to form the $2n$ zygote. Different organisms elaborate many variations on this scheme.

If a genome has sustained a transposon insertion, recombination during meiosis may allow the production of gametes containing different numbers of transposons. For example, in a genome with five transposons, recombination could redistribute the elements so that one genome had four transposons and the other six. Meiosis followed by fertilization yields daughter organisms with different numbers of elements. If selection against high numbers of transposons is strong, offspring with fewer elements would leave more descendants and so expand in the population.

Models for the purpose of sex may be testable by studying asexual organisms. Examples of eukaryotic species that do not engage in sex are rare, and proposed asexual organisms have often been found, with more study, to engage in sex after all. However, there does seem to be

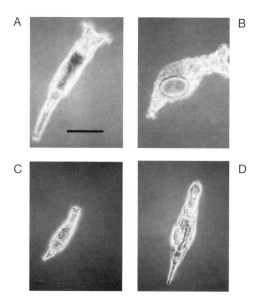

FIGURE 14.1. Four species of bdelloid rotifers. (A) *Philodina roseola*, (B) *Macrotrachela quadricornifera* (the oval is an egg), (C) *Adineta vaga*, (D) *Habrotrocha constricta*. The scale bar in A is 100 μm. (Reprinted, with permission, from Welch and Meselson 2000 [© American Association for the Advancement of Science]. [49])

a group of organisms, the bdelloid class of rotifers, that really do live without sexual reproduction (Fig. 14.1). The bdelloids are diploid, allowing comparison of the two copies of each gene present in homologous chromosomes. Studies by Matt Meselson and colleagues revealed that for those genes sequenced, the two copies within a single organism were much more different from each other than in sexual rotifers or other sexual organisms. These findings strengthen the idea that a lack of meiotic recombination accompanying sex has allowed the two chromosomes to evolve independently. The rates of divergence were as expected if the genes on each chromosome have been evolving independently since the divergence of asexual bdelloids from sexual rotifers some 35–40 million years ago (the age of the oldest specimens found fossilized in amber). In addition, in two bdelloid rotifer species, some of the chromosomes have no obvious homologs, suggesting that the differences have accumulated to the point that some previously homologous chromosome pairs are no longer recognizable by morphological criteria. These data strongly support the proposal that bdelloid rotifers multiply without even cryptic or rare sexual reproduction.

The bdelloid rotifers thus offer an opportunity to test the relationship between sex and mobile DNA. Arkhipova and Meselson surveyed the genomes of bdelloids, sexual rotifers, and myriad other species for their content of mobile elements. Stunningly, they found that retrotransposons are selectively absent in bdelloid rotifers but present in all other species examined (Table 14.1). *Mariner/Tc1* elements, in contrast, were present in bdelloids and many other species. Why retrotransposons

TABLE 14.1. LINE-like and *gypsy*-like RTase Sequences and *mariner/Tc1*-like Sequences in Diverse Species

Phylum	Species	DNA transposons		Retrotransposons	
		Mariner	Tc1	LINE	Gypsy
Sarcomastigophora	*Giardia lamblia*			+	
Porifera	*Halichondria bowerbanki*	+		+	+
	Spongilla sp.	+		+	+
Cnidaria (L,M)*	*Hydra littoralis*	+	+	+	−
	Aurelia aurita	+		+	
Ctenophora	*Condylactus sp.*	−		+	+
Platyhelminthes (M)	*Dugesia tigrina*	+		+	+
Rotifera (Acanthocephala)	*Moniliformis moniliformis*	−	−	+	+
Rotifera (Monogononta)	*Brachionus plicatilis*	+	−	+	+
	Brachionus calyciflorus	−	−	+	
	Sinantherina socialis	−	+	+	+
	Monostyla sp.	−	−	+	
Rotifera (Bdelloidea)	*Philodina roseola*	+	−	−	−
	Philodina rapida	+	−	−	−
	Habrotrocha constricta	+	−	−	−
	Adineta vaga	+	−	−	−
	Macrotrachela quadricornifera	+	−	−	−
Gastrotricha	*Lepidodermella sp.*	+	+	+	−
Nemertea	*Lineus sp.*	+	+	+	+
Priapulida	*Priapulus caudatus*	−		+	+
Sipuncula	*Themiste alutacea*	−		+	+
Annelida	*Glycera sp.*	−		+	+
Echiura	*Lissomyema mellita*	+	−	+	+
Mollusca (L)	*Chione cancellata*	+		+	+
Brachiopoda	*Glottidea pyramidata*	−		+	+
Bryozoa	*Amathia convoluta*	+	+	+	+
Phoronida	*Phoronis architecta*	+		+	+
Nematoda (L,G,M,T)	*Caenorhabditis elegans*	+	+	+	+
Onychophora	*Euperipatoides rowelli*	+	+	+	+
Arthropoda (L,G,M,T)	*Drosophila melanogaster*	−	+	+	+
	Drosophila pseudoobscura	+	+	+	+
	Drosophila virilis	+	+	+	+
	Lasius niger			+	+
	Formica polyctenum			+	+
	Aphis sp.			+	+
Tarigrada	*Milnesium sp.*	+		+	+
Chaetognatha	*Sagitta sp.*	+		+	+
Echinodermata (G)	*Echinometra mathaei*			+	+
	Strongylocentrotus purpuratus			+	+
Hemichordata	*Saccoglossus kowalevskii*	+		+	+
Chordata (L,G,M,T)	*Branchiostoma floridae*			+	+
	Danio rerio			+	+
	Onchorhynchus keta			+	+
	Xenopus laevis			+	+
	Mus musculus			+	+
	Bos Taurus			+	+

(Reprinted, with permission, from Arkhipova and Meselson 2000 [© National Academy of Sciences]. [2])
  * The designations L (LINE), G (gypsy), M (mariner), and T (Tc1) indicate that the transposon family was detected in the indicated organism in previous studies.

should be selectively absent in bdelloids is unclear, but affords fertile territory for further research. Perhaps retroelements are more genotoxic in some way than DNA transposons, or maybe they induce different host cell defenses. Another unanswered question is whether sex or retrotransposons were lost first in the bdelloids. Maybe retrotransposons were somehow lost by the bdelloids, thereby permitting the organisms to survive without sex to control proliferation of retrotransposons. Alternatively, maybe sex is required to maintain retrotransposons in a population. According to this idea, if an occasional individual loses active retrotransposons, their offspring could reacquire them from their other parent, thanks to sexual reproduction. Although many mysteries remain, Meselson's discovery suggests that there is a deep link between sex and retrotransposons, and the bdelloids provide an experimental system for investigating these questions.

DNA Transfer and the Early Evolution of Life (3, 15, 21–23, 30, 37, 40, 52)

Leaving the realm of experimental science entirely, it is intriguing to speculate on the role of lateral DNA transfer in the early evolution of life. The RNA World idea of Orgel, Crick, and Woese posits that RNA was the original informational biopolymer, since it can both store information and serve as an enzyme. This idea has been greatly strengthened in recent years with the discovery of RNA catalysis in mobile introns, RNase P, and, more recently, the ribosome. In another line of study, many laboratory experiments have succeeded in evolving new RNA molecules with diverse catalytic activities, documenting the ability of RNA to act both as an informational molecule and as an enzyme. David Bartel and coworkers have even succeeded in evolving a ribozyme that acts as an RNA polymerase, providing proof of principle for the idea that RNA molecules could be self-propagating. These findings strongly support the idea that RNA was central in the early evolution of life.

Modern cellular organisms, however, have genomes composed of DNA. This is quite understandable, since DNA has much greater chemical stability in water than does RNA. This raises the question of how the transition was made from the RNA World to the modern DNA World.

One clue is the homogeneity of modern life. With a few minor exceptions, the genetic code is universal. Similarly, the enzymes of central metabolism, the protein synthetic machinery, and the transcriptional machinery are related in all organisms, as are many other cellular systems. These observations suggest quite strongly that there was some force acting early in the DNA World to homogenize emerging organisms at the molecular level.

Lateral DNA transfer provides a strong candidate. Early RNA segments would presumably have been short, no more than a few hundred nucleotides, since longer molecules would have been too unstable. Early reverse transcriptase enzymes would have copied these into short DNA strands. Lateral transfer of strands among early organisms could have led to the observed similarities among modern genomes. According to this idea, lateral DNA transfer was one of the main forces shaping the modern DNA World.

Another line of reasoning also suggests that there was extensive lateral transfer in the early evolution of the DNA World. If you look at all the gene types present today in all organisms, the number is so high that it seems unlikely that they would have resided in a single organism early in the DNA World. Certainly new genes have been created since the origin of the DNA World, but very old gene types can be identified by their wide phylogenetic distribution and wide sequence divergence. The collection of genes expected to have been present in the founder organism seems much too large—instead, it is more likely that a pool of genes existed that was actively exchanged between individuals. According to this idea, modern organisms would have inherited particular subsets from the early pool, leading to the pattern of distribution seen today.

THE IMPACT OF LATERAL DNA TRANSFER: LOOKING FORWARD

Looking forward, it seems likely that lateral DNA transfer will be increasingly in the news. The effects of gene transfer on our daily lives are only starting to be appreciated. The transfer of sequence information among pathogens strongly influences the diseases they cause. Antibiotic-resistant infection, mobile pathogenic islands, and drug-resistant HIV are all examples of lateral transfer dictating the course of human disease.

Studies of lateral DNA transfer are increasingly important in assessing the use of genetically modified organisms in agriculture. Is it safe to sow fields with genetically modified plants? Will the genes spread to nearby plants and animals? If so, does it matter? At this writing there have been a few press reports on possible spread of engineered genes from genetically modified organisms to indigenous plants, but relatively little solid information. For example, it may be that pollen from genetically modified plants can fertilize related species growing nearby. In rare cases where the hybrids were fertile, this might have resulted in the transfer of engineered genes. Insects feeding on genetically modified plants might also provide a means for gene transfer, as with P elements and mites. Any engineered prokaryote introduced into the environment is likely to transfer genes to native species, at least occasionally. Further routes of gene transfer are not hard to imagine.

Given the frequency of lateral transfer generally, it seems likely that if large numbers of genetically modified organisms are introduced into the environment, at least a low rate of gene transfer will take place. However, this needs to be considered against the background of high rates of lateral transfer generally. If organisms are exchanging genes at a high rate, does it matter whether engineered genes are transferred occasionally? Obviously, the nature of the genes involved is a major determinant, and care must be taken in any setting where laboratory DNA constructions can be introduced into the environment. Still, in the seas, genes move between cells some 20 million billion times per second, leaving any human contribution vanishingly small by comparison. Looking forward, it will be crucial to carry out the debate on uses of genetically modified organisms against a background of solid information on natural rates of lateral gene transfer.

Last, a consequence of understanding lateral gene transfer is to understand ourselves. We were formed from a composite of the myriad creatures previously inhabiting earth. In our genes, we are closely related to all the life around us, even insects and plants and bacteria. Although evolutionary lineages have diverged, they are rejoined at least on occasion by lateral DNA transfer. Understanding lateral transfer emphasizes our connection to the rest of life on earth.

REFERENCES

1. Alm R.A., Ling L.S., Moir D.T., King B.L., Brown E.D., Doig P.C., Smith D.R., Noonan B., Guild B.C., deJonge B.L., Carmel G., Tummino P.J., Caruso A., Uria-Nickelsen M., Mills D.M., Ives C., Gibson R., Merberg D., Mills S.D., Jiang Q., Taylor D.E., Vovis G.F., and Trust T.J. 1999. Genomic-sequence comparison of two unrelated isolates of the human gastric pathogen *Helicobacter pylori*. *Nature* **397:** 176–180.
2. Arkhipova I. and Meselson M. 2000. Transposable elements in sexual and ancient asexual taxa. *Proc. Natl. Acad. Sci.* **97:** 14473–14477.
3. Ban N., Nissen P., Hansen J., Moore P.B., and Steitz T.A. 2000. The complete atomic structure of the ribosomal subunit at 2.4 Å resolution. *Science* **11:** 905–920.
4. Bennetzen J.L. and Kumar A. 1999. Plant retrotransposons. *Annu. Rev. Genet.* **33:** 479–532.
5. Berry W.B.N. and Hartman H. 1998. Graptolite parallel evolution and lateral gene transfer. In *Horizontal gene transfer* (ed. M. Syvanen and C.I. Kado), pp. 425–434. Chapman & Hall, London.
6. Bestor T.H. 2000. The DNA methyltransferases of mammals. *Hum. Mol. Genet.* **9:** 2395–2402.
7. Chaconas G. 1999. Studies on a "jumping gene machine": Higher-order nucleoprotein complexes in Mu DNA transposition. *Biochem. Cell Biol.* **77:** 487–491.
8. Courvalin P., Goussard S., and Grillot-Courvalin C. 1998. Gene transfer from bacteria to mammalian cells. In *Horizontal gene transfer* (ed. M. Syvanen and C.I. Kado), pp. 107–116. Chapman & Hall, London.
9. Covacci A., Telford J.L., Del Giudice G., Parsonnet J., and Rappuoli R.

1999. *Helicobacter pylori* virulence and genetic geography. *Science* **284:** 1328–1333.

10. Crick F.H.C. 1979. Split genes and RNA splicing. *Science* **204:** 264–271.

11. Dawkins R. 1976. *The selfish gene.* Oxford University Press, Oxford, United Kingdom.

12. Doolittle W.F. and Sapienza C. 1980. Selfish genes, the phenotype paradigm and genome evolution. *Nature* **284:** 601–603.

13. Fink G.R. and Boeke J.D. 1999. Yeast, an experimental organism for all times. *ASM News* **65:** 351–357.

14. Fisher G., James S.A., Roberts I.N., Oliver S.G., and Louis E.J. 2000. Chromosomal evolution in *Saccharomyces. Nature* **405:** 451–454.

15. Gesteland R.F. and Atkins J.F. 1993. *The RNA world.* Cold Spring Harbor Laboratory Press, Cold Spring Harbor, New York.

16. Gregory T.R. 2001. Coincidence, coevolution, or causation? DNA content, cell size, and the C-value enigma. *Biol. Rev. Camb. Philos Soc.* **76:** 65–101.

17. Grillot-Courvalin C., Goussard S., Huetz F., Ojcius D., and Courvalin P. 1998. Functional gene transfer from intracellular bacteria to mammalian cells. *Nat. Biotechnol.* **16:** 862–866.

18. Hendrix R.W., Roberts J.W., Stahl F.W., and Weisberg R.A. 1983. *Lambda* II. Cold Spring Harbor Laboratory, Cold Spring Harbor, New York.

19. Houck M.A., Clark J.B., Peterson K.R., and Kidwell M.G. 1991. Possible horizontal transfer of *Drosophila* genes by the mite *Proctolaelaps regalis. Science* **253:** 1125–1128.

20. Janeway C.A., Travers P., Walport M., and Capra J.D. 1999. *Immunobiology: The immune system in health and disease,* 4th edition. Elsevier, London.

21. Johnston W.K., Unrau P.J., Lawrence M.S., Glasner M.E., and Bartel D.P. 2001. RNA-catalyzed RNA polymerization: Accurate and general RNA-template primer extension. *Science* **292:** 1319–1325.

22. Joyce G.F. 1996. Building the RNA world. *Curr. Biol.* **6:** 965–967.

23. Joyce G.F. 2000. RNA structure. *Science* **289:** 401–402.

24. Judson O.P. and Normark B.B. 2000. Evolutionary genetics. Sinless originals. *Science* **288:** 1185–1186.

25. Kalendar R., Tanskanen J., Immonen S., Nevo E., and Schulman A.H. 2000. Genome evolution of wild barley (*Hordeum spontaneum*) by BARE-1 retrotransposon dynamics in response to sharp microclimatic divergence. *Proc. Natl. Acad. Sci.* **97:** 6603–6607.

26. Karpen G.H. and Alshire R.C. 1997. The case for epigenetic effects on centromere identity and function. *Trends Genet.* **13:** 317–320.

27. Kinsey J.A., Garrett-Engele P.W., Cambareri E.B., and Selker E.U. 1994. The Neurospora transposon Tad is sensitive to repeat-induced point mutation (RIP). *Genetics* **138:** 657–664.

28. Kobayashi I., Nobusato A., Kobayashi-Takahasi N., and Uchiyama I. 1999. Shaping the genome—Restriction modification systems as mobile genetic elements. *Curr. Opin. Genet. Dev.* **9:** 649–656.

29. Krassilov V.A. 1998. Character parallelism and reticulation in the origin of angiosperms. In *Horizontal gene transfer* (ed. M. Syvanen and C.I. Kado), pp. 409–422. Chapman & Hall, London.

30. Kruger K., Grabowski P.J., Zaug A.J., Sands J., Gottschling D.E., and Cech T.R. 1982. Self-splicing RNA: Autoexcision and autocyclization of the ribosomal RNA intervening sequence of *Tetrahymena. Cell* **31:** 147–157.

31. Lambowitz A., Caprara M.G., Zimmerly S., and Perlman P.S. 1999. Group

I and group II ribozymes as RNPs: Clues to the past and guides to the future. In *The RNA world*, 2nd edition (ed. R.F. Gesteland et al.), pp. 451–485. Cold Spring Harbor Laboratory Press, Cold Spring Harbor, New York.

32. Lander E., Linton L.M., Birren B., Nusbaum C., Zody M.C., Baldwin J., Devon K., Dewar K., Doyle M., FitzHugh W., et al. 2001. Initial sequencing and analysis of the human genome. *Nature* **409:** 860–921.

33. Lawrence J.G. and Ochman H. 1998. Molecular archaeology of the *Escherichia coli* genome. *Proc. Natl. Acad. Sci.* **95:** 9413–9417.

34. Levis R., Ganesan R., Houtchens K., Tolar L.A., and Sheen F.-M. 1993. Transposons in place of telomeric repeats at a *Drosophila* telomere. *Cell* **75:** 1083–1093.

35. Li L., Olvera J.M., Yoder K., Mitchell R.S., Butler S.L., Lieber M.R., Martin S.L., and Bushman F.D. 2001. Role of the non-homologous DNA end joining pathway in retroviral infection. *EMBO J.* **20:** 3272–3281.

36. Matic I., Rayssiguier C., and Radman M. 1995. Interspecies gene exchange in bacteria: The role of SOS and mismatch repair systems in evolution of species. *Cell* **80:** 507–515.

37. Nissen P., Hansen J., Ban N., Moore P.B., and Steitz T.A. 2000. The structural basis of ribosome activity in peptide bond synthesis. *Science* **289:** 920–930.

38. Ochman H., Lawrence J.G., and Groisman E.A. 2000. Lateral gene transfer and the nature of bacterial innovation. *Nature* **405:** 299–304.

39. Oettinger M., Schatz D., Gorka C., and Baltimore D. 1990. RAG-1 and RAG-2, adjacent genes that synergistically activate V(D)J recombination. *Science* **248:** 1517–1523.

40. Orgel L.E. and Crick F.H.C. 1980. Selfish DNA: The ultimate parasite. *Nature* **284:** 604–607.

41. Orgel L.E., Crick F.H.C., and Sapienza C. 1980. Selfish DNA. *Nature* **288:** 645–646.

42. Ptashne M. 1992. *A genetic switch*, 2nd edition. Cell Press and Blackwell Scientific, Cambridge, Massachusetts.

43. Rayssiguier C., Thaler D.S., and Radman M. 1989. The barrier to recombination between *Escherichia coli* and *Salmonella typhimurium* is disrupted in mismatch-repair mutants. *Nature* **342:** 396–401.

44. Schubbert R., Renz D., Schmitz B., and Doerfler W. 1997. Foreign (M13) DNA ingested by mice reaches peripheral leukocytes, spleen, and liver via the intestinal wall mucosa and can be covalently linked to mouse DNA. *Proc. Natl. Acad. Sci.* **94:** 961–966.

45. Shapiro J. 1979. A molecular model for the transposition and replication of bacteriophage Mu and other transposable elements. *Proc. Natl. Acad. Sci.* **76:** 1933–1937.

46. Stellwagen A.E. and Craig N.L. 1998. Mobile DNA elements: Controlling transposition with ATP-dependent molecular switches. *Trends Biochem. Sci.* **23:** 487–490.

47. Sun X., Wahlstrom J., and Karpen G. 1997. Molecular structure of a functional *Drosophila* centromere. *Cell* **91:** 1007–1019.

48. Thomas C.L., Jones L., Baulcombe D.C., and Maule A.J. 2001. Size constraints for targeting post-transcriptional gene silencing and for RNA-directed methylation of *Nicotiana benthamiana* using a potato virus X vector. *Plant J.* **25:** 417–425.

49. Welch D.M. and Meselson M. 2000. Evidence for the evolution of bdelloid rotifers without sexual reproduction or genetic exchange. *Science* **288:** 1211–1215.

50. Wendel J.F. and Wessler S.R. 2000. Retrotransposon-mediated genome evolution on a local ecological scale. *Proc. Natl. Acad. Sci.* **97:** 6250–6252.

51. Williamson D.I. 1998. Larval transfer in evolution. In *Horizontal gene transfer* (ed. M. Syvanen and C.I. Kado), pp. 436–452. Chapman & Hall, London.

52. Woese C.R. 2000. Interpreting the universal phylogenetic tree. *Proc. Natl. Acad. Sci.* **97:** 8392–8396.

53. Xiong Y. and Eickbush T. 1990. Origin and evolution of retroelements based upon their reverse transcriptase sequences. *EMBO J.* **9:** 3353–3362.

Index